AF802451

Özge Dolunay

# "Electricity is Life"
# Exploring the *Process* of Modernization in Off-Grid Energy Governance in Namibia

Geographie
Geography

Band 30

LIT

Özge Dolunay

# “Electricity is Life”
# Exploring the *Process* of Modernization in Off-Grid Energy Governance in Namibia

LIT

Funded by the German Federal Ministry of Education and Research (BMBF) – Client II Initiative (PROCEED Project).
Funded by the Chair of Social and Population Geography (Prof. Dr. Eberhard Rothfuß) of the University of Bayreuth.

Cover:
Photos: Özge Dolunay
Visual Design: Özge Dolunay
Illustration and graphic support by Elbie Bentley

This book is printed on acid-free paper.

**Bibliographic information published by the Deutsche Nationalbibliothek**
The Deutsche Nationalbibliothek lists this publication in the Deutsche Nationalbibliografie; detailed bibliographic data are available on the Internet at https://dnb.dnb.de.

ISBN 978-3-643-91697-6 (pb)
ISBN 978-3-643-96697-1 (PDF)
Zugl.: Bayreuth, Univ., Diss., 2024

© LIT VERLAG Dr. W. Hopf Berlin 2025
Contact:
Fresnostr. 2 D-48159 Münster
Tel. +49 (0) 2 51-62 03 20 Fax +49 (0) 2 51-23 19 72
e-Mail: lit@lit-verlag.de https://www.lit-verlag.de
**Distribution:**
In the UK: Global Book Marketing, e-mail: mo@centralbooks.com

*Dedicated to the communities living in the "off-grid" areas whose voices have not yet been heard...*

## Acknowledgments

It has been a long and adventurous research journey for someone who was never in Sub-Saharan Africa. I cannot imagine this without the support of many people whom I would like to express my deepest gratitude. My supervisor and mentor, Eberhard Rothfuß, who introduced me to the universe of social sciences, opened the door of phenomenology, and has always encouraged me with his invaluable guidance and constructive feedback for my thesis and projects. Special thanks to my second mentor Thomas Dörfler, who motivated me from the very beginning and encouraged me throughout the doctoral process with his guidance and comments and has been key in shaping this dissertation. It has been a pleasure for me to work with both of you, exploring human geography with your insights, and learning firsthand. I am grateful to Tanja Märkisch, who supported me in navigating the web of administrations in Germany and helped me all the way through this adventure. Thank you Michael Wegener for your maps. And, Elbie Bentley, your timing was simply perfect, much grateful for your support, your help, giving life to visual designs, your maps, and your patience with my thesis, proofreading, and your invaluable comments.

The field research in Namibia would not have been possible without the support of Helvi Ileka, who introduced me to Namibia, her culture, renewable energy in Namibia, and the communities. I would like to thank the community members in Tsumkwe and Gam who contributed to this research. I am especially grateful for the field support of Charlie Nqeisji, Likoro Masheshe, Virginia Karumbu, and Lukas Kandjii.

I am thankful for the opportunity provided by the German Federal Ministry of Education and Research within the PROCEED project and project partners who were involved. Thank you, Wilfried Zörner, for proposing PROCEED and SEED and leading these wonderful research projects. The COVID-19 outbreak made us all an unexpected and unpleasant surprise in the initial years, in which I am thankful to the local support of former Turkish Ambassador to the Republic of Namibia Berin Makbule Tulun, and consular officer Zeynep Bahar Bildik whose help was invaluable during the lockdowns in Namibia as well as Clemens von Doderer who supported throughout the project, including the lockdowns.

GEO colleagues Matthias, Claudia, Andrew, Veronika, Valerie, Axel, Susan, Heide, Festus, Behnaz, John, Meng, Harald, and Julia, thank you all for enriching this doctoral journey and supporting me in the difficult times.

Last but not least, I am grateful to scholars I have never met before, such as Norbert Elias, whose works, ideas, and recorded interviews inspired me during this dissertation; van der Loo and van Reijen; Edmund Husserl; Alfred Schütz; Edward Relph; and David Seamon, who enlightened my way.

My dear family in Türkiye, Luna, Lale, Burcu, Cemal, Göker, and the members who are not with us anymore, Aysel, Ahmet, and Çiğdem. I am grateful for your infinite support, encouragement, patience, and love. Özlem, words cannot fully express my gratitude - thank you for accompanying me on this remarkable Ph.D. journey but also being my bestie in life. All my friends, whom I consider family in Türkiye and worldwide. Thanks for your cordial, genuine support throughout this period. I would not have been able to come so far without you.

*Hepinize gönülden teşekkürler…*

## Summary

Despite the wide availability of energy-related technologies, 568 million people in Africa still lack electricity access (IRENA, 2022). The absence of progress in energy expansion stems from two sources, according to the literature: First, the need to acquire more knowledge of off-grid systems, which remains despite the passage and implementation of several national RE energy initiatives in recent years, and second, the neglect of interdisciplinary and social science research practices in energy and policy studies, which could provide in-depth insights to better understand and provide innovative ways to advance off-grid RE energy development. Furthermore, research in the Namibian context tends to focus on the more technical, economic, and ownership aspects of energy policy and implementation – while neglecting individual, local perspectives. Thus, although energy policies and rural electrification plans have been developed in Namibia over the past 30 years, the optimal level of electrification has yet to be met.

Currently, Namibia´s overall electricity access is approximately 50%, with rural access at 33% (WB, 2022). The vast and sparsely populated geography has also added complexities to the expansion of off-grid electrification. Considering these challenges, this study explored renewable energy governance from a bottom-up social sciences perspective using two case studies in Tsumkwe and Gam, two rural communities served exclusively by mini-grids. The goal was to better understand issues related to off-grid electrification in Namibia. This study continued with an investigation of the multi-level governance framework at regional, national, and supranational levels. This study used specified research questions to investigate the current state of renewable energy governance in Namibia from a multi-governance perspective, paying special attention to off-grid areas' energy practices and local community members' visions.

The research was conducted inductively, using a mixed methods, qualitative research methodology: This included data collection and analysis using participatory observation, ethnomethodology, and the phenomenology of life-worlds combined with hermeneutic interpretations as well as semi-structured interviews, surveys with open-ended questions, and secondary data sources. MAXQDA and SPSS software were used to analyze qualitative and quantitative data. The exploratory research findings suggest a modernization process occurring at all levels.

The process of modernization (van der Loo & van Reijen, 1992) has been explained through the processes of differentiation, individualization, rationalization, and domestication, within which the contributing factors were the differentiation between local actors, public, private, and community-based businesses, as well as differentiated tariffs. These processes work with their re-enchanting worldviews through electricity, energy practices, electric appliances, and their interlinkage within everyday lives and futures. Furthermore, in both settlements, it was found that community members viewed electricity as: "electricity is light," "electricity is life," and "electricity is like a source of life," and it was essential for everyday life. Here, lighting, cooking, cooling, heating, sewing, laundry, welding, communication, media, and leisure were found to be the primary energy practices. Both communities confirmed that they were satisfied with the mini-grid providing electricity for their appliances and its reliability and availability 24/7, enabling them to carry out energy practices. Their dissatisfaction was related to technical or weather interruptions and the prohibitive cost of the related tariffs. Despite the communities´ different ethnic and historical backgrounds, both individually and as a collective, they envisioned their future in a "town" characterized by services and infrastructure such as shops, malls, streetlights, banks, hospitals, tar roads, and vocational training centers. Another key finding of the study was the estimation of the electricity access, with approximately 49% in Tsumkwe and 73% in Gam.

At the national and regional levels, the differentiation of energy market actors, their collaboration with supranational actors and the differentiation within the RE and off-grid-related policy and regulatory frameworks with goal-oriented, planned, specific measures contributed to the differentiation and rationalization processes. The use of digital tools and the incorporation of varied energy sources with their specific technologies, such as RE, contributed to domestication at the national and regional levels. Electrification and RE development are recognized as key components of the National Vision of an industrialized, socio-economically developed Namibia. The current state of RE governance was identified as a multi-level meta-governance consisting of subnational, national, and supra-national actors, led primarily through government/state authority but with the involvement of both public and private entities.

In conclusion, capacity building, capacity development, and public awareness using digital tools, elaborated by the Participatory Action (Research) methodology, and transparency in the off-grid electrification schedule were recommended for their ability to include the local and national perspectives needed to improve off-grid electrification and access to the residents of Namibia..

## Zusammenfassung

Trotz der breiten Verfügbarkeit von Energietechnologien haben 568 Millionen Menschen in Afrika immer noch keinen Zugang zu Elektrizität (IRENA, 2022). In der Literatur wird darauf hingewiesen, dass dieser mangelnde Fortschritt auf zwei Ursachen zurückzuführen ist: Erstens die Notwendigkeit, netzunabhängige Implementierungen besser zu verstehen, die trotz der Verabschiedung und Umsetzung mehrerer nationaler Erneuebare-Energie-Initiativen (EE) in den letzten Jahren immer noch besteht, und zweitens die Vernachlässigung sozialwissenschaftlicher Forschungspraktiken, mit denen interdisziplinäre Energie- und Politikstudien zu einem besseren Verständnis und zur Förderung der EE-Energieentwicklung beitragen könnten. Darüber hinaus befasst sich die Forschung im namibischen Kontext eher mit technischen, wirtschaftlichen und eigentumsrechtlichen Fragen in Bezug auf die Politik und ihre Umsetzung, während die individuelle, lokale Perspektive vernachlässigt wird. Obwohl in Namibia in den letzten 30 Jahren Energiepolitiken und Pläne zur Elektrifizierung des ländlichen Raums entwickelt wurden, konnte der gewünschte Grad der Elektrifizierung noch nicht erreicht werden.

Derzeit liegt der Zugang zu Elektrizität in Namibia bei etwa 50 %, der Zugang auf dem Land bei 33 % (WB, 2023). Die weitläufige und dünn besiedelte Geografie hat die Ausweitung der netzunabhängigen Elektrifizierung zusätzlich erschwert. Angesichts dieser Herausforderungen untersuchte diese Studie die Governance im Bereich der erneuerbaren Energien aus einer sozialwissenschaftlichen Bottom-up-Perspektive anhand von zwei Fallstudien in Tsumkwe und Gam, zwei ländlichen Gemeinden, die ausschließlich durch Mini-Netze versorgt werden. Ziel war es, ein besseres Verständnis der Probleme im Zusammenhang mit der netzunabhängigen Elektrifizierung in Namibia zu entwickeln. Die Studie wurde mit einer Untersuchung in einem Multi-Level-Governance-Rahmen auf regionaler, nationaler und supranationaler Ebene fortgesetzt. Die Studie untersuchte anhand von spezifizierten Forschungsfragen den aktuellen Stand der Governance im Bereich der erneuerbaren Energien in Namibia aus einer Multi-Governance-Perspektive, wobei den lokalen Energiepraktiken und den Visionen der Gemeindemitglieder in den netzunabhängigen Gebieten besondere Aufmerksamkeit geschenkt wurde.

Die Forschung wurde induktiv durchgeführt, unter Verwendung einer gemischten, qualitativen Forschungsmethodik: Dazu gehörten die Datenerhebung und -analyse mittels teilnehmender Beobachtung, Ethnomethodologie und Phänomenologie der Lebenswelten mit hermeneutischen Interpretationen sowie halbstrukturierte Interviews, Umfragen mit offenen Fragen und Sekundärdatenquellen. Für die Analyse der qualitativen und quantitativen Daten wurden die Software MAXQDA und SPSS verwendet. Die Ergebnisse der explorativen Forschung deuten darauf hin, dass ein Modernisierungsprozess auf allen Ebenen stattgefunden hat, und die Ergebnisse wurden entsprechend erklärt.

Der Prozess der Modernisierung (van der Loo & van Reijen, 1992) wurde durch Prozesse der Differenzierung, Individualisierung, Rationalisierung und Domestizierung mit der Differenzierung zwischen lokalen Akteuren, öffentlichen, privaten und kommunalen Unternehmen sowie differenzierten Tarifen erklärt, die alle zum Prozess der Differenzierung beitragen. Diese arbeiten mit der Wiederverzauberung ihrer Weltanschauung durch Elektrizität; mit ihren Energiepraktiken, mit elektrischen Geräten und wie diese mit ihrem Alltag und ihrer Zukunft verbunden sind. Darüber hinaus wurde in beiden Siedlungen festgestellt, dass die Gemeindemitglieder Elektrizität wie folgt betrachten: "Elektrizität ist Licht", "Elektrizität ist Leben" und "Elektrizität ist wie eine Lebensquelle" und ist für das tägliche Leben unerlässlich. Hier wurden Beleuchtung, Kochen, Kühlen, Heizen, Nähen, Wäschewaschen, Schweißen, Kommunikation, Medien und Freizeit als die wichtigsten Energiepraktiken genannt. Beide Gemeinschaften bestätigten, dass sie mit der Stromversorgung ihrer Geräte durch das Mini-Netz zufrieden sind und dass diese zuverlässig und rund um die Uhr verfügbar ist, um die Energiepraktiken auszuführen. Ihre Unzufriedenheit bezog sich auf technische oder wetterbedingte Unterbrechungen und die unerschwinglichen Kosten der entsprechenden Tarife. Trotz der unterschiedlichen ethnischen und historischen Hintergründe der Gemeinschaften, sowohl individuell als auch kollektiv, stellten sie sich ihre Zukunft in einer "Stadt" vor, die durch Dienstleistungen und Infrastruktur wie Geschäfte, Einkaufszentren, Straßenbeleuchtung, Banken, Krankenhäuser, Asphaltstraßen, Berufsbildungszentren usw. gekennzeichnet ist. Ein weiteres wichtiges Ergebnis der Studie war die Einschätzung des Stromanschlusses, der in Tsumkwe bei 49 % und in Gam bei 73 % liegt.

Auf nationaler und regionaler Ebene trugen die Differenzierung der Energiemarktakteure und ihre Zusammenarbeit mit supranationalen Akteuren sowie die Differenzierung innerhalb des EE- und netzfernen Politik- und Regulierungsrahmens mit zielgerichteten, geplanten, spezifischen Maßnahmen zu den Differenzierungs- und Rationalisierungsprozessen bei. Der Einsatz digitaler Werkzeuge und die Einbeziehung verschiedener Energiequellen mit ihren spezifischen Technologien, wie z. B. EE, trugen zur Domestizierung auf nationaler und regionaler Ebene bei. Elektrifizierung und EE-Entwicklung werden als Schlüsselkomponenten der nationalen Vision eines industrialisierten, sozioökonomisch entwickelten Namibia anerkannt. Der gegenwärtige Stand der EE-Governance wurde als eine Mehrebenen-Meta-Governance identifiziert, die aus subnationalen, nationalen und supranationalen Akteuren besteht und in erster Linie durch die Regierung/Staatsgewalt geführt wird, an der aber sowohl öffentliche als auch private Stellen beteiligt sind.

Abschließend wurde empfohlen, den Aufbau von Kapazitäten, die Entwicklung von Kapazitäten und die Sensibilisierung der Öffentlichkeit mit Hilfe digitaler Werkzeuge, die durch die Methodik der partizipativen Aktionsforschung (Participatory Action Research) entwickelt wurden, sowie Transparenz im Zeitplan für die netzunabhängige Elektrifizierung zu fördern. Infolgedessen wurde dieser

Ansatz empfohlen, da er die lokalen und nationalen Perspektiven einbezieht, die zur Verbesserung der netzunabhängigen Elektrifizierung und des Zugangs für die Bewohner Namibias erforderlich sind.

## Table of Contents

## List of Figures

## List of Tables

## List of Boxes

## List of Abbreviations

| | |
|---|---|
| AfDB | African Development Bank |
| AMDA | Africa Mini-Grid Developers´ Association |
| BESS | Battery energy Storage System |
| BMBF | Federal Ministry of Education and Research |
| CBNRM | Community Based Natural Resource Management |
| CENORED | Central North Region Electricity Distributor |
| DM | Deputy Minister |
| ECB | Electricity Control Board |
| EIA | Energy Information Administration |
| Erongo-RED | Erongo Regional Electricity Distributor |
| ESI | Electricity Supply Industry |
| ESMAP | Energy Sector Management Assistance Programme |
| FNB | First National Bank Namibia |
| GCF | Green Climate Fund |
| GEF | Green Environmental Facility |
| GIS | Geographic Information System |
| GLCEP | Geospatial Least Cost Electrification Plan |
| GRN | Government |
| GWh | Gigawatt hours |
| HPP | Harambee Prosperity Plan |
| IEA | International Energy Agency |
| IFC | International Finance Corporation |
| IPPP | Independent Power Producer Policy |
| IPPs | Independent Power Producers |
| IRENA | International Renewable Energy Agency |
| IRDP | Integrated Rural Development Programme |
| JICA | Japanese International Cooperation Agency |
| JUWA | Juwa Farmers´ Union |
| KfW | Kreditanstalt für Wiederaufbau – KfW Development Bank |
| kWh | kilo Watt hours |
| LA | Local Authority |
| LAC | Legal Assistance Center |
| MLG | Multilevel Governance |

| | |
|---|---|
| MLRR | Ministry of Lands, Resettlement and Rehabilitation |
| MME | Ministry of Mines and Energy |
| MPE | Ministry of Public Enterprises |
| MSB | Modified Single Buyer Model |
| MTF | Multitier Framework ESMAP |
| MTC | Mobile Telecommunications Company Namibia |
| MW | Megawatt |
| NACSO | Namibian Association of CBNRM Support Organisation |
| NAD | Namibian Dollars |
| NamPower | Namibian Power Corporation |
| NAMREP | Namibian Renewable Energy Program |
| NBC | Namibian Broadcasting Corporation |
| NDP | National Development Plan |
| NEEP | National Energy Efficiency Programme in Buildings |
| NEF | National Energy Fund |
| NEI | Namibia Energy Institute |
| NELF | National Electrification Fund |
| NELFP | National Electrification Funding Portfolio |
| NELP | National Electrification Policy |
| NEP | National Energy Policy |
| NERA | Namibia Energy Regulatory Authority |
| NES | National Electrification Strategy |
| NHIES | Namibia Household Income and Expenditure Survey |
| NIDS | Namibia Inter-censal Demographic Survey |
| NIRP | National Integrated Resource Plan |
| NORED | Northern Region Electricity Distributor |
| NPC | National Planning Commission |
| NREP | National Renewable Energy Policy |
| NSA | National Statistics Agency |
| NUST | Namibia University of Science and Technology |
| NNC | Nyae Nyae Conservancy |
| NNDF | Nyae Nyae Development Foundation |
| NNFC | Nyae Nyae Farmers´ Cooperative |
| O&M | Operations and maintenance |

| | |
|---|---|
| OGEMP | Off-Grid Energization Master Plan |
| OTRC | Otjozondjupa Regional Council |
| PoM | Process of Modernization (van der Loo & van Reijen, 1992) |
| PPA | Power Purchase Agreement |
| PROCEED | Pathway to Renewable Off-Grid Community energy for Development |
| PVP | Photovoltaic Pumps |
| QOQ | Quotes from Open-ended Questions |
| RC | Regional Council |
| RE | Renewable Energy |
| REEESAP | Southern Africa Renewable Energy and Energy Efficiency Strategy and Action |
| REDMP | Rural Electrification Distribution Master Plan |
| REDs | Regional Electricity Distributors |
| REFIT | Renewable Energy Feed-In Tariff |
| RERA | Regional Energy Regulators Association |
| SACREEE | SADC Centre for Renewable Energy and Energy Efficiency |
| SADC | Southern African Development Community |
| SADCC | Southern African Development Coordination Conference |
| SAPP | Southern African Power Pool |
| SDG | Sustainable Development Goal |
| SHS | Solar Home Systems |
| SOLTRAIN | Southern African Solar Thermal Training and Demonstration Initiative |
| Southern-RED | Southern Regional Electricity Distributor |
| SRF | Solar Revolving Fund |
| SWA | South West Africa |
| SWAPO | South West Africa People's Organisation |
| SWAWEK | South West Africa Water and Electricity Corporation |
| SWH | Solar Water Heaters |
| TA | Traditional Authority |
| TAA | Traditional Authorities Act |
| UN | United Nations |
| UNAM | University of Namibia |
| UNFCCC | United Nations Framework Convention on Climate Change |

| | |
|---|---|
| WB | World Bank |
| WBG | World Bank Group |
| VSP | Village Schools Project |

## Glossary and Definitions

**Appliances (also called end-use devices):** Equipment powered by electricity or other energy sources that accomplish some function or task to deliver an energy service (e.g., light bulb, electric fan, cookstove, refrigerator, radio, washing machine, x-ray machine, drilling machine, etc.) (ESMAP).

**Capacity of energy supply:** An attribute of energy supply that relates to the quantity of energy made available to the user. It can be measured as a combination of total energy available over a period of time and the maximum power (rate of energy delivery) that can be used. For example, for electricity, capacity can be measured as the maximum power available (in watts) or the total energy available each day (in Watthours).

**Demand-side management.** The application of economic incentives and technology in the pursuit of cost-effective energy efficiency measures and load-shifting on the customer side, to achieve least-cost overall energy system optimization.

**End user:** The ultimate consumer who requires energy for desired energy services at any locale—a household, productive enterprise, or community institution.

**Energy access:** The ability of the end user to utilize an energy supply that is usable for the desired energy services. Improvement in energy access is achieved through enhancement of the usability of the energy supply with improvement in attributes.

**Energy supply:** The provision of energy regardless of the availability of end-use equipment.

**Independent Power Producer:** A privately-owned generation company and not part of a utility company that sells power to the grid under a long-term contract.

**Installed Capacity:** Actual installed capacity of the power plant in kWp or MWp.

**Kilowatt:** A unit of energy equal to 1,000 watts.

**Kilowatt-Hour:** A unit of energy equal to 1,000 watts over 1 hour; unit of electricity production and consumption.

**Mini-grid/Micro-grid.** For distributed renewable energy systems for energy access, a mini-grid/micro-grid typically refers to an independent grid network operating on a scale of less than 10 MW (with most at very small scale) that distributes electricity to a limited number of customers (NES).

**Multi-Tier Framework:** It is a framework for measuring access to electricity adopted by the WBG, Energy Sector Management Assistance Programme, ESMAP, and Climate Investment Funds (Please see Box. 3 for more details).

**Peak:** Hours with the highest power demand during the day.

**Prepaid Meter:** A device that monitors energy use after a customer has paid for electricity in advance.

**Productive Use:** An activity that uses energy to earn income or generate other non-leisure benefits.

**Reliability of energy supply:** An attribute of energy supply that entails the absence of unpredictable outages of energy supply. It is measured by the frequency and length of unpredictable outages.

**Off-grid:** An electricity supply system or distribution network that is not connected to the national electricity grid, irrespective of the location or proximity to the grid.

**Off-grid electrification:** Provision of electricity services using electricity systems that are not connected to the national electricity grid, such as a mini-grid or solar home systems.

**Solar photovoltaics (PV):** A technology used for converting solar light directly into electricity. Solar PV cells are constructed from semiconducting materials that use sunlight to separate electrons from atoms to create an electric current.

**Utility:** An entity that generates, transmits, and/or distributes power from facilities that it owns and operates.

## Energy Units and Abbreviations

Units of Power, Capacity of Energy Power Plants

1 watt

1 kilowatt (kW) = 1,000 Watt

1 megawatt (MW) = 1,000 kW

1 gigawatt (GW) = 1,000 MW

Units of electricity production and consumption

1 watt-hour (Wh)

1 kilowatt-hour (kWh) = 1,000 Wh

1 megawatt-hour (MWh) = 1,000 kWh

1 gigawatt-hour (GWh) = 1,000 MWh

# 1 Introduction

´Marco Polo describes a bridge, stone by stone.
*"But which is the stone that supports the bridge?"*
Kublai Khan asks.
*"The bridge is not supported by one stone or another,"*
Marco answers, *"but by the line of the arch that they form."*
Kublai Khan remains silent, reflecting.
Then he adds: *"Why do you speak to me of the stones? It is only the arch that matters to me."*
Polo answers: *"Without stones, there is no arch*[1]*.* ´

Despite rapid technological advances and the wide availability of renewable and conventional energy technologies, seven hundred thirty-three million people globally, of which 568 million are on the African continent, still do not have access to reliable energy sources (International Renewable Energy Agency [IRENA], 2022. As of 2021, the overall electricity access rate in Sub-Saharan Africa was 50.6% (World Bank [WB], 2023), while the rural access rate was even lower at 30.4% of the rural population. Economic, technical, social, and policy factors pose off-grid electrification's primary challenges (Hoeck et al., 2022). “Energy” has diverse meanings: people with technical backgrounds interpret it according to its physical characteristics; traders and developers interpret it according to its monetary value; many people live with it, unable to imagine a life without it; others lack it entirely and face many inconveniences. Pasqualetti (2021) characterizes energy as a thread weaving through the activities of daily life; Perkins (2022) sees it as ubiquitous to modern life. Under the umbrella of Sustainable Development Goals (SDG) number 7, all multilateral organizations, all countries, and all supporters of renewable energy follow the motto of SDG7 “ensure access to affordable, reliable, sustainable and *modern* energy for all.” Energy is intertwined within many contexts and interpreted as a magical (electrical) touch or a cure for all.

Current trends show that an estimated 670 million people globally will remain without energy access by 2030 (IRENA, 2022). Perhaps it would be helpful to look at the story of Italo Calvino (1972/1985) in The *Invisible Cities*[2] to reflect on the bridge made of many stones forming an arch, as people, communities, and energy actors create an arch for the establishment of necessary measures to

---

[1] *Invisible Cities* by Italo Calvino.

[2] Also cited in van Heerikhuizen (2015, p. 79).

reach electricity access globally. Furthermore, to look and see from a different perspective, as Pasqualetti (2021, p. 1) describes:

> "[Energy] weaves its way through all activities, from the most uncomplicated daily chore to the most complex international agreement, and through all ideas, from the memory of the past to every dream of the future…[I]ts role… is fundamental as it is overlooked"

in order to reflect on its governance.

The research and development of new technologies in all energy-related disciplines have contributed to different energy acquisition methods. Although conventional resources such as fossil fuels are still in use, renewable resources are increasingly used in the current energy mix. Increasing the ratio of renewables to expand energy access, reducing energy poverty, contributing to carbon-neutral technologies to combat climate change, and many more directly and indirectly energy-related goals have been declared by the United Nations (UN) under the seventeen SDGs. Unequal distribution of energy and growing populations highlight the importance of adequate models of energy governance, including renewable energy, to reach a wider array of people across diverse geographies. In this regard, off-grid renewable energy systems with mini-grids or stand-alone systems are sustainable and cost-effective for expanding electricity access (IRENA, 2018). The declining cost of off-grid system components since 2009, along with technological advances, has enabled the timely deployment of renewable off-grid systems, of which 83% have taken place on commercial and public services and the rest for household electrification (Dolunay, 2021; IRENA, 2019). So, although these expansion efforts have contributed to SDG 7 directly with the aim of "*ensuring access to affordable, reliable, sustainable, and modern energy for all by 2030*" (UN, 2015), they have also contributed to poverty reduction (SDG 1), the reduction of inequalities (SDG 10). by electricity/energy being as a condition for economic development (UN, 2018).

Namibia is one of the countries that has adopted SDG7 and ratified the Paris Climate Agreement to UNFCCC, which commits to increase its share of renewables in electricity production by 70% as well as increase energy efficiency and demand side measures (Ministry of Mines and Energy [MME], 2017a, National Energy Policy [NEP], 2017). Thus, Namibia is one of the Sub-Saharan countries that contributes with its efforts to bring energy to its vast geography, and nearly one million residents who lack access to energy. Namibia´s overall access to electricity was 36.5% in 2000 (WB, 2023), reaching 55.2% in 2021, above the Sub-Saharan African access rate of 50.6% (WB, 2023). Furthermore, the urban population's access to electricity increased from 73.2% (2000) to 74.7% (2021) by remaining below the Sub-Saharan Africa ratio of 80.7% (WB, 2023). The ratio of the rural population with access to electricity increased from 18.9% (2000) to 33.2% (2021), situating itself above the Sub-Saharan average of 30.4% (2021) (WB, 2023). As a result, these access rates present the necessity for more action to reach the national goals and SDG 7 targets and maintain the commitment to the Paris Climate Agreement.

## 1.1 Off-Grid Renewable Energy Governance in Namibia: Problem Statement

Governance describes a shift in the process of moving from the traditional hierarchical government to network governance, which includes actors and networks in governing society and economy (Lo, 2018). Given the background of Namibia in terms of electricity access and off-grid electrification in urban and rural areas, and electricity/energy belonging to the infrastructure of the countries, RE governance becomes prominent in advancing electricity access, reaching the national and international targets, as well as addressing the challenges of off-grid electrification in rural areas (Hoeck et al., 2022). Furthermore, the multi-level governance framework involving supra-national, national, and sub-national actors enables the assessment of the RE governance concept and process(es) for better, faster, and more sustainable ways of off-grid electricity expansion in Namibia. Thus, framing the research at different levels of governance, starting from the local, regional, national, and supra-national levels, brings forward a perspective that has not been previously elaborated. This perspective addresses the challenges in off-grid areas by focusing on RE, and more precisely solar energy, for the Namibia case, as it is already an accepted solution for off-grid electrification.

Within this context, Namibia´s RE consideration started in 1998 with the White Paper on Energy Policy recognizing the prominence of RE and energy efficiency (EE) (MME, 1998) after the independence of Namibia in 1990. Subsequently, the Electricity Act of 2000 (Republic of Namibia, 2000) reorganized the electricity sector, enabled the establishment of the Electricity Control Board (ECB), and provided the basis for the formation of the Regional Electricity Distributors (REDs). Thus, NamPower´s sole responsibility for electricity generation, transmission, distribution, and bulk supply was partially decentralized to five regions. The ECB was established to control the Electricity Supply Industry by issuing licenses regulating electricity generation, transmission, distribution, supply, import, and export (Dolunay, 2021). Furthermore, during the following years, three REDs were established according to the Companies Act with government-owned and public shareholding in Namibia's North, Central North, and West regions, supplying electricity to the residents of their jurisdiction areas (Dolunay, 2021).

Namibia Vision 2030 (Republic of Namibia, 2004, p. 15) aims for a “prosperous and industrialized Namibia, developed by her human resources, enjoying peace, harmony, and political stability,” and presents the long-term policy framework for national development for the industrialization of Namibia. The principle of sustainable development constituted the foundation for reaching the objectives in education, science, technology, health and development, sustainable agriculture, peace, social justice, and gender equality (Namibia Vision 2030, 2004; Dolunay, 2021). Even though minimizing the disparities between urban and rural areas was one of the main themes of Namibia Vision 2030, rural electrification of on-grid RE and its policies remained the focus. (Republic of Namibia, 2004). The Rural Electrification Distribution Master Plan (REDMP) was adopted in 2000 (MME, 2000a; MME, 2000b; MME, 2000c) and updated in 2005 and 2010 by the MME (MME, 2010a; MME, 2010b; MME, 2010c; MME, 2010d). This plan recognized the role of RE sources and their technologies as an integral part of the rural electrification strategy. It aimed to improve access to long-distance localities, which bear the burden of disproportioned cost and technical constraints for electrification (Dolunay, 2021). Although REDMP focused on rural electrification through grid extension with a priority in the government buildings, it also identified off-grid areas thoroughly, enabling the planning with Geographical Information Systems (GIS).

The Off-Grid Energization Master Plan (OGEMP) for Namibia, aligned with the REDMP 2005 to provide access to energy technologies in rural areas, was issued in 2007 (Schultz & Schumann, 2007, published by MME, 2007). To create an electrification plan, the OGEMP determined, according to the REDMP, the off-grid, pre-grid, and grey areas that would not have access to electricity within the next 20 years. The National Energy Policy (NEP/MME, 2017a) also adopted specified policy conditions for rural and urban households (Policy 29), and for rural institutional and commercial (P30) access to energy, policies for integrating energy into the planning of national development (P32). Additionally, NEP adopted conditions towards transparency and good governance (P34), energy research and development (P39), as well as other policy statements addressing economic and social development (Dolunay, 2021). The adoption of the policy on Independent Power Producers (IPP) in 2018 enhanced the implementation of grid-connected RE generation projects by involving private actors. On the other hand, the Off-Grid Electrification Policy (MME, 2019), which remains in the draft, was later integrated into the National Electrification Strategy and Implementation Plan (NES) (MME, 2021) and was still awaiting approval by the Cabinet at the time of this research.

At the beginning of the 2020s, the Electricity Bill, the establishment of the National Energy Regulatory Authority (NERA), the Electrification Fund, the full adoption of the Modified Single Buyer Model, and the new electrification conditions and strategies for the off-grid and on-grid electrification were in the pipeline of energy regulatory plans of the energy market actors (MME, 2021; MME, 2022a; MME 2022b). Despite the past 30 years of policy and planning efforts to expand electricity access, Namibia's access rates have remained behind the planned and desired rates. So far, only two solar hybrid mini-grids have been implemented in the two off-grid settlements in Tsumkwe and Gam (Figure 1.1), and some stand-alone systems have been implemented under the scope of some national and international projects, primarily for government buildings (Dolunay, 2021). Thus, even though policy development has taken place in the past, the postponement of the Off-Grid Energy Policy has negatively impacted off-grid electrification progress and contributed to the adversities facing electricity access (Hoeck et al., 2022). Further, the lack of energy policy implementation has also hindered the socioeconomic development of rural populations whose basic energy needs are still not being met (Bhattacharyya & Palit, 2013).

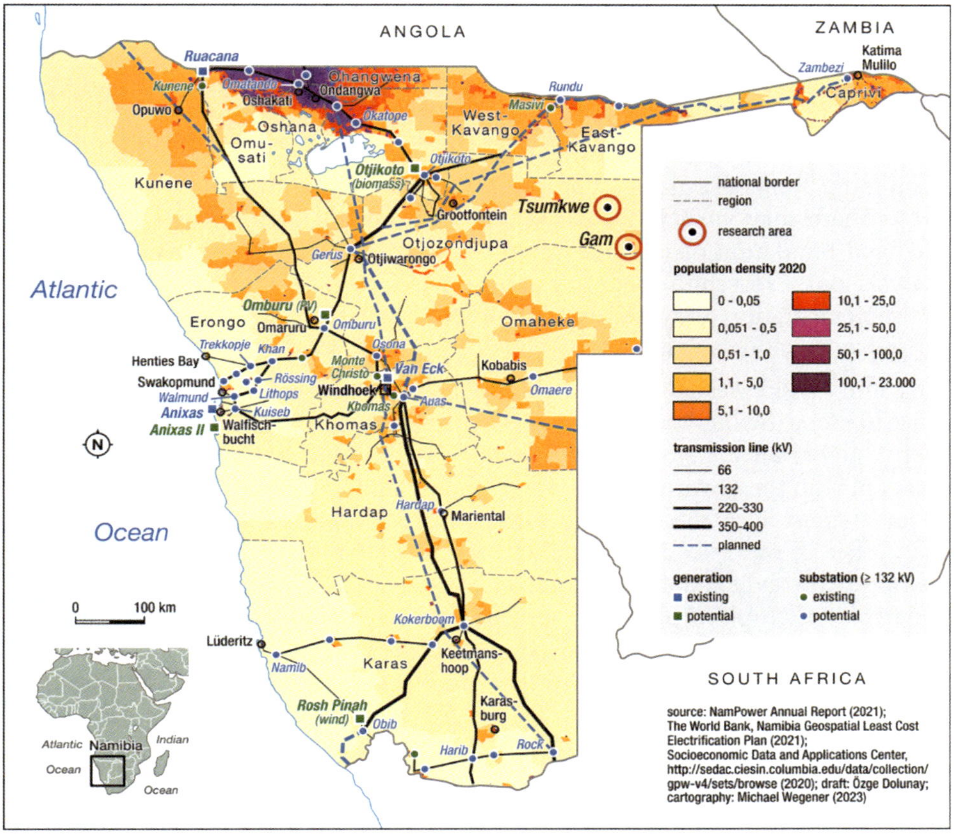


*Figure 1.1. Population Density and the Electricity Transmission Lines in Namibia*

## 1.2 Research Aim and Research Questions

Namibia´s policy strategies and frameworks support RE and off-grid electrification, but they have not been sufficient to reach 100% electricity access over the past thirty years. Furthermore, the (renewable) energy studies carried out in Namibia focused on the technical, economic, policy, and ownership issues (Asemota, 2013; von Oertzen, 2015; Chiguvare & Chiguvare, 2016; Chiguvare & Ileka, 2016; Schütt, 2016; von Oertzen, 2016; Ileka et al., 2017; von Oertzen, 2018; von Oertzen, 2019); representing a techno-economic bias when trying to resolve the issues related to off-grid energy access. In their analysis of rural electrification through off-grid systems in developing countries, Bhattacharyya and Palit (2013, p. 292) argue that despite numerous initiatives and suggested further research on case studies with systematic analysis, the development of a comprehensive analytical framework that encompasses multiple dimensions and levels, facilitating the assessment of off-grid interventions integrating social, economic, technical and environmental aspects into design and evaluation, there remains limited understanding of off-grid projects.

Furthermore, Bhattacharyya and Palit (2013) argue that techno-economic electrification interventions are insufficient to ensure viable and socially acceptable outcomes. They also identify the optimization of an off-grid system could include social and governance dimensions with a need for practical tools to support decision-making processes on the ground, especially with regard to capacity building for future needs that emerge in the development of off-grid initiatives (Bhattacharyya & Palit, 2013). In a last remark, they propose the governance and regulation of off-grid initiatives receive more attention with in-depth analysis and research in order to establish conducive conditions for the successful implementation and operation of off-grid electrification endeavors offering participatory models and capacity development (Bhattacharyya & Palit, 2013).

This study addresses Namibia's RE governance using a bottom-up approach to off-grid electrification from a social science perspective. Namibia's thirty-year energy governance history shows the roles that public, private, and civil society actors play in the governing process of energy and its development. It also helps to understand how actors at the different levels and institutions interact with each other in order to coordinate action and collaborate authoritative decision making that produce collective outcomes (Ansell & Torfing, 2016, p. 1). Thus, understanding multilevel energy governance (MLG) in Namibia will contribute to improving this governance to obtain the desirable outcomes for future off-grid (renewable) energy scenarios that are clean, affordable, and accessible for all.

This research implements an exploratory research design, taking existing energy plans, strategies, and policies into account, that uses phenomenology to explore life-worlds, energy practices, and visions of the communities to investigate off-grid RE governance, This research is guided by three primary questions: "What is the current state of off-grid RE governance in Namibia?", "What are local energy practices within the context of Namibia's current off-grid areas?" and "What visions do individual community members have?".

## 1.3 Relevance of the Research

This study makes multiple scientific contributions with its practical place-based recommendations. It explores off-grid renewable energy governance by investigating the life-worlds in Tsumkwe and Gam along the different levels of MLG from a bottom-up approach and explains the findings with the emerging process of modernization as a result of the findings. As Seamon (1979) suggested, the things and events of the daily experience are explored by phenomenology. A phenomenological perspective in this understanding, to look and see everyday life, to the unquestioned realities of the life-worlds, provides a tool for the policy-makers and serves as a framework for the people who live in this specific place (Seamon, 1979). With this phenomenological life-world approach, this research uncovers everyday life energy practices. In this way, it brings an understanding of the energy visions to be included and/or considered in the energy planning and policy-making of RE off-grid electrification, which would foster socioeconomic development. Thus, it would also fill in the gap in the understanding of RE governance (Bhattacharyya & Palit, 2013) from a human-centered, bottom-up social science perspective rather than the technical, economic, or techno-economic top-down perspectives.

On the other hand, the life-world methodology contributes to the research on phenomenology, human geography, and energy geographies. Subsequently, the emergence of the process of modernization through exploration introduces a different approach to off-grid RE governance research and implementation. Furthermore, as an unexpected contribution, it also contributes to the theory of the process of modernization by including digitization and digitalization in the process of domestication.

The results of the research, in addition to scientific contributions, provide place-based recommendations for the settlements as well as policy and regulatory framework recommendations in Namibia for multi-level RE governance. Thus, it aims to transfer the scientific outcomes into practical recommendations within the nexus of people, energy, and technology in off-grid electrification.

## 1.4 Background of Namibia

Namibia is located in the Southeast of the African continent, bordering South Africa in the south, Botswana in the east, Angola in the north, and Zambia in the northeast of the country. Its western coast lies in the Atlantic Ocean with a surface area of 824,292 km2. Since 2013, Namibia has been divided into fourteen administrative regions. The population of Namibia grew from 1.83 million in 2001 (Namibia Inter-censal Demographic Survey [NIDS], 2017, published by the Namibia Statistics Agency [NSA]), to 2,604,172 (as of 2023) (World Population Review, 2023). The population density rose from 2.01 to 3.16 in 2001 and 2023, with a growth rate of 1.6%, placing Namibia in the 227th rank among one of the most sparsely populated countries in the world (World Population Review, 2023). Furthermore, 60.3% of the population lives in urban areas and 39.1% in rural, accounting for 1,034,095 people (Worldometers, 2023). Regarding demography, Ovambo presents the largest ethnic group, with 49.8% mainly residing in the north (NIDS, 2017). They are followed by other ethnic groups such as Kavango (9.3%), Damara (7.5%), Herero (7.5%), white (6.4%), Nama (4.8%), Caprivian (3.7%), San (2.9%), and Basters (2.5%) (NIDS, 2017). In addition, 6.4% of the population comprises whites (Portuguese, British, German, and Afrikaner). The median age in Namibia is 21.4, and the fertility rate is 3.21 children, with a life expectancy of 59.5 years (Worldometers, 2023).

The settlement of Namibia began with the arrival of the San people, the hunter and gatherers, the oldest ethnic group, around 25,000 BC, who settled in the southern mountains (Hackl, n.d.). Around the 1$^{st}$ century BC, the Nama arrived and subsisted from livestock farming. It was only around the 15th and 16th centuries that the nomadic Bantu tribes, mainly Herero, reached Namibia from the northeast, expelling San people from their territories (Hackl, n.d.). They settled in the central part of the country around the 18$^{th}$ century, leaving the north to the Himba. In the 17$^{th}$ century, the Ovambo came from Central Africa to the north, followed by the Orlaam in the 19$^{th}$ century, and shortly after the Boers, white farmers, came from the south. The descendants of Nama women and Boer men, the Basters, were the last indigenous settlers to arrive in Namibia (Hackl, n.d.). These first settlers were already in conflict, whether for weapons or land, before the non-Africans arrived.

Following explorations of the coastline of Namibia by two Portuguese explorers, Diego Cão and Bartholomew Diaz, in 1486 and 1487, Dutch settlers took over the Cape Colony and Walvis Bay in 1793 (Hackl, n.d.). The United Kingdom soon followed, taking control of the region in 1797. The UK's occupation penetrated the country, which worked to pre-empt German ambitions in this area and ensured sole power over the deep-water harbor, Walvis Bay; this marked the first official land claim (Green, 2023). The German trader Adolf Lüderitz arrived in 1883 and discovered the nation´s large diamond deposits; he subsequently made a deal with Nama chief Joseph Frederick of Bethanie, securing him the rights over Angra Pequena, which was soon renamed "Lüderitz" (Katjavivi, 1990). In the meantime, these events had taken Germany's political and economic interest. In 1884, the German Chancellor Otto von Bismarck claimed the region before the British. That same year, at the Berlin Congo Conference in Germany, "German South-West Africa" colony borders were established with the British and the Portuguese. In 1890, Windhoek was made the capital of the German administration, and the Germans remained in Namibia until 1915 (Hackl, n.d.). The bovine plague outbreak, the resistance of Nama, the great Herero Rising under Samuel Maharero, the battle in Waterberg followed by the Herero´s forced relocation into the desert resulting thousands of casualties, the rise of the Nama under the leadership of Hendrik Witbooi and Jacob Morenga against Germans, and other more minor riots were just a few of the tribulations and upheavals that took place during this period (Hackl, n.d.). The indigenous traditions and community structures were affected by these events; cattle raising was forbidden, and their access to education was stopped (Katjavivi, 1990).

In 1915, South Africa defeated the German forces and took control (Dugard, 2018). Following the peace treaty after the Conference of Peace in Paris in 1919, the "League of Nations" was formed, and a mandate system was introduced (Katjavivi, 1990). German South-West Africa was then under the South African administration as South West Africa (SWA) (Dugard, 2018). However, South Africa continued colonial behavior. At the end of World War II in 1945, the United Nations (UN) Charter replaced the League of Nations´ mandate system. In 1946, South African Prime Minister Jan Smuts called Namibia the fifth South African province. South Africa refused to acknowledge the UN as the successor of the League of Nations and, in 1951, applied its South African Apartheid laws to SWA (Hackl, n.d.). The resistance and call for independence grew in those years, and the Owamboland People´s Organization (OPO), which later changed its name to SWAPO (South-West African People´s Organization), was formed. In 1964, South Africa introduced the Odendaal Plan, a development plan for South West Africa (Hackl, n.d.). The UN withdrew South Africa´s mandate to administer SWA, but South Africa rejected it again. SWAPO called for an armed fight, which led to a declaration of emergency in 1966 (Hackl, n.d.). In 1968, the UN General Assembly officially renamed SWA "Namibia." People´s Liberation Army of Namibia (PLAN) was founded by the SWAPO in 1971 and recognized by the UN in 1973 for being the first authentic representation of Namibia (Hackl, n.d.). From 1975 until 1983, several attempts at discussions over a new Namibian constitution and a ruling party took place.

Finally, in 1988, South Africa agreed to have new dialogues, which resulted in a ceasefire and a UN resolution. In 1989, with the new elections, SWAPO was elected with a total majority, naming the leader Sam Nujoma as its first president (Green, 2023). The new constitution was effective as of March 21st, 1990, and it marked the day of independence of Namibia. South African troops left the country (Green, 2023). In 1993, the Namibian Dollar was introduced as the new currency. Sam Nujoma stayed in power for three legislative periods. He was followed by another SWAPO member, Hifikepunye Pohamba, for the successive two terms (Green, 2023). In 2014, Hage Geingob won 85% of the votes and elected the next president (Green, 2023). Today, the Parliament of the Republic of Namibia has a bicameral legislature, which consists of the National Council (upper house) and the National Assembly (lower house). There are 146 seats in both houses, where eleven political parties are presented, with SWAPO being the governing party. Twenty ministries are currently in operation.

## 1.5 Research Sites: Local Setting

This research was conducted within the PROCEED Project, which investigated the sustainability of mini-grids in Namibia from their social, economic, and technical perspectives between 2019 to 2023. The project was funded by the Federal Ministry of Research and Education (BMBF) and included research project partners from Namibia and Germany. The settlements of Tsumkwe and Gam were selected for this project because they are home to the only two existing mini-grids in Namibia, which partially electrify two settlements. Additionally, these two communities contrasting histories and social backgrounds enriched the study through comparative perspectives of everyday lives of community members and the mini-grids. This study explored community experience with mini-grids between the time of their initial operation, 2011 in Tsumkwe and 2014 in Gam, to the completion of this research in 2022. No further mini-grids were established in subsequent years, nor have any additional studies exploring local communities and mini-grids from a qualitative, social sciences perspective been conducted (to the author's knowledge). Therefore, the findings of this research are expected to contribute to social and interdisciplinary sciences as well as place-based local and national implementations in new and unique ways.

Tsumkwe and Gam settlements are situated in the Otjozondjupa Region, which is the fourth largest of the fourteen regions, towards the Botswana border. The region's administrative capital is Otjiwarongo. According to the 2011 census, the population was 142,000 inhabitants. Tsumkwe Constituency is situated in the northeast of the Otjozondjupa region, where Tsumkwe Settlement and Gam are located. Tsumkwe is 304 km from Grootfontein, the nearest town, connected by a gravel road, and 735 km from Windhoek (Figure 1.1). Gam is about 100 km from Tsumkwe in the southeast and 100 km from the Botswana border.

Link et al.'s (2011) definition of a community is "a shared locality with homogenous interests." MacQueen et al. (2001) define community as "a group of people with diverse characteristics who are linked by social ties, share common perspectives, and engage in joint action in geographical locations or settings." McMillan (1986) discusses two major uses of the term community: (1) territorial and geographical notion of community (neighborhood, town, city), and (2) relational notion, which is concerned with the "quality of character of human relationship, without reference to location." McMillan (1986) also states what Durheim´s observation on how modern society develops community around interests and skills more than the locality. It is important to note that within this research, the term *community* is used in its territorial and relational context. Therefore, in each case study area, the local communities´ energy practices and visions have been considered from household and business (public and private) perspectives.

### 1.5.1 Tsumkwe

Tsumkwe is one of the off-grid settlements identified in the REDMP (2000), where San, Kavango, Herero, Damara/Nama, Ovambo, and Zambezi ethnic groups reside (Ileka et al., 2017). In particular, it is known as the capital of the San people in Namibia, considered to be the most vulnerable and marginalized ethnic group in the country (Suzman, 2001). The population of the Tsumkwe Constituency was 9,000 people, of which Tsumkwe Settlement hosted an estimated 3,800 people (Boyd et al., 2012). During the South African apartheid, Tsumkwe Settlement served as a military post (Ileka et al. 2017). After independence in 1990, the government established a school, a clinic, and a police station for the community, where the diesel generators and the electrical infrastructure left by the South African Government were used (Boyd et al., 2012). In 2005, the Councilor of Tsumkwe requested an enhancement of the electricity, and NamPower realized a feasibility study for Tsumkwe (Dolunay, 2021). The grid connection cost resulted in more than N$ 150 million for a 270 km distance from the national grid (Ileka et al., 2017).

The Ministry of Mines and Energy (MME) initiated an investigation into the current energy situation in Tsumkwe with experts from the Desert Research Foundation of Namibia (DRFN) who had the experience from the solar-diesel hybrid pilot project of the Gobabeb Training and Research Center (Dolunay, 2021). Based on the investigation, a similar system was decided for Tsumkwe (Dolunay, 2021). Thus, the existing electrical infrastructure was upgraded into a solar-diesel hybrid mini-grid with a total budget of 2.9 million Euro, which was funded by the European Commission (75%), the Otjozondjupa Regional Council (OTRC) (11%), and NamPower (14%) (Boyd et al., 2012). The mini-grid consisted of a 202kWp solar PV capacity with a 650kVA diesel generator and 766 kWh battery capacity (Ileka et al., 2017), and the system was upgraded to 303 $kW_p$ in the following years (Boamah, 2020). The ownership was transferred to the OTRC, while the operation and maintenance fell under the responsibility of the Department of Works. In 2017, the mini-grid's ownership, operation, and maintenance were transferred to the Central North Regional Electricity Distributor (CENORED).

### 1.5.2 Gam

The second case study area, Gam, is situated 100 km from Tsumkwe, located on limestone rock with a 20cm soil depth, which is also not a very fertile and suitable type for cultivating crops (Harring & Odendaal, 2002). The settlement is in the semi-arid zone which lacks water resources; there is no permanent surface water, and groundwater resources are scarce. It is tough to locate the aquifers; they also present technical difficulties for drilling and installing boreholes (Tarr, 1998).

In relation to the National Programme for Resettlement, conducted by the Ministry of Land, Reform, and Resettlement (MLRR), many resettlement projects have been realized following independence, and Gam Resettlement being one of them. According to the MLRR data, the project served approximately 423 families (Harring & Odendaal, 2002; Ileka et al., 2017). Although some San families lived in the area, the project took place exclusively for Herero people who were voluntarily repatriated to Namibia after independence in 1990. They were all the descendants of the people who fled to Botswana during the war in 1904-1907. The families arrived between 1993 and 1996, with priority given to the women and children in the early years. They are mainly cattle farmers, and state employees from the Ministry of Agriculture, Ministry of Water and Forestry, police station, clinic, and schools (Hoeck et al., 2022).

After the initial part of the resettlement was completed, the Gam Development Organization requested the MME for the electrification of Gam in order to enhance its development (Dolunay, 2021). In 2014, the central part of Gam, including the essential and the non-essential lines, was electrified through a stand-alone solar hybrid mini-grid of 292 $kW_p$ capacity and a 180 kW, 3-phase inverter/charger, a lead-acid battery bank with two diesel generators, all of which was commissioned by the MME (Ileka et al., 2017). Upon commissioning, the mini-grid was handed over to the Regional Council (RC), which was responsible for collecting the consumption fees. The Tsumkwe-based Department of Works was made responsible for its operation and maintenance. In 2017, the mini-grid's operation and maintenance were then handed over to CENORED. The two diesel generators were removed to Tsumkwe after CENORED took ownership since the mini-grid only used 28% of its power from the generator. There was a surplus of electricity since not all households and businesses in Gam were connected to the mini-grid (only 200 households, clinics, school, and police station were connected). This energy scenario changed drastically at the end of 2020 and the beginning of 2021, when an additional 400 connections were made (Hoeck et al. 2022).

## 1.6 Research Design and Methodology

This research was designed with an inductive exploratory research approach, adopting a qualitative methodology, including ethnomethodology, life-worlds, and hermeneutical interpretation with mixed methods. Data were collected through participant observation, semi-structured interviews, surveys, and secondary data and used to address the research questions. First, the field research and data collection in two case study areas, Tsumkwe and Gam, was conducted. Preliminary analysis shows that modernization emerged as a lived experience in the everyday life of the settlements, which was explained through the process of modernization theory (van der Loo & van Reijen, 1992). As a part of the MLG framework, interviews with the national and regional actors were conducted. and off-grid-focused policy analysis was carried out within the (temporal) hermeneutics, reflecting the policy developments in the past and present while moving to the future. The validation of the qualitative and quantitative data was carried out in the PROCEED Workshop (March 2023) and in the PROCEED Conference (September 2023), with the involvement of the energy market actors and the community members from two settlements.

This research employed a mixed methods inductive, exploratory research approach. First, field studies in the two case study areas, the communities of Tsumkwe and Gam, were conducted; data were collected through participant observation, semi-structured interviews, surveys, and secondary data. These datasets were then analyzed using ethnomethodology, life-worlds, and hermeneutical interpretation approaches to understand everyday energy practices and the intricate relationships between community members and the mini-grids from diverse perspectives. As part of inductive exploration, the process of modernization (van der Loo & van Reijen, 1992), as experienced through the everyday lives of community members in both settlements, emerged through initial analysis and served as a guiding framework in the explanation of findings and analysis. This step contributed to understanding energy practices, visions, and local energy governance. Second, interviews with national and regional actors were conducted within the MLG framework to understand regional and national perspectives regarding current off-grid electrification and future visions. Furthermore, off-grid-focused policy analyses were carried out to situate the off-grid energy governance and electrification within the MLG by adopting a (temporal) hermeneutic interpretation reflect past and present policy developments while also moving to the future. Finally, to triangulate the results, the datasets were validated in the PROCEED Workshop (March 2023) and the PROCEED Conference (September 2023), which involved energy market actors, including community members from the two settlements. As a result, the research questions could be answered, and policy recommendations could be elaborated based on them.

## 1.7 Limitations and Opportunities

The COVID-19 pandemic coincided with the expected start date for the research, which significantly impacted data collection in the field resulting in a smaller survey sample size than initially intended. However, this situation provided the opportunity to introduce digital tools to conduct interviews with assistance from the settlements. Likewise, the initial interviews with national and regional actors were conducted online, followed by on-site observations in the subsequent years after the travel restrictions had been lifted. Furthermore, the collected data and analysis have brought many place-based insights into the localities; however, due to the focus on off-grid RE, only key points could be highlighted.

The researcher had no previous contact with the communities or Namibia before this research. The researcher's curiosity toward a new continent, a new country, and new cultures offered its own uniqueness for the research. The field research has been conducted with an open mind, an open heart, and an eagerness to learn. The author has an environmental engineering background, with Master of Science degrees in sustainable resource management and renewable energy systems, and more than fifteen years of working experience in different geographies of the world.

## 1.8 Research Outline

This study is composed of eight sections, excluding references and appendixes. Following the introduction chapter, Chapter 2 explains the framework of perpetual theories used in the study. The methodology is detailed in Chapter 3. Chapter 4 provides the qualitative and quantitative research findings in Tsumkwe, while Chapter 5 presents the findings of the second case study area, Gam. Chapter 6 focuses on renewable energy governance in Namibia, including the policies and key inputs about the opportunities and limitations expressed by national and regional actors. This is followed by the discussion of the comparative findings in Chapter 7, and the overall conclusion, along with policy and regulatory framework recommendations, is presented in Chapter 8.

## 2 Framing and Linking Perpetual Theories

This section will focus on the three central bodies of literature that provide this study's foundation: phenomenology, multilevel governance, and the process of modernization (Figure 2.1). A phenomenological epistemology provided the inductive foundation for the life-world approach employed in this research. This epistemology followed an *ex ante* multilevel governance (MLG) theoretical framework implemented at the local, regional, national, and supra-national levels to address off-grid energy governance; the definitions of governance and multilevel governance will thus be reviewed to situate off-grid energy governance in Namibia. In this context, as a part of phenomenological sociology (Heap & Roth, 1973), the life-world approach constituted the methodological anchor of energy practices central to establishing bridges between the major theories and frameworks. The preliminary analysis of the inductive research exposed modernization as an *ex post* theory. Therefore, at the intersection of human geography, MLG of energy, and macro social processes, Norbert Elias´ *Civilising Process* (Elias, 1976/2000; Elias, 1976/2004) will be revisited in order to understand the figuration concept and the basis of the *processes*. After reviewing modernization's history and critiques, the process of modernization theory by van der Loo and van Reijen (1992) that is utilized to explain the processes within the multilevel off-grid energy governance will be detailed.

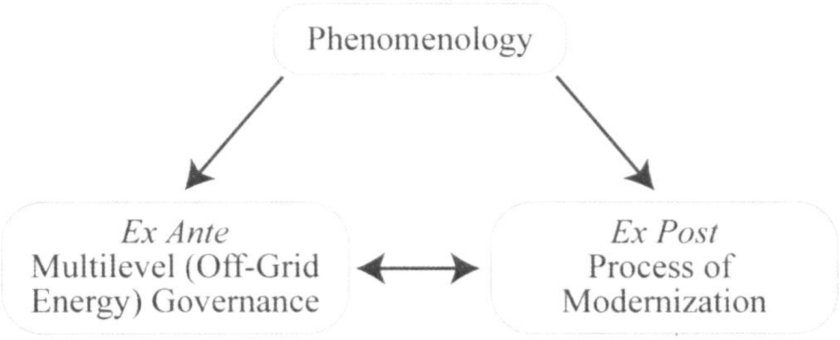


*Figure 2.1. Framing Perpetual Theories*

*Source: Author*

### 2.1 Multilevel Energy Governance

This section frames the governance and multi-level governance definitions investigated within this study to better situate off-grid energy governance in Namibia and to understand the issue(s) regarding off-grid electrification, RE, and energy access. Fundamentally, governance is defined as "the act or process of governing or overseeing the control and direction of something" (Webster Dictionary, 2024). The concept of governance is used as a "new mode of governing" (Mayntz, 2001; Rhodes, 2007, p. 4) and as a "political steering in modern societies" (Mayntz, 2001, p.1). Governance is also described as a "shift in the process of public decision-making, governing of society and economy" (Lo,

2018, p. 1) from the hierarchical models towards the non-hierarchical models of governing with the involvement of the private actors in the formulation and implementation of public policy (Mayntz, 2001, p. 1). Furthermore, Rhodes (2007, p.4) views governance as "an interdependence among different organizations and actors, continuing game-like interactions between networks, regulated by rules that are negotiated and agreed by network actors, and a significant authority of the state." Under these definitions, governance shifts from the traditional state hierarchy to governance with the participation of private, semi-public, and public actors and networks (Lo, 2018, p. 1).

Based on this governance concept, through the emergence of the plurality of actors, networks, and their coordination, Hooghe et al. (2020, p. 197) defined MLG as the "dispersion of authority within and beyond national states," whereby state and non-state actors have been are involved at the national, subnational, and supranational levels. In this way, the dispersing authority of the nation-state has been spread bidirectionally within supranational and subnational levels, linking different participants from different organizations, corporations, and civil society in the decision-making process (such as in the case of the European Union) (Hooghe et al., 2020, p. 194; Mayntz, 2001, p. 2). Thus, the MLG suggests a decentralization of decision-making processes at different levels and between different actors (Saito-Jensen, 2015, p. 2). At the same time, the power of government has been modified and shared with the involvement of these actors during the process (Saito-Jensen, 2015, p. 2). As a consequence, the differentiation between the actors and emerging networks sets forth a "series of reconfigurations of the relationships" (Saito-Jensen, 2015, p. 2), whereby "the role of the state is being transformed" (Bache & Flinders, 2004).

Hooghe and Marks (2001, p. 5) differentiate a "type I governance that is federalism, which deals with the power-sharing between a limited number of governments at just few levels" within the context of the national and sub-national state governments that do not overlap in their relationships, have "multi-task jurisdictions," and present a "hierarchical governance" (Saito-Jensen, 2015, p. 2). On the other hand, type II governance has "vast jurisdictions" and operates at "diverse territorial scales," whereby "the jurisdictions are specific and flexible" (Hooghe & Marks, 2001, p. 6) – also termed "polycentric governance" (Ostrom, 2010; Saito-Jensen, 2015, p. 3). Thus, through the interactions between differentiating structures and actors, the hierarchical levels and types of governance are "blurred" (Saito-Jensen, 2015, p. 3), allowing refigurations to occur. Hooghe et al. (2020, p. 194) define the two logics behind these two types of MLG as "functionalist logic" that situates governance as "an instrument for the efficient delivery of goods that individuals cannot provide for themselves" and the second logic of "the demand for self-rule by those living in distinctive communities," which is linked with an identity logic. Hooghe et al. (2020, p. 194) thus argue that "the larger a state´s population grows, the greater the number of different governance levels" become in bidirectional ways.

### *Polycentric (Energy) Governance*

The change in structuring within government or institutional operations, such as in the case of energy and electricity provision and new governance alternatives, have also been considered within energy governance research. Among many diverse governance approaches, polycentric governance has emerged through discussions surrounding common pool resources, public goods, and collective action problems (Sovacool, 2011). Polycentrism has been considered a strategy for gathering the power of diverse perspectives from multiple organizations, promoting cooperation, and involving stakeholders outside of the governments (Sovacool & Drupady, 2012; Sovacool & Martiskainen, 2020). Although Sovacool (2011) stated that the success of polycentrism has not been guaranteed, it provides "benefits for local action without compromising many of the benefits of global action" (Sovacool & Martiskainen, 2020, p. 4). Sovacool (2011) and Sovacool and Martiskainen (2020) investigated the polycentric governance approaches to energy and climate change, claiming that when implemented correctly, it has the potential to provide an "equitable, inclusive, informative, accountable, protective and adaptable framework for promoting renewable energy and energy efficiency, fighting energy poverty, reducing greenhouse gas emission and improving energy security" (Sovacool, 2011, p. 3842).

Sovacool (2011, pp. 3840-3841) elaborated polycentric governance under six categories each with its own dimension: (1) "equity or proportional equivalence of costs and benefits," preventing some users from having all/more benefits at little cost (e.g. subsidies or cost sharing); (2) "inclusivity and local involvement, or the active involvement of a diverse number of stakeholders," that usually bring participatory approaches of local actors (households, users, rural areas, private sector) in the planning processes; (3) "information, demonstration, and innovation," that include the distribution of data about benefits, monitoring programs encouraging feedback and knowledge sharing; (4) accountability and ownership, that users accept some of the cost of governance themselves, they are responsible for their own actions (eg. actions or rules); (5) "organizational multiplicity," that enables the involvement of multiple actors and offers parallel systems of governance (beside government actors at all scales, industry, civil society, transnational actors); (6) "experimentation and flexibility," or adaptability. Thus, this would bring several mechanisms in place to cope with the unexpected challenges and to take an open approach to management (Sovacool & Martiskainen, 2020, p. 10).

As a result, based on the provided definitions of governance, MLG, hierarchical governance, and polycentric governance, the off-grid (renewable) energy governance in Namibia has been approached by looking at various levels within the MLG framework. In this sense, the adoption of a bottom-up perspective by framing energy as a social practice at a local level, enables us to understand how social practices develop, change, or intersect to understand energy demand and supply (Shove & Walker, 2014, p. 47). In addition to the different levels of governance, energy practices provide an understanding of energy use and networks in multiple ways, from individuals to communities, from users to suppliers. Thus, it would be possible to contribute to off-grid energy governance and policy strategies from a broader perspective often only addressed as a part of the techno-economical-(political) debates. In doing so, the epistemological ground of phenomenology would open different insights in looking at and seeing different perspectives within this framework.

## 2.2 From the Epistemological Ground of Phenomenology to the "Life-world"

Epistemology plays a critical role in guiding and shaping research and the investigation of research questions and informs knowledge production, interpretation, and dissemination. In this study, interpretive epistemology, with its approach to understanding the lived experiences of individuals and phenomena, was adopted among positivist, post-positivist, and critical epistemologies (Cal & Tehmarn, 2016). Edmund Husserl's (1859-1938) philosophy of phenomenology is a branch of modern philosophy (Relph, 1970) that arose at the beginning of the twentieth century and was characterized by four epistemologies: First, transcendental phenomenology, that of Husserl himself, which was followed by Martin Heidegger's hermeneutic phenomenology, Maurice Merleau-Ponty's perceptual phenomenology, and finally Alfred Schütz's sociological phenomenology (Cal & Tehmarn, 2016). Edmund Husserl's transcendental phenomenology introduced the concepts of *noema* and *noesis* in the quest to understand consciousness, its intentionality, and the relationship between subject and object (Cal & Tehmarn, 2016), with the intentionality of the consciousness projecting light, like a torch (of Husserl) into the object-subject relation. The term "*noema*" refers to the meaning of a mental act, while "*noesis*" refers to the mental process of thinking and pondering through which the noema is experienced (Shahabi & Rassi, 2015). Shahabi & Rassi (2015) state that according to Husserl, *noema* and *noesis* are two poles of any intentional act, which consist of an "I-pole" (*noesis*) and "object-pole" (*noema*) that are always in relation to each other (p. 30).

Following the birth of phenomenology, Alfred Schütz, rejecting parts of Husserl´s transcendental philosophy and its reduction method but maintaining its general form (Costelloe, 1996), proposed the idea of intersubjectivity as a given of everyday life. He explained that the "intersubjective world is one consisting of bunded subjects and that the possible constitution of others is possible" (Schutz 1962, p. 126-127); these ideas were developed for application in the social sciences and went on to influence the field of sociology. Relph (1970) discussed phenomenology and the phenomenological method in relation to geography by conceptualizing it under two categories: (1) the objects of nature, which are utensils for [hu]man exist only through their utility or disutility, and they are discovered only in connection with the practical pursuits of [hu]man;

(2) the anthropocentric stand-point in the understanding of "[hu]man and nature as a single source or system" (p. 196). He stated that human´s relationship to the world is not only cognitive but permeates the [hu]man´s whole being (Relph, 1970, p. 197). The overall human-world system consists of multiple world views that are unified through the common intentions of people; these create synergy, resulting in valuable results in exploring geographical issues (Relph, 1970). Rothfuß and Dörfler (2023) extend the application of phenomenology within human geography by drawing attention to an overlooked aspect of Alfred Schütz´s social theory. Rothfuß & Dörfler (2023) suggest a `*Leib*-based` approach to studying the social and physical world. They proposed a new interpretation focused on spatial and bodily experience to enhance explanations of social phenomena by incorporating a bodily perspective in exploring social and spatial interrelations in human geography (Rothfuß & Dörfler, 2023).

## 2.3 Civilizing Process: Norbert Elias revisited for Human Geography

Norbert Elias (1897-1990), the "last classical sociologist" of the 20th century (van Heerikhuizen, 2015), explored the interconnectedness of the individual and their connections/relations to society in his Magnum *opus*, *Über den Prozess der Zivilisation (1939)*[3]. Elias´ *magnum opus* went through the somewhat lengthy *process of understanding* by scholars in the social sciences over the next thirty-seven years. He worked on the basis that the "individual" and "society" concepts do not consist of two separate independent objects from each other, but rather as two aspects of humans, which are inseparable from each other, and that those aspects commonly can be comprehended only when they are considered within a structural transformation (Elias, 1976/2000).

In *The Civilising Process*, Elias first had a closer look at the concept of "civilization" in France and Germany. He compared the concepts of "culture" and "civilization" in those countries, including his wish to bring a better understanding from a historical perspective among French, German, and British (Elias, 1976/2000). One shall not forget that this magnum opus was written after World War I and right before World War II. He utilized a figurational concept instead of a social system or constellation while avoiding evolutionary determinism (Evers, 2022). Elias explained the concept of figuration with an analogy to dance or dancing, where the dancers constantly change positions and partners, but the type of dance (figuration) remains visibly the same (Evers, 2022). How does this process happen? According to Elias, this occurs "through the civilizing process of increasing control of feelings, sophistication of manners, control of violence and the institutionalization of these refined manners, of inner control, in the form of the state" (Evers, 2022, p. 18).

[3] It was later on published in Germany in 1969 and the latest edition The Civilising Process (1976) in English, after the Second World War.

Evers (2022) aimed to explain the typical figures related to the process of civilization in his analysis. To better understand the fundamental problems, he saw it as essential to lay out how the behaviors and emotions of Western subjects changed after the Middle Ages (Elias, 2000). He claimed it is necessary to confirm the psychological and mental changes observed within Western history only by investigating the materials that belong to these historical experiences. In this regard, he shed light on how human behaviors and emotions developed in a specific direction over centuries, from the Middle Ages until the 19th century. He considered table manners, sleeping or fighting manners from the written materials of Erasmus, Tannhäuser, de la Salle, John Russel, Della Casa, Caxton, de Courtin, and de Callieres, but not limited to time (Elias, 1976/2000). He also kept an eye on the depiction of the paintings as well as the literary materials and clothing. He claimed that the historical experience shows the important role that the change in the sense of shame and embarrassment plays in the process of "civilization." Societal wishes and the standard of forbidden things change, and consequently, the border of fear defined by society changes. Thus, the problematic of the social formation of the fears that belong to the people becomes one of the central problems of the process of civilization.

Debates surrounding civilization and the civilizing process occupied sociologists of this period. According to Elias (cited in Bogner & Mennell, 2022, p. 1), one of the tasks of sociologists was to look at societies over a longer period, accordingly Elias criticized contemporary sociologists for investigating over shorter time frames, which they deemed as present-centered or ahistorical approaches; Elias claimed these were a "retreat into the present." Bogner and Mennell (2022) noted that Elias preferred to use the term "evolution" for biological processes and the term "development" for social processes; because social processes are generally reversible, and biological processes are not (Bogner & Mennell, 2022). Further definitions for the civilizing process focused on long-term historical developments. Alfred Weber, for example, defined the process as "a directed, irreversible process of increasing consciousness of the surrounding world (*Bewusstseinsentwicklung und Durchreflektion des Daseins*)" and that it can be transferred from one ´civilization´ *(Geschichtskörper*) to another without transcending the whole world" (Bogner & Mennell, 2022, p. 17). The earlier opposition between culture and nature, between the cultivated world and the uncultivated, has been replaced by the opposition between system and lifeworld/*Lebenswel*t, between affluence and poverty, and between center and periphery (Bogner & Mennell, 2022, p. 21). Alfred Weber and Norbert Elias indicated that social differentiation accompanies the process of civilization, in which new figurations arise. However, their opposition in the field should also be noted (Bogner & Mennell, 2022): In his later work published in 1995, Elias defined both the process of technization and the civilizing process as "both unplanned long-term processes, moving in a discernible direction – with spurts and counter-spurts to and from– but without any long-term aim. They are unplanned since they arise from the interweaving, the conjunction, co-operation, and confrontation of many planned activities" (Elias, 1995, p. 7) and provided a perspective on the process of technization.

Elias strongly argued that "there is no group of humans on Earth without civilizing processes" (cited Bogner & Mennell, 2022, p. 25). Elias claimed that one must return to the roots to understand the place it has reached so far (cited Bogner & Mennell, 2022, p. 25). The process of "civilization," within which one finds oneself or as part of, was seen as a meta, and how one obtained it has never been questioned before (Elias, 1976/2000). All the things associated with this process, machines, scientific inventions, and state figurations, are all human relations, social formations, and human behaviors (Elias, 1995). According to him, the main issue was whether it was possible to observe behavioral changes and the social process called "civilization" over a specific period as close as possible to reality by a reflecting consciousness (Elias, 1976/2000). Sociological debates surrounding "civilizations," the "civilizing process," and "modernity" also maintained interest; Elias convened a conference in 1984 in Bielefeld with 30 international scholars who were interested in the investigation of long-term processes of social development (Bogner & Mennell, 2022). One could also conclude that Elias´ wish to work on the long-term processes from a figurational approach also included the transferability of this knowledge toward a future level or future generations.

Elias´ figurational concept and view of processes in a continuous and evolving way shaping societies provided a different perspective adopted by future scholars and elaborated further to the modernization process. Thus, inspired by the figurational concept and the characteristics of processes defined by Elias, van der Loo and van Reijen (1992) investigated the process of modernization (PoM) and framed it with four macro processes and their paradoxes, which are explored in this study and will be elaborated in sub-section 2.6. However, the history of modernization and its critiques have been considered before continuing this thorough elaboration of PoM.

## 2.4 The Genesis of "Modern," "Modernity," and the History of "Modernization"

Wootton (2015) stated that the word ´modern´ or *modernus* dates to the sixth century with the dismissal of Rome by the Visigoths and the establishment of a new Christian order. The meaning of "modern" has also changed from century to century, from discipline to discipline. Throughout the millennia there were ancients and moderns, pagans, and Christians. The terms "ancient, middle, and modern times` were used by Filippo Villani (1382). The terms "the Renaissance" and "the Enlightenment" of the nineteenth century have been the basis for the early modern discussions of the last fifty years (Wootton, 2015). Whereas modern had been used by Shakespeare to mean both "ordinary" and "contemporary" without really emphasizing the features of the modern world but acknowledging in a way "the Reformation" (Wootton, 2015). Indeed, "modern" has been discussed for centuries in many disciplines, including arts and science.

In the middle of the twentieth century, amidst the post-war-driven conditions in the 1950s, "modernization" emerged as a Western, European, and North American concept, including the processes of industrialization, democratization, bureaucratization, and secularization in the social sciences (Mergel, 2012). Through this classical approach, the development logic of modern societies has been investigated (Mergel, 2012). Talcott Parsons also joined in the theoretical development of the concept with a Parsonian model of the development of modern societies (Parsons, 1985). The approaches combining the long-term historical dimension of social class formation with economic development and contemporary political cultures tended to consider terms of development stages and assumed the development processes were progressive with some temporary setbacks (Mergel, 2012). According to Mergel (2012), this theory also claimed a superiority of the North American/Western European model of capitalist democracy over the Eastern European and Fascist authoritarianist models, which were regulated and autocratic (Mergel, 2012). There was also an expectation that Third World countries catch up, following the path of "developing countries" and develop along the path of modernity of the European/North American countries, which was adopted by the development agencies (Mergel, 2012; MacNeill, 2020).

Unlike the somewhat optimistic, progressive, functionalist, and future-oriented connotations of modernization, Shmuel Eisenstadt presented modernization as a conflict model in the 1960s that elucidated the de-differentiation, disorganization, and protest and focused on the role played by imperialism (Mergel, 2012). Barrington Moore also discussed the possibility of progress on different paths of modernity related to societies with their specific history, traditions, and how they were processed (Wiener, 1975). The key concepts of modernization theory included "developmental paths" and "path-dependency" (Mergel, 2012). After 1990, with the end of communism, there was an unexpected revitalization of the theory of modernization, which had been previously criticized. As a result, Eastern Europe started proceeding in Western Europe's direction regarding reorganizing politics, the economy, and concepts of civil society (Mergel, 2012). Moreover, different development paths emerged with varying goals in China and Russia, where structural economic transformations with a certain degree of capitalism were not necessarily accompanied by democratization but rather by authoritarian control (Mergel, 2012).

Edward William Soja investigated *The Geography of Modernization in Kenya* (1967) as a spatial phenomenon in the transition from traditional to modern forms after the independence of Kenya. Here, Soja sought to understand the complex interactions between social, economic, and political organization and behavior through the areal variations and existing data at an inter-state level to draw out the patterns of the nation-state Kenya. He claimed that changes occur in space as well as in time, and thus spatial dimensions must be part of the theoretical processes, which occur alongside the temporal dimensions (Soja, 1967, p. 5).

Furthermore, Olúfémi Táíwò asked Africa in *Africa Must Be Modern:* A *Manifesto (2014)* to "embrace modernity" to make Africa better, akin to countries currently regarded as modern (Taiwo, 2014). Taiwo bases the concept of modernity on these three tenets: The principle of subjectivity, the centrality of reason, and the idea of progress (Taiwo, 2014, p. 41; Sillah, 2015). His call rested on the idea that "modernity is life" and is usually taken for granted by those who live it as natural (Taiwo, 2014, p. xii).

> "Africa must be modern. Put simply, Africa must embrace individualism as a principle of social ordering; make reason central in its relation to, activity upon, understanding of, and producing knowledge about the world, both physical and social, that it inhabits; and adopt progress as its motto in all things" (Taiwo, 2014, p. xxv).

Thus, discussions and debates surrounding the concept and terminology of "modern" have progressed within the African context as well and developed across various disciplines and among various scholars until present.

## 2.5 Critiques of Modernization

The terms "modernity" and "modernization" have been criticized historically and theoretically through critics of colonial and dictatorial pasts (Mergel, 2012; Bhambra, 2011). Dipesh Chakrabarty (2000) critiqued the "enlightenment concepts of universal human experience and secular modernity," rejected "the temporal totality of history," opposed to Marx´s history of capital, while in opposition to "two histories of capital concepts: "histories posited by capital" and histories that exist outside of "capital's life process" (Borah, 2015). Thus, opened the gates for further research on "modernization" after 2000 (Mergel, 2012). Furthermore, Chakrabarty also discussed the terms "modern," "modernity," "modernization," and "early modern" with his reflections on the historian´s perspective based on the history of India with comparisons of European, Western, Chinese, African, or South Asian modernities (Chakrabarty, 2011). He defined being modern as "to judge one´s experience of time and space and, thus, create new possibilities for oneself" (Chakrabarty, 2011, p. 674). Thus, Chakrabarty contributed to the reflexivity in the process of modernization in India through a historical lens in the *Muddle of Modernity* (Chakrabarty, 2011).

Eisenstadt and Schluchter (1998, p. 2) considered the reassessment of the convergence-based "older visions of modernity and modernization" of the fifties and sixties, which would, theoretically, eliminate differences in the world and "lead to a uniform modern world." Thus, Shmuel Eisenstadt, who already criticized the theoretical framework of modernization from the outset, introduced the concept of "multiple modernities," which countered the general discourse of the classical theories on modernization (Eisenstadt, 2000), arguing that European modernity was merely one model of many (Mergel, 2012). Eisenstadt explicitly states that modernity and Westernization are not identical; however, they shared a primary historical reference point: "Western patterns of modernity are not the only ′authentic′ modernity" (Eisenstadt, 2000). From his viewpoint, multiple modernities enabled one to understand the contemporary world by "seeing it as a story of continual constitution and reconstitution of cultural programs" (Eisenstadt, 2000, p. 2). According to this description, social actors from different social, political, and intellectual backgrounds carried multiple institutional and ideological patterns; they constituted different perspectives on the modernity of society(ies) (Eisenstadt, 2000).

Furthermore, the degree of reflexivity of modernity emphasized different interpretations of core transcendental visions and the related institutional patterns while increasing the awareness of multiple visions (Eisenstadt, 2000, p. 4). The autonomy of humans played an essential role within this cultural program of modernity, creating emancipation from cultural, traditional, and political authorities (Eisenstadt, 2000, p. 5). When the legitimacy of multiple individuals, the goals of a group, and interests were recognized, various interpretations of the common good were enabled (Eisenstadt, 2000). Modernity also made way for the construction of collective identities and their boundaries, and social movements of protest transformed modern settings (Eisenstadt, 2000).

Another theoretical advance followed by Ulrich Beck, Anthony Giddens, and Scott Cash that considered reflexivity as a characteristic of modern society and people posited the concept of cognitively and politically connected social change (Chang, 2017). Chang (2017) suggests that modern individuals or groups cannot be reflexive self-consciously; instead, they are forced to choose among the multiplicity of options in the name of rationality. Reflexive modernity is termed the "second modernity" by Ulrich Beck (it follows the first modernity that appeared due to the linear industrial phase). This, in turn, reflected a transformation from an industrial to a risk society (Sørensen, 2019).

Post-colonial critiques began with the psychoanalyst Frantz Fanon's (1925–1961) *The Wretched of the Earth* (1968/2002b) and *Black Skin, White Masks* (1986/2002a); in both works, he challenged Europeans to reflect on the historical experiences of the colonies (Peet & Hartwick, 2009). Jinadu (1978, p. 121) further suggested that African political thinkers like Fanon, Nyerere, Cabral, and Nkrumah were concerned with the modernization process regarding their societies' ethical issues and ordering. Moreover, he stated that "every society, whether industrial, non-industrial or post-precolonial, colonial, neo-colonial, or post-colonial, is a modernizing society" (Jinadu, 1978, p. 123). Ranajit Guha and Gayatri Spivak (1988) also extended post-colonial critiques to the "subaltern" groups, those who are subordinate in terms of class, caste, race, gender, and culture (cited in Peet & Hartwick, 2009, p. 210). Edward Said dramatically influenced critiques in post-colonial thought by introducing the concept of "Orientalism," which originated from European culture in the post-Enlightenment period as a complex system of "discourse supported by various institutions, vocabulary, scholarship, imagery, doctrines, and even colonial bureaucracies and styles" (Peet & Hartwick, 2009, p. 211).

These scholars, among others, contributed to a critical re-evaluation of modernization theory from decolonial and post-colonial perspectives by highlighting its limitations and proposing alternative frameworks for understanding social change and development in a global context. In contrast, van der Loo and van Reijen (1992) interpreted the PoM from a universal perspective by embracing four macro processes and their simultaneous paradoxes. Thus, their holistic understanding provided a macro framework with various intersections and intertwinedness that incorporated an alternative approach to multi- and interdisciplinary issues.

## 2.6 *Ex Post*: Process of Modernization by van der Loo and van Reijen

The PoM defined by van der Loo and van Reijen (1992) defines modernization using a macro-sociological perspective. Their definition is based on the four macro processes of differentiation, rationalization, domestication, and individualization, each of which includes their own paradoxical characteristics, which might also be observed as a figurational concept. Van der Loo and van Reijen (1992) define modernization as:

> "A complex of interrelated structural, cultural, psychological, and physical changes that have crystallized over the past several centuries, shaping and still shaping the world in which we currently live" (p. 11).

Van der Loo and van Reijen's (1992) work explored the long-term developments that introduced changes and transformations in social realities. It follows that these are reflected in the works of sociologists, such as Durkheim, Simmel, and Weber (van der Loo & van Reijen, 1992): Émile Durkheim's work explored the increasing division of labor; Georg Simmel looked at the rising individualization above everything else; while Max Weber emphasized the growing rationalization (van der Loo & van Reijen, 1992, p. 18). The modernization discussed by these classical sociologists was not only a historical process, but also a process that exists in the world where it is also nonlinear and practically irreversible (van der Loo & van Reijen, 1992, p. 18).

Van der Loo & van Reijen (1992, p. 18) criticized classical thinkers´ one-sided "evolutionist" standpoint claiming that they do not consider the different results of the process of modernization across different cultural and temporal settings enough and they also underestimate the modern and traditional mixtures of social realities. Thus, the objections to modernization thought have been gathered in general under the evolution thoughts, total development, characteristics with harmony, monocausal explanations, structure, and actors (van der Loo & van Reijen, 1992, p. 18). Their evolution-based concept had two general assumptions: the social reality consisted of phenomena that could be ranked from lower to higher forms of life and that this process that takes over time in which lower forms evolve into higher ones (van der Loo & van Reijen, 1992, pp. 18-19).

However, modernization is not a linear evolution, and it differentiates from country to country and sector to sector (van der Loo & van Reijen, 1992, p. 19). Additionally, due to critiques from anthropologists, the idea of multilinear development in response to replacing the image of nonlinear evolution has been attempted (van der Loo & van Reijen, 1992, p. 20). Different types of modernization were possible, depending on the time and the country where they occurred (van der Loo & van Reijen, 1992, p. 20). Furthermore, van der Loo and van Reijen (1992, p. 23) agreed on modernization's continuous and uniform development, but disagree with the modernization process's harmonious course. From this point onward, they acknowledged the contradictory characteristics of modernization and term them "paradoxes." They considered them instead in a dialectic, two sides of the same coin, rather than oppositional characters (van der Loo & van Reijen, 1992, p. 23).

Van der Loo & van Reijen's standpoint on the relationship between the material "substructure" and the cultural "superstructure" shall not be considered monocausal but rather multicausal, an explanation investigating the interconnectedness of various modernization processes (van der Loo & van Reijen, 1992, p. 24). They argued that monodisciplinary approaches, in the search for monocausal explanations, have not yielded fruitful results: the tendency for economists was to view the economic factors; the demographers to focus on the rapid population growth; the political scientists to investigate power and authority structures; and the psychologists to look at the psychological changes at the cradle of modernization (van der Loo & van Reijen, 1992, p. 25). Therefore, the modernizing society could not be diminished into a single denominator; it would be more plausible to have a multidisciplinary and interdisciplinary approach and to pursue the events that have stimulated or hindered each other in various subprocesses (van der Loo & van Reijen, 1992, p. 25). Hence, it became necessary to investigate concrete phenomena without relying only on abstract impressions and observations to discern different processes that construct the swirling stream of modernization (van der Loo & van Reijen, 1992, p. 25).

Further, van der Loo & van Reijen (1992) observed from the perspective of the sociology of knowledge that the phenomenon of modernization itself was also a product of modernization, in which the network of human activities/agency became more and more complicated and extended, where more groups of people became dependent on each other to meet each other's needs (van der Loo & van Reijen, 1992, p. 25). They considered human beings as culture-producing beings; who could defend themselves against existing institutions and structures, and after a particular time, they could even replace them (van der Loo & van Reijen, 1992, p. 26). In this way, people may maintain and change structures through their actions; on the contrary, structures make actions possible (van der Loo & van Reijen, 1992, p. 26).

In conclusion, van der Loo and van Reijen (1992, p. 27) reflected the function of various theories on a different level, providing a much-necessary orientation in a very complex reality. They agreed that simplifying the facts is necessary for clarity and usability, enabling them to discuss them and use them as instruments for specific use (van der Loo & van Reijen, 1992, p. 28). Thus, they started working with an already existing scheme, initially designed by Parsons (1975, 1985) and later generalized by Adriaansens (1980), which collected the social reality and the human action from four perspectives or dimensions: reciprocal combinations of structure, culture, person, and nature (van der Loo & van Reijen, 1992, p. 29). Based on this general scheme, they interpreted modernization as the processes of differentiation, rationalization, domestication, individualization, and all possible combinations (Figure 2.2) as well as the associated paradoxes – these form the swirling continuum of modernization (van der Loo & van Reijen, 1992, p.30).

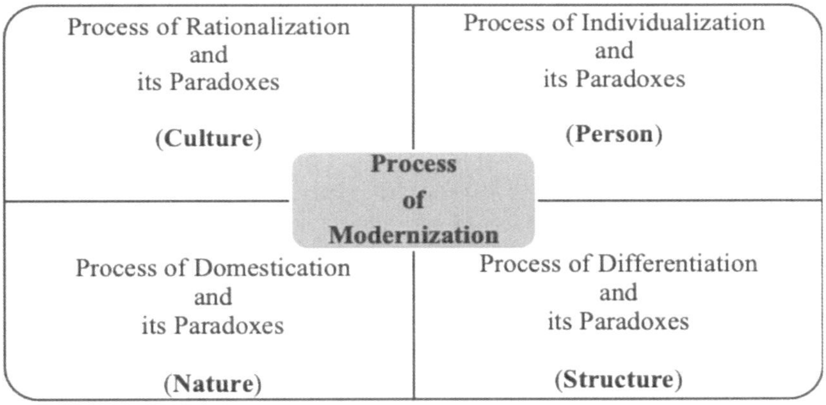


*Figure 2.2. Four Perspectives of Modernization*
*Source: Author, (based on the PoM by van der Loo & van Reijen, 1992)*

### 2.6.1 Process of Differentiation

According to van der Loo and van Reijen (1992, p. 31), differentiation indicates "the splitting of an originally homogenous whole into parts, each with their character and composition," where each becomes more specialized in certain functions. From a structural perspective, people fulfill certain social roles (i.e., father, mother, teacher, or student) within their social reality, and they constantly interact with others, from which certain patterns emerge (van der Loo & van Reijen, 1992, p. 31). Through the differentiation process, the actions and interactions of people's daily lives become independent; new organizations and institutions form from their orientations (van der Loo & van Reijen, 1992, p. 31).

Van der Loo and van Reijen (1992) investigated the differentiation from three aspects: division of labor and solidarity, social stratification, and self-control/mastery. They claimed that modern relationships are distant, complex, and changeable and that modern society is formed from individuals and specialized entities dedicated to subtasks while maintaining an interconnection to each other (van der Loo & van Reijen, 1992, p. 83). "Solidarity" refers to the connection and interconnectedness of individuals that constitute a group or a society together and makes it solid, whereby in the absence of this bond, a society cannot exist (van der Loo & van Reijen, 1992, p. 83). The starting point was the (inter-)dependency of people/human beings (van der Loo & van Reijen, 1992, p. 84). In contrast with the liberal intellectuals of the era who viewed society as the result of social contracts, Durkheim claimed that human relationships were set in a series of existing values, norms, and rules of the game and that human beings were first and foremost social beings (van der Loo & van Reijen, 1992, p. 84). When they are part of a shared social group, living together with others, one is included in a supra-individual community and forms a collective consciousness that existed before the individual and will exist after the extinction of the individual from Earth (van der Loo & van Reijen, 1992, p. 84). Durkheim suggested that the social relationships and feelings of belonging that shaped the traditional society changed during the modernization process, and new social bonds and a new type of solidarity appeared, characterizing traditional and modern societies, which will contribute to mechanical and organic solidarity (van der Loo & van Reijen, 1992, p. 85).

Furthermore, traditional society was characterized by a low division of labor and a relatively simple social structure, where people were naturally and mechanically bonded to each other, leaving no room for individuality (van der Loo & van Reijen, 1992, p. 85). Within this tightly connected network of relationships and obligations, each individual knew their place. In this sense, Durkheim referred to mechanical solidarity, suggesting that people's relationships with one another occurred mechanically or automatically, whereby the individual and collective consciousness coincided (van der Loo & van Reijen, 1992, p. 85). With the increasing population and increasing contact with others, the closedness of traditional society came under pressure, and another type of economic production emerged based on specialization and sharing tasks (van der Loo & van Reijen, 1992). Thus, people started distinguishing themselves more from one another while becoming increasingly dependent on one another.

However, as indicated by Durkheim, the division of labor is far more advanced in modern society (van der Loo & van Reijen, 1992). Overall, that was characterized by an extensive division of labor; each had its own place in a highly complicated system, whereby all parts would need to cooperate (van der Loo & van Reijen, 1992, p. 86). For such a case, van der Loo and van Reijen (1992, p. 87) indicate that Durkheim suggested "organic solidarity," referring to the functioning of a biological system. The organs constituting this organism were more or less autonomous in their internal functioning, but they could only continue existing if every individual organ was coordinated and worked together (van der Loo & van Reijen, 1992, p. 87).

Similarly, the successful coordination and adaptation of these parts, such as the division of labor, would form modern solidarity (van der Loo & van Reijen, 1992, p. 87). However, modernization, within this context, did not lead much to a reduction in dependence but rather a change/shift in the form of dependence. In such a society, more and more people became less and less dependent (van der Loo & van Reijen, 1992, p. 87). Hence, the individual and collective got on the same track, although individuals became increasingly occupied with their subtasks (van der Loo & van Reijen, 1992, p. 87). Individual responsibility and social solidarity became two sides of the same coin (van der Loo & van Reijen, 1992, p. 87). Nevertheless, Durkheim also considered that one day, both sides could develop very far apart (van der Loo & van Reijen, 1992, p. 87).

The term stratification refers to social inequality as "strata," in which the differentiation originating nearly automatically from the division of labor leads to hierarchical positions (van der Loo & van Reijen, 1992, p. 97). As mentioned earlier, people were interdependent and worked together. However, this rarely happened on an "equal" basis (van der Loo & van Reijen, 1992, p. 97). Thus, cooperation among people did not lead to the equal distribution of goods, prestige, and power, which created social inequalities between women and men and the young and old (van der Loo & van Reijen, 1992). Social stratification in modern society was a fairly open system that allowed people to move up or down the "social ladder" (van der Loo & van Reijen, 1992, p. 97). On the other hand, the anthropologist Dumont differentiated the traditional with *Homo hierarchicus* (1967) and the modern humans with *Homo aequlis* (1977) (van der Loo & van Reijen, 1992, p. 97). In return, modern aspirations tended to play a more significant role in social equality while maintaining existing inequalities (p. 98). Van der Loo and van Reijen (1992, p. 110) claimed that a rising dichotomy of society did not accompany the increasing differentiation as Marx presupposed but instead by increasing fragmentation and individualization of social differences.

Social differentiation was related to social changes and psychological changes (van der Loo & van Reijen, 1992, p. 110). According to Elias, progressive differentiation leads to both an adaptation of social relations and a more comprehensive and differentiated complex of behavioral standards (cited, van der Loo & van Reijen, 1992, p. 111). As a result of this progressive differentiation, the more complicated the social structure becomes, the better people keep their emotions under control, refraining from their desires, and planning for the long term (van der Loo & van Reijen, 1992, p. 111). This observation was based on the idea that progressive differentiation was accompanied by a growing awareness of mutual dependencies and the necessity of better coordinating behavior (van der Loo & van Reijen, 1992). Furthermore, the coercion for self-control was initially exerted on each other by the people who were themselves concerned (van der Loo & van Reijen, 1992, p. 111). At another stage, this external coercion became self-coercion. Through upbringing, one learns how to keep oneself more and more under control, behavioral contrasts disappear, and one starts thinking more long-term about one´s intentions and those attributed to others (van der Loo & van Reijen, 1992, p. 112). By growing differentiation, there was a growing dependency (van der Loo & van Reijen, 1992). Thus, there was a higher demand for controlling emotions, which was essential for the individual to emerge as a personality in social life. As a result, the more an individual became part of the dependency networks, the more one had to control one´s emotions and affections (van der Loo & van Reijen, 1992).

#### *Paradoxes of Differentiation*

Differentiation is also accompanied by its paradoxes within the process of modernization. From the point of differentiation described by van der Loo and Van Reijen (1992, p. 115), "modernization is a development in which society becomes organizationally complex without falling into chaos," and solidarity was adapted to this complexity. Through the progressive division of labor, the differences among them gave rise to new forms of cooperation, groups, or associations while not limiting one´s group but rather extending far beyond it (van der Loo & van Reijen, 1992). This process diminished the differences between members of different groups. However, "if the differences within the groups increase and those between the groups decrease, the boundaries of the original group become blurred, and new social groups emerge, which may go beyond the group's boundaries" (van der Loo & van Reijen, 1992, p. 116). In this sense, the differentiation process enabled specialization, scale reduction, and scale enlargement.

Van der Loo and van Reijen (1992) have indicated that, according to Elias, the processes of scale reduction are continuous. There is progressive individualization from one side, where "I-identity" rises in importance versus "we-identity." Additionally, the scale of enlargement continues to develop among national, international, and worldwide organizations.

### 2.6.2 Process of Rationalization

Van der Loo and van Reijen (1992, p. 31) also discussed modernization from the cultural perspective of differentiation, as found in rationalization. They consider the lived reality as cultural reality, including the perception, values, norms, meanings, ideas, and symbols that give meaning to our actions and have a role in our culture. They interpreted rationalization as "the ordering and systemizing of reality to make it calculable and controllable, which happens on different and interrelated levels" (van der Loo & van Reijen, 1992, p. 31). They specifically discussed the rationalization process from the worldview level, collective action, and individual action.

Van der Loo and van Reijen (1992, p. 119) considered the worldview focusing on Weber´s “disenchantment of the world” at one level, which referred to the disappearance of the magical, enchanted life toward a rational, calculable, and controllable future. As Weber referred to fate's domination and unanticipated moments were beyond humans' power in pre-modern times (van der Loo & van Reijen, 1992). Therefore, this pre-modern world was depicted as a magical garden with pleasant surprises. They argue that over the centuries, through modernization, the idea of the mysterious forces of nature faded away and left its place to a controllable life by calculation and rational action (van der Loo & van Reijen, 1992, p. 118). At the collective action level, this rationalization presented an ability to target specific goals systematically using appropriate means of precise calculations (van der Loo & van Reijen, 1992). When rationalization arrived at a level of individual action, a tendency toward the reign of reason appeared in all aspects of life (van der Loo & van Reijen, 1992). As a result, from a cultural perspective, people practiced more and more control over each other, themselves, and their environment (van der Loo & van Reijen, 1992, p. 119).

By exploring rationalization through cultural lenses, they investigated the worldview that characterizes every society, which gives meaning to life and makes understandable the natural and social order around us (van der Loo & van Reijen, 1992). They highlighted that this worldview, passed from generation to generation, also meant something about how the members of these groups (must) think about reality (van der Loo & van Reijen, 1992). In traditional societies, worldviews that were strongly shaped by mythical or religious influences had an underlying characteristic of the mythical world whereby “life is governed by unpredictable and uncontrollable forces and powers” (van der Loo & van Reijen, 1992). Moreover, this worldview made a sharp differentiation between humans and the world, subject and object. However, it considered humans a part of nature, a form of consciousness still independent of nature (van der Loo & van Reijen, 1992).

During the Renaissance, more attention also shifted, giving greater significance to individual experience, and empiricism, obtained through our senses as observation, measurement, and experimentation (van der Loo & van Reijen, 1992, p. 121). This movement turned society into a knowledge society, where people relied on scientific knowledge to forecast and control developments (van der Loo & van Reijen, 1992, p. 121). Governments, political parties, and companies invested according to the scientific results. In this way, the modern-scientific worldview created a link between knowledge and feasibility (van der Loo & van Reijen, 1992). The modern worldview embarked on the perceivable world instead of the contemplative world, and rationalization gradually enabled the liberalization of the diverse scientific fields from the control of religion and myth (van der Loo & van Reijen, 1992). Consequently, the detachment of human activity from the mysterious, unpredictable, and mythical increased the assessment according to the criteria of expediency and usefulness (van der Loo & van Reijen, 1992).

Some Enlightenment thinkers such as Voltaire, D'Alembert, Diderot, and Montesquieu (Bristow, 2023) presented rationality itself as an article of faith and supported the idea that it was also possible to improve the life of humankind and promote general happiness based on scientific discoveries; this led to the emergence of a social elite composed of scientists and technicians (Bristow, 2023; van der Loo & van Reijen, 1992). According to van der Loo and van Reijen (1992), Weber considered this a rationality of purpose, referring to the ability to search for concrete goals most effectively and efficiently without being connected to moral or religious considerations. In this way, technique/technology could be interpreted as the embodiment of the rationality of purpose, providing autonomy and modernization expansion (van der Loo & van Reijen, 1992). Once more, Weber pointed out the close connection between differentiation and rationality of purpose, which was creating the development of various sectors such as science, arts, or politics, looking for ways to achieve their goals in their way, without relying on moral or religious values (van der Loo & van Reijen, 1992). Weber saw the world as the culmination of rationality and irrationality, where "the unimagined ability to proceed purposefully and efficiently is matched by the inability to critically examine the goals themselves against certain fundamental and overarching values" (van der Loo & van Reijen, 1992, p. 129). As a result, moral problems tended to be reduced to a simple question of effectiveness, while the meaning of "good" turned out to be the one that "worked out or functioned" (van der Loo & van Reijen, 1992).

In this process, the organization was the most visible sign in collective actions because organizations were deliberately planned and not grown (van der Loo & van Reijen, 1992, p. 135). It meant that people equipped with adequate means of action consciously created a network to achieve specific goals, whereby each individual's contribution to the whole group became controllable and calculable (van der Loo & van Reijen, 1992). New organizations appeared in different sectors of society for specific purposes and to meet diverse social functions, such as schools, hospitals, administrations, factories, prisons, institutions, etc. (van der Loo & van Reijen, 1992); the bureaucracy within the modern organizational concept was the main line of rationalizing collective action. This concept consisted of three main characteristics: functions, interconnected hierarchical structures, and the impersonality of bureaucratic organizations (van der Loo & van Reijen, 1992, p. 137). The core was subdivided into individual areas, essential functions, and everything else shaped around them (van der Loo & van Reijen, 1992). Officials carried out tasks and supervised within the procedural written rules and levels of the hierarchy without allowing personal tendencies or interests (van der Loo & van Reijen, 1992). Thus, people happened to be "numbers" to be dealt with or "cases" "to be settled" (van der Loo & van Reijen, 1992). This process also led to the formation of a risk-averse person, the modern bureaucrat or "employee," as a new type of personality in the institutionalized settings also concerned with their career, climbing the stairs was no longer merely working in an organization but also belonging to it (van der Loo & van Reijen, 1992).

Van der Loo and van Reijen (1992) argue that Weber saw the rationalization of collective actions, namely the emergence of modern bureaucracy, as an irreversible process. This impacted a planned lifestyle independent from other social processes (van der Loo & van Reijen, 1992). Consequently, the modern individual would be confronted with bureaucratic mechanisms at work, during leisure activities, or other private activities in family life. Moreover, modern individuals would also become omnipotent planners. Thus, life planners would order and systemize their existence to integrate their individual actions by where to spend time, money, and energy to make sense of life (van der Loo & van Reijen, 1992). This would eventually have a role in modern consumerism (van der Loo & van Reijen, 1992).

***Paradoxes of Rationalization***

From the cultural perspective, modernization appeared to be a process of progressive self-control/mastery on the levels of the worldview, collective action, and individual action by the increasing ordering and systematization of natural and social reality to make it predictable and calculable (van der Loo & van Reijen, 1992, p. 157). The opposing developments resulted from this process in the "narrowing of vision" and the "widening of vision" from pluralization to generalization. While different, independent entities, organizations, and institutions achieved their own goals, they also derived their legitimacy from specific values and norms with a function-bound character (van der Loo & van Reijen, 1992). In contrast, all sorts of previous separate cultural systems became attached from the original foundations and blended, which brought he generalization of cultural systems a greater overarching power (van der Loo & van Reijen, 1992). These processes of pluralization and generalization did not oppose each other, but rather, they were two sides of the same coin (van der Loo & van Reijen, 1992). Modern people did not have the same counts for individuals; more options/aspects of personality were subsumed under one idea of a consistent life (but now `different´ or `diverse`) (van der Loo & van Reijen, 1992).

### 2.6.3 Process of Individualization

Considering the changes that took place during the processes of differentiation and rationalization the "self-experience" of the individual, as well as the individual "identity," were also affected (van der Loo & van Reijen, 1992, p. 159). According to van der Loo and van Reijen (1992, p. 32), individualization refers to "the growing importance of the individual who broke away from the collectivity of his or her immediate environment." They claimed that this happened because people had to deal with several social units scattered around, whereby the loyalty of the individual that made up each unit was reduced (van der Loo & van Reijen, 1992, p.32). As a result of belonging to different units, the individual obtained certain personal independence concerning each of these units (van der Loo & van Reijen, 1992, p. 32). Scholars of different backgrounds from all over the world noted that "more and more people preferred self-development, autonomy, and privacy, while the docile life based on traditions was becoming increasingly rare" (van der Loo & van Reijen, 1992, p. 160). Moreover, people endeavored to achieve something in life, to decide for themselves, develop personally, live an exciting life, and enjoy life as much as possible (van der Loo & van Reijen, 1992). These terminologies also brought negative and positive assessments of individualization (van der Loo & van Reijen, 1992).

From a more pessimistic perspective, some authors claimed that increasing freedom and individualization would cause "hyperindividualism," egoism, and hedonism; others viewed it as a transition from self-denial to self-development, enabling modern individuals to make their own decisions and live a rational and independent life (van der Loo & van Reijen, 1992). As van der Loo and van Reijen (1992) suggested, these different assessments view individualization as a detachment from traditional social ties, being the first side of the coin. However, the other side of the coin demonstrated the "decision to create new forms of social relationships" (van der Loo & van Reijen, 1992). Along the modernization process, personal social ties became external or anonymous, leading to a process of individualization whereby the "dependence of the individual shifted from the immediate environment to remote networks" (van der Loo & van Reijen, 1992). Individuals in traditional societies were indissolubly linked to the collectivity they were born into. In contrast, people in modern societies experience themselves as individuals whose lives are determined independently of others (van der Loo & van Reijen, 1992). Within the social sciences, individualization deals with people socially liberating themselves from traditional social ties and, on a cognitive level, withdrawing themselves from traditional beliefs and certainties (van der Loo & van Reijen, 1992, p. 162). With this, the world is seen as abundant with unlimited possibilities from which people can choose for themselves (van der Loo & van Reijen, 1992, p. 162).

Until the Renaissance, individuals were seen as part of a collective, and collective identity was more important than individual identity (van der Loo & van Reijen, 1992). The individual was born and raised in an environment where community precedes personal choice (van der Loo & van Reijen, 1992). Although free individuals existed during these times, they were seen from the traditional view as outliers who had placed themselves outside the social order (van der Loo & van Reijen, 1992). Among middle-class individuals during the Reformation, the idea of a "social contract" appeared as individuals joined together in the collectivities of their own free will rather than bound together for reasons of tradition (van der Loo & van Reijen, 1992, p. 163). However, with the Enlightenment, new valuation of the individual, individualism, developed within the middle class (van der Loo & van Reijen, 1992, p. 165). According to van der Loo and van Reijen (1992, p. 164), individualism considers every human being a personality. Thus, individualism started with self-confident and self-reliant individuals who took their fate into their own hands (van der Loo & van Reijen, 1992, p. 164). Furthermore, Romanticism also prompted individualism to take on a new dimension (van der Loo & van Reijen, 1992); the romantics made a short process of progressive optimism of enlightenment thinking and middle-class utilitarianism, where they described modern society as inhuman, alienating, and repressive and considered individual freedom an illusion (van der Loo & van Reijen, 1992, p. 165).

Georg Simmel noted that Romanticism contrasted the quantitative individualism of the $18^{th}$ Century with the qualitative individualism (van der Loo & van Reijen, 1992, p. 166). Quantitative individualism assumes equality of people, whereas qualitative individualism assumes each person is unique and thus people are not equal (van der Loo & van Reijen, 1992). Modern society addressed this, among other things, in the equal treatment of its citizens by the authorities (van der Loo & van Reijen, 1992, p. 166). Thus, the interpretation and the evaluation of the increasing individualization within the social sciences had two different approaches: one assumed the increasing importance of underlying social structures, and the other came to the opposite conclusion (van der Loo & van Reijen, 1992, p. 167). As a result, although modern people appeared to be independent, human freedom of movement was restricted by various bureaucratic structures (van der Loo & van Reijen, 1992, p. 167).

In contrast to the skeptical views, the optimistic view of modern individuals places them in the category of learning to determine their own destiny and responsibility (van der Loo & van Reijen, 1992, p. 171). In the pursuit of shaping their identities, modern individuals stood up for their interests more strongly than those of previous eras (van der Loo & van Reijen, 1992, p. 171). The advancement of individualization also improved the possibility of choice, while the question of "whether people want to choose and can choose" remained in discussion (van der Loo & van Reijen, 1992, p. 171). According to van der Loo and van Reijen (1992, pp. 171-172), Erich Fromm was the first in this area who wanted to give the psychoanalytic theory a "sociological basis," claiming that "it is not the biological conditions that determine individual and social processes, but the social has primacy."

In addition, van der Loo and van Reijen (1992, p. 185) suggested that the "self-denial" of the past could be replaced by "self-development" when it comes to individualization; people could decide more and more for themselves, without being forced to strictly follow social values and norms. Among different perspectives on self-development, the discovery of one´s own "inner world" and the development of one´s personality have been linked (van der Loo & van Reijen, 1992, p. 186). The personal growth aspect was mainly focused on humanistic psychology, with Carl R. Rogers and Abraham Maslow introducing the concept of a growing and self-actualizing individual (van der Loo & van Reijen,

1992, p. 186). Within the humanistic conceptual framework, individuals live in a process of permanent change and growth, in which the human "self" develops gradually as a result of a series of experiences (bodily sensations, learning experiences, inner reflection) (van der Loo & van Reijen, 1992, p. 187). According to Rogers, the most crucial driving force in life was the quest for self-actualization: realizing one´s possibility (van der Loo & van Reijen, 1992, p. 187). Rogers saw this process as very similar to that of the genetic program of a seed which grows into a tree; accordingly, the potential of human beings was also innate in the human being. (van der Loo & van Reijen, 1992, p. 187).

Maslow (1980) explored a model that reflected human needs on a hierarchical level, which assumed that individuals focused their energy on spiritual development when the needs of a lower order – needs for security, food, and social respect – were satisfied (van der Loo & van Reijen, 1992, p. 188). The first level was related to the individual's physical and species related needs for biological survival, such as food, oxygen, and sexuality. The second level deals with the safety needs on a physical/bodily (to have a roof over one's head) and psychological basis (the feeling of stability and order) (van der Loo & van Reijen, 1992, p. 188). On a higher level, the need for social belonging and love represents a demand for warmth and affection. According to Maslow, feelings of alienation and isolation arise if these needs are left unsatisfied (van der Loo & van Reijen, 1992, p. 189). The fourth level consists of self-esteem and admiration or esteem from others; the self-esteem manifests itself in feelings of self-confidence and competence, whereas from others, it is expressed as recognition, prestige, power, and status (van der Loo & van Reijen, 1992, p. 189). At the top of the pyramid lies the need for self-actualization, where, according to Maslow, the satisfaction of this need is equivalent to "being fully human," and these "healthy" or "complete" individuals (only a small fraction of society) aim above all to develop in spiritual terms (van der Loo & van Reijen, 1992, p. 189).

### *Paradoxes of Individualization*

Van der Loo and van Reijen (1992, p. 192) also commented on the challenge of pluralism faced by modern culture; in their view, modern culture resembles a patchwork carpet composed of many meanings, norms, and values, some complementing, and some contradicting, each other. Furthermore, in modern cultures, individuals are confronted with divergent values, norms, meanings, lifestyles, and choices (van der Loo & van Reijen, 1992, p. 192). They also saw the relationship between the individual and the collective ambiguities. In traditional societies, the individual is subordinate to the collective, whereas in a modern society, the individual comes first (van der Loo & van Reijen, 1992, p. 194). The individual's dependency on the collective also changed through the course of modernization: previously, the dependency of the individual was on social units into which they were born and raised, whereas modern individuals depended on others in a more abstract, indirect, and non-transparent way (van der

Loo & van Reijen, 1992, p. 194). This abstractness and the pluralistic situation caused modern individuals to feel increasingly autonomous in their relationships with others (van der Loo & van Reijen, 1992, p. 194). When van der Loo and van Reijen (1992, p. 194) assessed optimistic and pessimistic views on individualization, they noted the paradox of individualization, where either view focused unilaterally on only one part of modern development and neglected the other. As a result, they claimed that the construction of an individual identity runs parallel with the construction of ever more comprehensive collectives (van der Loo & van Reijen, 1992).

### 2.6.4 Process of Domestication

Domestication, as defined by van der Loo and van Reijen (1992, p. 196), is "the taming of the natural and biological forces and, thus, the optimization of the possibilities, which nature and body offer." In pre-modern society, human beings depended on nature; with some technical support, they could adapt to the natural environment (van der Loo & van Reijen, 1992, p. 196). Throughout history, humans have used technical means to escape the limitations imposed by their bodies and the physical environment (van der Loo & van Reijen, 1992, p. 196). In modern society, due to progressive domestication, the relationship between humans and nature has changed radically; people began living with less connection to the natural environment and more in a created or fabricated places (van der Loo & van Reijen, 1992, p. 196). Thus, people became less dependent on nature's fluctuations by manufacturing artificial products to meet their primary life needs (van der Loo & van Reijen, 1992).

Concurrently, average life expectancy drastically increased due to improved techniques/technologies in hygiene, nutrition, sewage, water purification, and medicine (van der Loo & van Reijen, 1992, p. 196). Indeed, this did not bring any liberation from nature to people, but it became somewhat less directly perceptible due to the disappearance of direct contact (van der Loo & van Reijen, 1992, p. 196). As a result, becoming more independent from natural and physical forces also increased the dependence on technology, other people, and control over one´s behavior (van der Loo & van Reijen, 1992, p. 197). Some changes also occurred in human relationships and how humans controlled their drives and emotions, leading to changes in social networks and new thought patterns. (van der Loo & van Reijen, 1992, p. 197). Thus, technology not only covered the human deficit, but the human itself as a social and spiritual being also changed with the adoption of modern technology (van der Loo & van Reijen, 1992, p. 198). Productive, dynamic, effectively functional human beings were gradually needed to operate the technical equipment that functions precisely and intensely (van der Loo & van Reijen, 1992, p. 198).

According to van der Loo and van Reijen (1992, p. 200), a prerequisite for extensive domestication was a change in humans' rational and distant attitude towards nature, too. In a traditional society, the sense of connection and kinship with the living and inanimate would determine the attitude (van der Loo & van Reijen, 1992, p. 200). There was a feeling of connectedness where people depended on the developments and events in nature, in the sacred order (van der Loo & van Reijen, 1992). In this traditional view, human beings were confronted with nature but experienced it as if embedded in it; they were allowed to use natural resources under the supervision of Gods and spirits (van der Loo & van Reijen, 1992, p. 201). Society, nature, and the sacred were united (van der Loo & van Reijen, 1992, p. 201). Through time, the dependence on nature became more abstract and indirect, and the intensity of production started exceeding nature's capacity (van der Loo & van Reijen, 1992, p. 205). A new awareness had to be developed in the further modernization steps to reveal the interdependencies between humans and nature (van der Loo & van Reijen, 1992, p. 205). Thus, nature was perceived as a highly complex system and not regarded as a reservoir of exploitable resources (van der Loo & van Reijen, 1992, p. 205).

With the development of science and technique towards the end of the 18$^{th}$ century and the beginning of the 19$^{th}$, both merged into "technology" and had a significant influence on people and society in the taming of the forces of nature (van der Loo & van Reijen, 1992, p. 205). To some extent, some observers even call modern society "technological society" (van der Loo & van Reijen, 1992, p. 205). Van der Loo and van Reijen (1992, p. 205) suggested that most social scientists shared the opinion that "technology" was not only about tools or instruments but also about the social networks that formed around them as well as the mentality that was required to operate these tools/instruments. Furthermore, there were two opposing views in assessing technological developments: Technology as a blessing and technology as a curse (van der Loo & van Reijen, 1992, p. 205).

The optimistic viewpoint, which considers technology a blessing, assumes that it has liberated people from the daily survival struggles, thus freeing them to focus on satisfying higher needs and pursuing higher goals without being at the mercy of the whims of nature (van der Loo & van Reijen, 1992, p. 205). Modern culture saw technology as a cure-all for problems such as economic stagnation, environmental issues, physical disorders, or defense matters (van der Loo & van Reijen, 1992, p. 206). Contrarily, the pessimistic approach considered advancing technological development as a curse, this viewpoint focuses on the increasing limitations imposed on people in a society dominated by technology (van der Loo & van Reijen, 1992, p. 206). Modern human beings were increasingly subject to the "technological regime," (van der Loo & van Reijen, 1992, p. 206) and the complex values, norms, rules, and procedures necessary for the optimal functioning of the technological tools of control. At this point, modern human beings were at the mercy of technological dogma instead of the constraints of political and religious dogma (van der Loo & van Reijen, 1992, p. 206), and the ability of what a person could and could not do remained to be determined by technology in modern society. However, each of these views only took one side of the paradox of domestication into account (van der Loo &

van Reijen, 1992). While one considered freedom, the other focused exclusively on the dependencies caused by domestication (van der Loo & van Reijen, 1992, p. 207). Therefore, domestication brought forward liberation and dependence, not an either/or situation (van der Loo & van Reijen, 1992, p. 207).

Moreover, as van der Loo and van Reijen (1992, p. 207) highlighted, those views represented the outdated view of "technology" being a "thing" outside of society, a thing penetrating human life, whereas they argue the contrary, technology, people, and society are inextricably linked one another. They do not exist in isolation from each other but coexist (van der Loo & van Reijen, 1992). Thus, the domestication of nature led to a dependence on technology and other people (van der Loo & van Reijen, 1992). With the advanced division of labor and the highly sensitive technological control devices the goal was that modern individuals did not get out of control (van der Loo & van Reijen, 1992, p. 209).

Van der Loo and van Reijen (1992, p. 209) described social control as "the way people make or force others to comply with norms and rules." The forms this social control took within the process of modernization was through "disciplining" and "normalization" (terms borrowed from Michel Foucault). From Foucault´s perspective, modern individuals were not free and independent; on the contrary, they were trapped and surrounded by a structure of power relations (van der Loo & van Reijen, 1992, p. 209). The establishment of norms and their admissibility or inadmissibility were not defined by mythical or religious ideas but were based on science (van der Loo & van Reijen, 1992, p. 210). Modern norms, being more precise and abstract, could, thus, be applied to larger groups of people (van der Loo & van Reijen, 1992, p. 210). Within Foucault's framework, modern power was more productive than repressive, and modern power techniques aimed less at excluding people and limiting their freedom of action; they tried to integrate everyone into the society and let them participate (van der Loo & van Reijen, 1992, p. 215). Participation in social life was the core of this modern power and it also facilitated teaching the individual self-discipline, self-control and general norms (van der Loo & van Reijen, 1992, p. 215). According to Foucault, this modern situation created a paradox, where individuals believed they had free will but were actually making pre-determined choices (van der Loo & van Reijen, 1992, p. 215).

This self-control brought a systematic control to the drives and emotions over time that were also in line with Elias´ theory of civilization (van der Loo & van Reijen, 1992, p. 227). This transition affected all societies as well as individual lives (van der Loo & van Reijen, 1992, p. 228). Thus, this "process by which knowledge, symbols, habits, attitudes, skills, rules and regulations – in short, "culture" – are transmitted from generation to generation and from individual to individual" was what they called "socialization" (van der Loo & van Reijen, 1992, p. 228). Socialization involves learning specific skills and knowledge and coming into one´s person, whereby a home-grown individual forms an image of oneself and one´s own identity due to socialization (van der Loo & van Reijen, 1992, p. 229).

This “self” or “I-consciousness” developed through interaction with others and is not given by birth (van der Loo & van Reijen, 1992, p. 229). Moreover, increasing domestication and the emergence of technological devices, which led to new dependency networks among people, also led to a more withdrawn attitude that took attention to psychological mechanisms (van der Loo & van Reijen, 1992, p. 229). Sigmund Freud revealed the psychological mechanisms, investigated human unconsciousness, explored the depths of the psyche, and shed light on the irrational side of human nature (van der Loo & van Reijen, 1992, p. 229). From Freud´s perspective, a human was a “natural being” part of living nature and was governed by drives and instincts (van der Loo & van Reijen, 1992, p. 229). He believed that the human being consisted of an unbalanced and extremely unstable compromise between three different psychic instances containing all inherited drives and emotions: the id/subconscious, the ego, and the superego (van der Loo & van Reijen, 1992, p. 229).

#### *Paradoxes of Domestication*

The paradox of domestication is that the more humans have become dependent on the technical infrastructure they create, the more they free themselves from natural barriers (van der Loo & van Reijen, 1992, p. 40). Increasing domestication required modern people's more systematic self-control in their behavior and dealing with each other (van der Loo & van Reijen, 1992, p. 233). According to Elias, the interdependencies between people increased during the modernization process, including people who confronted people in different situations or in interactions with institutions (van der Loo & van Reijen, 1992, p. 233). Some professional institutions took over daily tasks instead of self-controlling (such as daycare centers children) (van der Loo & van Reijen, 1992, p. 233). Lasch viewed this in contrast to the means of psychoanalysis, but somewhat functional, where individuals in modern society are continually confronted with changing people, situations, choices, and risks from a young age, where they can learn to adapt to this (van der Loo & van Reijen, 1992, p. 234). As van der Loo and van Reijen (1992, p. 234) suggested, the optimist or the pessimist views toward modern society did not recognize the paradoxical character of domestication. The increasing deconditioning of natural circumstances could not be seen in isolation from increasing psychological and social conditioning (van der Loo & van Reijen, 1992, p. 235).

## 2.7 Conclusion

The interdisciplinary aspects of off-grid renewable energy and its development are complex, involving policy, finance, technology, administration, and society among all sectors and disciplines in general. Based on the preliminary analysis of the data, the PoM theory (van der Loo & van Reijen, 1992) provided four macro processes to explain the collected data within the MLG framework in Namibia. The PoM has consisted of the processes of differentiation, domestication, individualization, and rationalization, along with their paradoxes, in a swirling continuum intertwined among the processes. It has proposed a nonlinear and irreversible process in addition to the historical process that is differentiated from country to country, thus enabling different types of modernization. In this sense, it has also contributed to the multiple modernities of Eisenstadt. However, among all other theories, one of the key distinctions of the PoM was the acknowledgment of the contradictory characteristics of modernization as the paradoxes of modernization in a dialectical way, resulting in a universal theory. Furthermore, PoM was considered an interdisciplinary and multidisciplinary approach in investigating the interconnectedness of the various modernization processes. Thus, recalling also the definition of Elias (1995, p. 7) for the characteristics of long-term processes being unplanned, moving forward into a recognizable direction without any long-term aim, as a result of many planned activities that are intertwining, operating, and confronting with one another; the PoM presents a functional multidisciplinary approach to tackle with interdisciplinary issues in the case of off-grid renewable energy governance, deployment and expansion.

Furthermore, Husserl's torch, used to describe the basics of phenomenology with noema and noesis, symbolizes the energy/electricity research within the off-grid areas of this study, with its enlightening optical characteristics in daily life. This torch, along with its analogy, has also guided a bottom-up research design to discover unresearched aspects of everyday life and their connections to the different levels of governance. Hence, the light of the intentionality has been focused on the off-grid settlements and the communities' visions with respect to the rest of the MLG. The case study chapters will elaborate more on the electricity that enlightens everyday life through the eyes of the respondents.

Thus, the MLG framework has shed light on off-grid areas in Namibia in the quest to understand off-grid governance with the PoM and within the PoM of energy in Namibia. Consequently, the theoretical framework suggests that an interdisciplinary approach to addressing off-grid renewable energy development and deployment in multilevel governance has the potential to unlock the issues faced in the off-grid energy sector, aiding in the expansion of off-grid electrification and the achievement of universal access goals by 2040 in Namibia.

## 3 Methodology

A sound methodology is key to guiding research design, methodological implementation, and results interpretation as well as critical to ensuring rigor, integrity, and transparency, all of which when combined contribute to scientific knowledge production and advances (Sovacool et al., 2018). In terms of energy and society, the "relationship between the two is co-productive" between the generation and consumption of energy, determining the lifestyles as well as economic and cultural value systems guiding the resources that the energy will be generated (Calvert, 2016, p. 105). The intertwined and multi-layered nature of off-grid RE governance with its qualitative and quantitative characteristics informing its micro and macro scalability, should be approached using a robust mixed methods methodological framework which captures and unpacks its multifaceted character. Mixed methods allow an enhanced and in depth understanding of research topics from both culturally contextualized and statistical patterns contributing to both expanding practical implementations and theoretical contributions.

High-quality, action-oriented research in energy studies, building construction, and sustainability studies make valuable contributions to solving real-world problems (Sovacool et al., 2018). The necessity of a more significant role for the social sciences would be enhanced in terms of rigor, interdisciplinary reach, policy relevance, and the communication of results (Sovacool et al., 2018). With regard to energy research, the philosophical and methodological diversities of energy geographies present privileged research in engaging with the complexities of energy realities (Calvert, 2016). Therefore, as highlighted in the energy geographies` cross-cutting subfields and borderlands, the material, immaterial, socio-spatial, and policy-related research contributes to the past, present, and future patterns of energy generation, distribution, and consumption of different geographies (Calvert, 2016, p. 109). As a result, this theoretical, conceptual, and methodological framework enables the researchers and RE actors to understand off-grid RE governance, contribute to potential solutions to existing issues, and provide an alternative perspective to future off-grid electrification scenarios.

Thus, to tackle the issue(s) of energy and society for off-grid RE governance research in Namibia, a trajectory from the micro level with human geography to the macro level to the boundary discipline of macro-sociology (Calvert, 2016, p. 109) has been explored as a general methodology in linking local levels to national levels within the MLG framework. Based on the literature review in Namibia, a lack of qualitative research design has been identified in the field of off-grid (renewable) energy and its governance. Thus, inductive research with mixed methods has been designed to realize this study, which started with a bottom-up approach by first investigating the energy practices in the local communities and their visions through the use of a life-world-centered phenomenology with hermeneutical interpretation in interdisciplinary studies. Then, it continued with the regional, national, and supranational levels of governance. As the emerging finding of the exploration, the findings were explained through modernization theory (van der Loo & van Reijen, 1992). As a result, asking socially relevant and impactful research questions with a well-constructed research design constituted a strong foundation for this study (Sovacool et al., 2018; Hancké, 2009).

## 3.1 Objectives and Research Question(s)

One of the primary objectives of this research was to employ a social sciences perspective in exploring interdisciplinary issues, projects, and research topics to provide more comprehensive understandings of energy topics than those provided by purely economic, technical/technological, or techno-economical perspectives. This entailed developing a humanistic and pragmatic approach that neither labeled the human as an "end-user" or a "consumer/client" but rather a human in action, a human in practice. Hancké (2009) defines the aim of social sciences as "to understand the world as it is made by human beings, the structures and institutions they produce, and the actions they take within those structures." This research was guided by this unique qualitative perspective, which had not previously been used to explore renewable energy topics in Namibia.

The main objective of this research was to explore and understand energy practices and visions of the communities within the case study areas; hereby aiming to contribute to general understandings of RE governance as well as the medium to long-term energy demand scenarios crucial for the mini-grids' technical, operational, and social sustainability. With these goals in mind, a multi-level, bottom-up approach was developed to better understand the relationships taking place within off-grid RE governance. These results could then potentially be implemented in the planning and future off-grid systems for the prospective unelectrified localities in Namibia. This research was guided by three primary research and sub-questions:

- What is the current state of off-grid RE governance in Namibia? This broad question seeks to explore the RE governance landscape from the micro to macro levels: Who are the main energy market actors? What are the primary elements of the policy and regulatory frameworks that are linked with off-grid electrification? What is the state of existing (off-grid) infrastructure? Additionally, what are the complexities and challenges facing mini-gids? What are the opportunities that will come with future off-grid infrastructures in Namibia? This inquiry seeks to understand and realize how to improve current and future (off-grid) electrification scenarios within the current multi-scalar structure of RE governance.
- Secondly, what are local energy practices within the context of Namibia's current off-grid areas? How does energy availability influence the

everyday practices of community members? What specific practices emerge? How do community members relate to energy on an individual level? How does energy take on different meanings within the multi-layered social and material contexts? This line of questioning seeks to better understand the role of electricity in everyday lives and how knowledge of these practices can help inform RE governance at the macro level in Namibia, primarily following the ten to thirteen years of commissioning of two mini-grids.

- And third, what are the visions of the individual community members? This broad question seeks to understand the visions of the communities embedded in their social realities in connection with electricity and how these visions inform RE governance at the macro level with the visions of the state with regard to (off-grid) electrification.

## 3.2 Research Design - Qualitative Research with Mixed Methods

A mixed methods research design was developed with the goal of "seeking new insights in assessing phenomena in a new light" (Robson, 2002, p.59; Saunders et al., 2007). This mixed methods approach emerged through the initial exploratory research phase and involved the collection of both qualitative and quantitative data guided by a phenomenological philosophical approach (Cameron, 2015). A mixed method approach allowed a deeper understanding of the research problem than a study that would use only a single method (Creswell & Plano Clark, 2007, p. 5, cited in Cameron, 2014); within this study qualitative approaches were developed to uncover the unquantifiable, like the daily lived experience of community members and their relationships with energy in the home, which were complemented by the quantitative measures of electricity, consumption, expenses, socio-economics that contribute to the micro or macro level planning and governance of mini-grids.

The exploratory portion of this research began with ethnomethodology to explore and better understand local energy practices, the meanings of energy/electricity, and the everyday life routines of the communities while experiencing and observing them firsthand in the field. Harold Garfinkel developed this ethnomethodology in the 1960s. Garfinkel's goal was "to study the taken-for-granted "methods" used by the members of the collectivities to maintain a local sense of social order" (The Sage Encyclopedia of Qualitative Research Methods, 2008, p. 293). This approach is also known as the "third generation of phenomenology" because it explores the emergence of life-worlds as a result of micro-processes (Alvesson & Sköldberg, 2018, p. 99). Furthermore, according to Alvesson and Sköldberg (2018, p. 110), ethnomethodology and grounded theory represent a "micro-sociological revolution." Ethnomethodological research uses ethnographic observations and audio/video recordings to learn the circumstances and the details of members´ practices within their cultural context by asking "how" questions rather than "why" questions (The Sage Encyclopedia of Qualitative Research Methods, 2008).

Seamon (2018, p. 8), who focused for many decades on (human) geography and the life-world approach, defined phenomenology as "the directed effort to see the things and experiences of the world more clearly and completely;" phenomenology focused primarily on understanding the human experience. The phenomena are the things and experiences directly experienced by people; this can be any experience, situation, event, or object that a person lives through, understands, sees, feels, tastes, hears, or smells; anything related to human experience can be researched within phenomenology because "human beings can experience them" (Seamon, 2018, p. 8). As Dörfler and Rothfuß (2018, p. 8) argue, "without body, no space, without bodily connection in the theoretical and methodological language, no coverage of the spatiality of the world understood as its own attribute." Moreover, in their view, "this social and bodily connection to space represents a *conditio humana* and a *mathesis universalis* of the social connection to space" (Dörfler & Rothfuß, 2018, p. 8). This is how the "subject and the body are the centers of all experience in relation to space" (Dörfler & Rothfuß, 2018, p. 9). A phenomenological approach, with a life-world focus and a focus on the subject and bodies as the center of all experience, guided this research when exploring energy/electricity as the phenomena of the study.

Hermeneutical interpretation, used in the empiricist and theoretical approaches across disciplines of the humanities and social sciences (Alvesson & Sköldberg, 2018, p. 115), was used in this study to interpret the empirical findings. Hermeneutical interpretation is a valuable technique to analyze texts in context, where the basic hermeneutic circle "part/whole" depicts the main theme of "the meaning of a part can only be understood if it is related to the whole;" "and the whole only from the parts" (Alvesson & Sköldberg, 2018, p. 116).

The objectivist hermeneutics that appeared at the end of the 19$^{th}$ century exemplified a sharp division between the researched object and subject; this was in opposition to alethic hermeneutics that dissolved this polarity and considered an understanding of the original situation with the "revelation of something hidden, rather than the correspondence between the subjective thinking and objective reality" (Alvesson & Sköldberg, 2018, pp. 120-122). Thus, although there have been internal debates surrounding the various hermeneutics methods, four primary aspects have been agreed upon as central to hermeneutic interpretation, and are described as follows: as a "pattern of interpretation, text, dialogue, and sub-interpretation" achieved by placing the "text" to be interpreted into the "context with its socio-historical weave of connections" (Alvesson & Sköldberg, 2018, p. 125); these four fundaments incorporate a preunderstanding/understanding in the hermeneutic circle.

In addition to this circle, Hoy (2009, p. 91) stated that "the temporality is hermeneutical, just as hermeneutics is always temporal." Hurmerinta et al. (2016, pp. 812-813) also explain the inseparable links between the past, present, and future; they argue that the observable behaviors of individuals could be considered as actions that accumulate through the events in their contexts and present themselves in total as processes. Therefore, a temporal hermeneutic cycle that collects actions and events contributes to a new cycle within the spiral (Hurmerinta et al. 2016, p. 813).

A hermeneutic interpretation, including the context of temporal hermeneutics for people's bodily experiences within their everyday lives and environments, provided an approach to understanding the essence of the aspects of energy/electricity at local and national levels. Furthermore, it uncovered and shed light on the locally generated, created, and produced perspectives on place with a temporal cycle. Thus, this methodology provided research from the micro levels investigated with the everyday life experiences to the macro(-sociological) processes in a broader sense from a bottom-up perspective.

## 3.3 Research Process using Comparative Case Studies

This research (Figure 3.1) began with an interpretivism philosophy of phenomenology that enabled an inductive research approach using qualitative methods and a single quantitative method in two case study settlements, namely Tsumkwe and Gam, constituting the only community-scale local mini-grids in Namibia. This inductive approach, combined with interviews and participatory observations through a hermeneutic process, enabled the researcher to understand the relevant structures of these life-worlds regarding energy, development, and daily life routines at a local level of the off-grid energy governance. Modernization emerged as a lived experience within the communities, which could be explained through the process of modernization (described by four macro processes of differentiation, rationalization, individualization, and domestication) (van der Loo & Van Reijen, 1992). Qualitative data was also extended with the quantitative surveys to demonstrate the basis in the field of energy/electricity that is related to the scale-related technical and economic dimensions of the mini-grids and their futures for the settlements as well as future planning and policy-related issues regarding the off-grid electrification. Thus, the qualitative and quantitative data were connected and analyzed and explained using hermeneutics to provide a comprehensive and in-depth result.

Further, a mixed methods approach was also applied to address the MLG at the sub-national, national, and supranational levels. Interviews were conducted with energy market actors at national and regional levels to address the visions, limitations, and possibilities in the off-grid electrification framework. In addition, the field notes from the workshop and conference participation (which took place in Namibia) were used to strengthen and fill out the qualitative datasets. Furthermore, existing and planned policies and regulatory frameworks were investigated, beginning with the independence of Namibia in 1990, and explained in the process of modernization. A quantified outlook on the policies was presented to see and understand the state of off-grid electrification and comprehensively assess the MLG for its future planning and implementation.

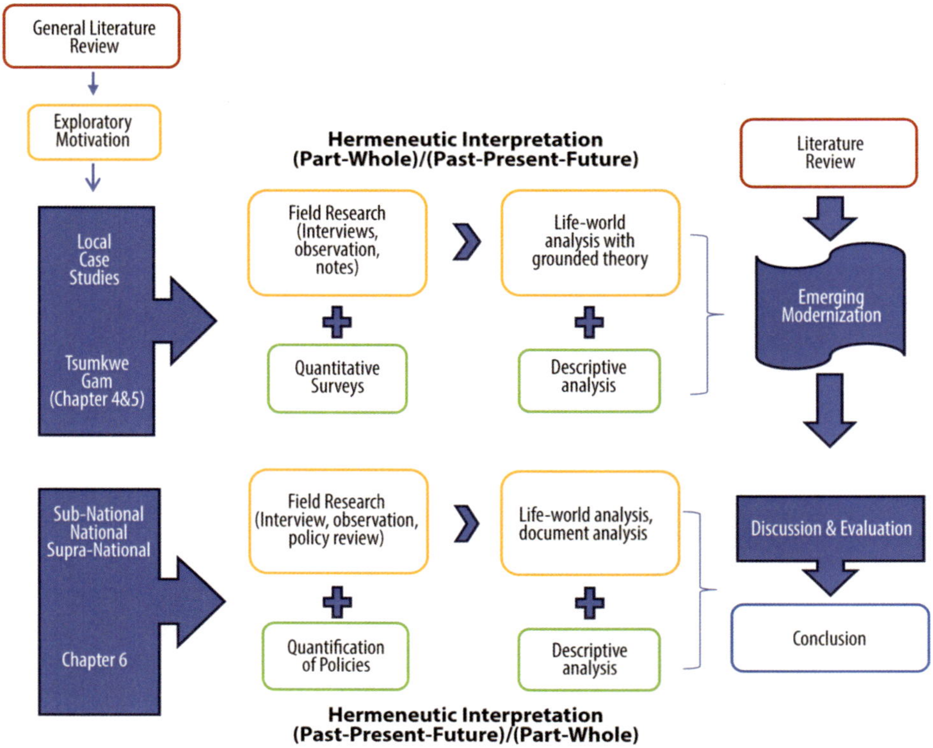


*Figure 3.1. Research Process*

*Source: Author*

As indicated earlier, the selected settlements have the only community electrifying mini-grids in Namibia, were chosen as the case study sites, these provided unique datasets for the assessment. As Eisenhardt (1989) stated, the *case study* is a research strategy that involves single or multiple cases that emphasize the understanding of dynamics in a particular setting. Additionally, by including a second case study settlement, the replication of the study can confirm and support a theory (Eisenhardt, 1989). Furthermore, the use of multiple data collection methods can strengthen the theoretical grounding with the triangulation of qualitative and quantitative data introducing a "synergistic view of evidence" (Eisenhardt, 1989). Therefore, in this case, insights from different ethnic backgrounds were used to enhance the validity of the research and their interpretation in an MLG framework. Thus, it brought an in-depth analysis of the situation with quantitative components that incorporated dimension and scalability of the analysis for future research or implementation. Finally, triangulation of the findings between two case studies was used to highlight crucial points for off-grid electrification in Namibia and the process of modernization in an MLG of energy.

## 3.4 Data Collection

### Qualitative Methods

Qualitative methods were used to shed light on the phenomena related to energy/electricity generated from the solar hybrid mini-grid in the daily life of the local communities. To this end, semi-structured interviews, participatory observations, and open-ended survey questions were utilized within two case study areas of Tsumkwe and Gam. The *case study* is a research strategy that enables realizing empirical research of "a particular phenomenon within its real-life context utilizing multiple sources of evidence" (Robson, 2002, p. 178; Saunders et al., 2007, p. 136) and is utilized in exploratory and explanatory research (Saunders et al. 200 7, p. 136). It provides the exploration of the context-based multiple data collected in the field and, thereby, a rich understanding of the processes (Saunders et al., 2007). The two case studies selected for this research represent unique cases in Namibia, with only two mini-grids providing electricity for each community. Furthermore, the communities´ histories and their different ethnic backgrounds enable a distinct comparison for the purpose of this study.

There were three primary phases of data collection: First, introductory meetings and interviews were performed, and preliminary field observations were completed first in Tsumkwe, then in Gam, and then ultimately with energy market actors/experts at the regional and national levels in 2020. The second field research phase in April 2022 coincided with the inauguration ceremony of the grid electrification in Mangetti Dune, a formerly off-grid settlement 85 km West of Tsumkwe, in which the author recorded the public speeches of the regional and national actors providing the visions in the off-grid electrification field. Third, an excerpt from the World Café discussion method organized at the PROCEED workshop was used to validate and triangulate the multi-level participatory data at the end of the research in March 2023.

The first phase of field research was initially intended to occur between 2020 and 2022 in both settlements of Tsumkwe and Gam. However, the COVID-19 pandemic interrupted the field research process in 2020, and alternative remote data collection methods needed to be adopted. The field research and data collection were thus shifted to an online format using the available means. Under these circumstances, interviews were conducted first in Tsumkwe with the support of a field assistant who visited the interviewees and made an online connection with the author to record the interview. Twelve online interviews (See Appendix B) were conducted with community members directly in Tsumkwe during August of 2020, under the fluctuating availability of internet connection., and two additional interviews were conducted in person in April 2022. The author recorded the interviews and transcribed them. In the necessary cases, the field assistant in Tsumkwe provided the translations.

During the COVID-19 pandemic, it was also impossible to conduct interviews in Gam as initially planned for 2020 due to the lack of field assistance and the lack of internet connection needed to conduct the interviews directly. Therefore, in-person interviews with community members could only occur after the COVID-19 travel restrictions had been loosened or abolished. Two years later, in April 2022, ten in-person interviews (See Appendix B) were conducted, recorded, and transcribed.

The COVID-19 restrictions halted in-person data collection, but September 2020 were used to collect data from energy market actors within the MLG of energy through online interviews using ZOOM. Six interviews (See Appendix B) were conducted, recorded, and transcribed. In the second field research in 2022, after the travel restrictions were lifted, the inauguration of the grid connection of a formerly off-grid settlement incorporated the views of the regional and national actors regarding the off-grid electrification and their visions. To complement these perspectives and to understand MLG policy and regulatory framework conditions, a policy and regulatory framework review was also realized to understand the policy landscape in relation to RE governance and off-grid electrification. Finally, secondary data such as policies and regulatory framework documents adopted or published by national authorities and institutions and research reports from international universities and organizations were collected between 2020 and 2023 from online databases, the Namibia National Archives and the Namibia National Library.

**Quantitative Method**

As a part of the exploratory, inductive research, surveys were used to extend the data regarding local energy practices with their material components in each settlement (Shove & Walker, 2014). Capturing the numeric values of demographics, structures, current and future electric appliances, and socio-economics in relation to energy consumption constituted technical and economic aspects at the local level to explain the processes that the communities have been going through, starting from the years of electrification. This component also contributed to the necessary technical and economic planning and forecasting, starting from the settlements and potentially reaching out to regional and national level planning since they have illustrated the patterns in which off-grid electrification might result over time for Namibia. Therefore, although surveys have been associated with a deductive approach (Saunders et al., 2007, p. 138), this study has benefited from the surveys as complementary data (Neal et al., 2015; cited in Cameron, 2015) to provide a comprehensive framework for the phenomena that were investigated.

Seventy-six semi-structured surveys, including open-ended questions, were conducted in Tsumkwe during the first field research in February/March 2020. This semi-structured survey (Appendix A) consisted of social, economic, and technical questions related to energy in order to provide an overall understanding of the locality and the community for each case study area. The survey area covered all existing settlement locations[4] in Tsumkwe. Although initially a probability random sampling method was to be employed to access interviewees, due to the time constraints in the field, convenience sampling was used by visiting the households and businesses and conducting the surveys based on their availability and the willingness to participate in the survey (Saunders et al., 2007, p. 207).

The first day of field research in Gam coincided with *the state of emergency* in Namibia due to the COVID-19 outbreak. This was followed by immediate restrictions across Namibia, which had a significant impact on foreigners. However, the investigation continued for three days, and eighteen surveys were conducted (See Appendix A). Increasing concern within the community, and the general concern about the COVID-19 outbreak, caused the field investigation in Gam to be halted. Considering the unpredictable conditions caused by the pandemic in 2020, surveys were conducted with the support of the field assistant in Gam. The assistant was instructed on interview procedures and processes, and daily communication via mobile phone, without reliance on internet, was maintained to ensure the field surveys were conducted appropriately and on schedule. The summertime (2020) in Namibia (which allowed social distancing measures) enabled 55 semi-structured surveys to be completed on paper form (See Appendix A). Once the surveys in the field were completed, the assistant compiled and photographed surveys; each survey was then transcribed into Excel format in preparation for further analysis.

## 3.5 Data Analysis and Triangulation

The qualitative and quantitative data analysis was conducted independently for each case study. First, F4 and MAXQDA software were used to transcribe the data. Next, the data were analyzed using grounded theory (Glaser & Strauss, 2006) In this way, in the analysis of the qualitative data, the basic principles of Grounded Theory were adopted to perform initial and in-vivo coding. The condensed meanings of the participants continued with focused coding, which allowed the categorial axial coding (Charmaz, 2006). All earlier coding was sorted and synthesized through axial coding with categories and subcategories to illustrate each dimension (Creswell, 2007; Charmaz, 2006, p. 60) and, construct abstractions, and reconstruct meanings, visions, and structures. Thus, the meaning of electricity, visions, and energy structures of life-worlds was reconstructed through the analysis of the qualitative data and interpreted hermeneutically.

Furthermore, during the PROCEED Energy Policy Recommendation Workshop (March 2023), both qualitative and quantitative data were presented and vali-

[4] Location is locally used word for "neighborhood" in Tsumkwe Settlement.

dated with the participation of the community representatives from Tsumkwe and Gam, as well as regional and national stakeholders. In this workshop, the *World Café* method was utilized to create a safe and creative environment for stakeholders from different levels of governance to engage in discussions intentionally through multiple ideas and perspectives (FSG Remaining Social Change, 2021). One discussion group specifically addressed *energy governance* and created vivid discussions of four groups in a row to reflect and provide input for the governance of off-grid electricity and off-electrification issues.

The collected policy and regulatory framework documents and national plans were analyzed to address governance-related research questions as supplementary research data, providing a means of tracking change and development over time and bringing background information with historical insights (Bowen, 2009). The secondary archival data leveraged existing data to produce a comprehensive and robust understanding of the off-grid RE governance, including the historical past of the settlements, which was not available through online sources. As a result, it also enhanced the life-world analysis and hermeneutical interpretation of the data in its temporal loop.

The quantitative datasets from Tsumkwe and Gam were cleaned and analyzed using SPSS software. The total processing time for the quantitative data transcription and preparation for the descriptive analysis was approximately one year. As an additional point in the collection of data from the field in Namibia, CENORED had the courtesy to share their customer numbers (this data also included the type of customer: residential, business, or institutional) in Tsumkwe and Gam, which brought new insights in the later stages of the research in dimensioning the population and consumption needs. This new dataset was used with descriptive statistics to analyze and compare the studies. The analysis allowed the author to present the data in a meaningful manner to describe the central tendency through means, median, and modes, the variability through the range and standard deviation, the frequency distribution, and the relations within the dataset with cross-tabulation to examine the differences in the datasets as well as in different case studies for their comparisons.

A final triangulation comparing both case studies focused on the relevant aspects of energy and min-grids (e.g., practices, appliances, prices, actors, and community visions), using a hermeneutical interpretation, was also undertaken. A comparison of the two case studies was conducted to uncover the similarities and differences between the communities using electricity from the mini-grid at a local level and within the regional and national context in Namibia. Furthermore, in the case of available data, selected findings were also compared with other mini-grids in Sub-Saharan Africa or the findings of the international context. This triangulation enabled both a contextual understanding of the life-worlds using power from the mini-grid as well as provided a temporal understanding of the phenomena at a local level for its future use and development. Thus, although the volume of the data was initially overwhelming, an iterative process was used to narrow down the research focus to solely off-grid electricity within the MLG context in Namibia, enabling emergent relationships that enhanced confidence, reliability, and validity (Eisenhardt, 1989).

## 3.6 Limitations

Amidst the changing times of the 21$^{st}$ century, field research activities occurred during the COVID-19 pandemic from the beginning of 2020 until the end of 2022. The number of instrumental, geographical, social, travel, and health limitations severely hindered the research process and data collection in the field; for example, research appointments for visits to the field had to be (re-)canceled over and over due to travel restrictions. This research is also a product of these mentioned restrictions, where a lot of emphasis had to be allocated in the reflection and evaluation cycles of the first field research data individually as well as in different scientific workshops, seminars, and in research group meetings in order to develop the core of the study. Additionally, the unexpected challenges and the limitations of reaching the field were also introduced spontaneously; digital methods were used in the data collection through the help of a field assistant in Tsumkwe. These digital instruments became the hands for the settlements that digital communication reaches. Finally, the local languages like Ju/'hoansi or Otjiherero were a barrier when English was not spoken by all the local people.

MLG research in the two mixed-methods case studies involved comprehensive data collection. However, it also introduced complex analysis cycles. On the other hand, the unexpected and unknown two-year interval between the first and second field research contributed to a temporal dimension to be observed within the fields and data analyzed, which was not initially planned. The challenges encountered with the limitations have been transformed into new opportunities, reflected, and discussed within this study. Unfortunately, not all findings could be discussed and elaborated accordingly; however, the focus was on the interdisciplinarity of off-grid energy based on the social sciences' perspective.

## 3.7 Ethical Considerations

The research was conducted within the PROCEED project, with the approval of the Ministry of Mines of Energy and the permission of the Otjozondjupa Regional Council. All interviewees and survey respondents were informed about the project and the scope of work. Upon their informed consent, the surveys and the interviews were conducted confidentially. Additionally, the interviews were recorded with anonymous titles to maintain privacy according to ethical considerations. The researcher observed that electricity is a topic of interest among all interviewees and respondents, which was openly discussed throughout the research process.

# 4 Everyday Life Matters in Tsumkwe: Energy Practices and Future Visions

> "Like this one, this one that is currently on, electricity is part of life, everywhere. It was actually implemented last, 2018. There was already a sort of, form of electricity which was being used. But it was not convenient to supply the whole area or with everyone. Even the current one is still not convenient to supply everyone in the area. So, not everyone is having electricity. So, it is for those who can afford and to the designated area. Where you situate yourself is where the electricity is, but most of the San people most of them are unemployed. Ninety-nine percent of the San people are unemployed. So, they are only relying on the tourism sectors. But they cannot get that fund to install electricity for everyone. Since you, whereby you install electricity to your place you need, there is a certain amount of money that you have to pay, is when they will connect you electricity in your house, but if you are unemployed, you'll not be able to get electricity. So, the benefit of electricity is only the side of government sectors, business areas employed people. [It] is only where you can find where people are having electricity. So, electricity was essential for everyone but not everyone has electricity" (Interview 6, Pos. 319-350).

This chapter sheds a phenomenological light on the everyday life and visions of the Tsumkwe people through qualitative and quantitative methods employing a hermeneutical interpretative approach. It aims to look at how "energy weaves the threads of life" (Pasqualetti, 2021, p. 1) of the past with the facts of the present to create a "better life" or a "normal life" for the future.

The chapter starts with the historical background of the Nyae Nyae region, beginning with the pre-colonial and post-colonial periods, through Namibia's independence, and the development of the Tsumkwe settlement to the present. The qualitative and the quantitative findings follow, respectively. The qualitative findings have been deliberately presented prior to the quantitative findings since the exploratory research portion was initiated through preliminary observation and talks before the surveys were conducted.

## 4.1 Pre-Colonial, Colonial, and Post-Colonial Tsumkwe and Nyae Nyae Region

The literature uses the terms "San," "Bushman," and "Basarwa[5]" to refer to the same group of people living in southern Africa (Suzman, 2001). The Namibian

---

[5] In Botswana

Association of CBNRM[6] Association Support Organization (NACSO) suggests that, although both words have different negative connotations, the Regional San Conference 1993 adopted "San" as the official term. Thus, in this chapter, Bushmen and San are used according to the period and the source of information within which they were cited. In the interviews, not every person used Ju/'hoansi specifically, but referred to San in general, irrespective of their ethnic backgrounds. It is also kept in that version. However, the author uses the term "San" in general, including all sub-groups, and uses "Ju/'hoansi" to refer to the group living in the region of Nyae Nyae.

### 4.1.1 Pre-colonial Period

The San are the indigenous people of Namibia and Southern Africa, who are the original hunter-gatherers, and the oldest (recorded) ethnic group to settle in southern Africa (Nyae Nyae Development Foundation of Namibia [NNDF], 2024)). The Ju/'hoansi, descendants of the San, have been living in the ancestral region of Nyae Nyae for thousands of years and are presently the last self-sufficient and independent hunter-gatherers in southern Africa (Ritchie & Marshall, 1991). Traditionally, within the San community, land rights and food and water resources were inherited from generation to generation (Marshall, 1995; Biesele et al., 1986). The San people lived in the *n!oresi*, their traditional land area, where a family group of 30 to 50 could be provided sufficient resources (Dieckmann et al., 2014). However, depending on the conditions, they could split into smaller groups of two or three families while remaining in the same land area. They were not dependent on each other (Marshall, 1995). An area of 45,000 km$^2$ was necessary to support 1,200 people (Ritchie & Marshall, 1991). Biesele and Hitchcock (2011) informed that the Ju/'hoansi encountered the Europeans for the first time in the 1850s.

### 4.1.2 Colonial Period in Tsumkwe and Namibia

In 1884, the German Chancellor Otto von Bismark claimed the region before the British. The same year, at the Berlin Congo Conference in Germany, "German South West Africa" colony borders were established with the British and the Portuguese. In 1890, Curt von Francois made Windhoek the capital of the German administration. Germany remained in Namibia until 1915 when South Africa defeated the German forces and took control (Dugard, 2019). Following the peace treaty after the Conference of Peace in Paris in 1919, the "League of Nations" was formed, and a mandate system was introduced (Katjavivi, 1989). Therefore, German South West Africa was put under the South African administration, as South West Africa (SWA) (Dugard, 2019). However, South Africa continued the colonial behavior; in 1949, a commission was appointed to pre-

[6] Community Based Natural Resource Management

serve the Bushmen (Biesele & Hitchcock, 2011). The Marshall family began conducting ethnographic studies between 1951 and 1958, and in 1959. Tsumkwe became an administrative center (Biesele & Hitchcock, 2011).

In 1970, “Bushmanland” was recommended by the Odendaal Commission and officially declared in 1976 under Proclamation 208 (Biesele & Hitchcock, 2011). Based on the homelands policy, the administration provided 11,000 $km^2$ of northern Nyae Nyae to Kavango, of which 5,000 $km^2$ was declared as the Kaudom Game Reserve, and 20,000 $km^2$ southern Nyae Nyae was given to the Herero ethnic group (Ritchie and Marshall, 1991); this marked the loss of 90% of Ju/'hoansi's traditional land, effectively ending hunting-gathering in Nyae Nyae as their large population could no longer subsist on the remaining land area (Ritchie & Marshall, 1991). According to their report (Ritchie & Marshall, 1991), 920 people moved into the “rural slum” in Tsumkwe in the mid-1970s, where there was a clinic and a school but, few jobs for men and no jobs for women; these changes resulted in a 5% population decline due to malnutrition and disease. The average per capita income was recorded as less than 42 Rand per year. Dieckmann et al. (2014, p. 96) indicate the difficulty of Ju/'hoansi, who could not cope with the social changes and the limitations of the Tsumkwe *n!ore* for hunting and gathering, resulted in social disintegration with increased dependency on alcohol, and an increase in crime and poverty caused by few wage earners. Thus, it became known as the “place of death” by Ju/'hoansi (Biesele & Hitchcock, 2011, p. 10).

Furthermore, the report of Ritchie and Marshall (1991) also mentioned that in 1981, the South African defense force was at war with Angola, recruited 100 Ju/'hoansi in the army and thus caused the Ju/'hoansi population of Tsumkwe to become dependent on their salaries, which was 600 Rand per month. In the meantime, 7,000 refugees from Angola also immigrated to Namibia, increasing the total population to 30,000. In Tsumkwe, deaths by alcohol consumption, violence, and homicide increased, taking over the birth rate. Tsamkxao[7] was one of the Ju/'hoansi who left Tsumkwe in 1983 because of sickness, fighting, alcohol, lack of work, and lack of governmental support (Ritchie & Marshall, 1991). He was convinced that keeping cattle and farming was the only way to survive on their land. Hence, they founded Ju/wa Farmers' Union in 1986 to help people start farming. Between 1982 and 1983, some Ju/'hoansi returned to their *n!oresi* for farming (Biesele & Hitchcock, 2011). By 1986, 500 people were supported in 11 Ju/'hoansi farming communities. They created a mixed economy by providing 40% of the products from farming, 30% from hunting, and 30% from military salaries. However, the challenges around the lions, the elephants, the waterholes, and the unsupportive measures of the government officials at the time did not make their lives easy (Ritchie & Marshall, 1991). In 1989, Ju/wa Farmers' Union became Nyae Nyae Farmers' Cooperative (NNFC)

[7] Present Chief at the Ju/'hoansi Traditional Authority in 2023.

led by Tsamkxao, who were busy drafting the statutes and NNFC′s drilling programs had helped fifteen farming communities to drill a well and to settle (Ritchie & Marshall, 1991).

> "Nothing starts or happens in Namibia without water. Every Namibian knows that water is life. From water flow the economy, development, and the future" (Ritchie & Marshall, 1991, 00:31:50).

### 4.1.3 Post-Colonial Period in Tsumkwe

The independence of Namibia introduced a land reform based on the apartheid homeland land tenure system, which divided the land into individual 5 ha plots (Dieckmann et al., 2014). This land reform did not fit the traditional *n!oresi* of !Xun and Ju/′hoan San of West Tsumkwe (Dieckmann et al. 2014). During the Conference on Land Reform and the Land Question in 1991, NNFC pioneered in defending their traditional land system against the proposed model and succeeded in keeping the traditional n!ore and became a pilot area on CBNRM in land use by resettling the natural resource rights to the Ju/′hoansi community. Thus, NNFC was registered as the first Namibian Conservancy in 1998, changed its name to Nyae Nyae Conservancy (NNC), and paved the way for Namibia's next conservancies and communities.

Following this official recognition and the new or traditional land tenure system, the NACSO was founded in 1999. As of 2023, they are assisting 86 registered conservancies, covering an area of 166,045 $km^2$ with approximately 238,071 people (approximately 9-10% of the overall Namibian population) in Namibia (NACSO, 2023a).

NACSO (2023a) stated that Ju/'hoansi are one of the only two out of the six San Traditional Authorities who received official recognition; and development initiatives have been contributing to schools, health facilities, permanent water points, livestock herding, and gardening. Accordingly, The Nyae Nyae Conservancy is situated in the Tsumkwe Constituency, whose area is outside of the settlement boundaries. There are 38 small villages based on the *n!ore*, usually inherited from the parents or through marital status (Dieckmann et al., 2014).

San groups live in southern Africa, which is divided across the national boundaries of Namibia, Botswana, Angola, Zimbabwe, and Zambia. Suzman (2001) stated that the population of San groups was affected by wars and migration, and governments were reluctant to include socio-linguistic census data. However, based on the data from Biesele and Hitchcock (2011, p. 4), the overall San population in southern Africa was 96,800 (11% of the overall population), with the majority of 48,000 in Botswana and 34,000 (1,6 %) in Namibia. On the other hand, a regional assessment suggests that the figure is approximately 32,000 and represents 1.8% of the population in Namibia (Suzman, 2001, p. 5), indicating a 0.2% decline in the San population in Namibia.

Furthermore, Dieckmann et al. (2014, p. 96) stated that before the 1960s, the Ju/'hoansi population living semi-permanently in today's Tsumkwe Constituency (West and East) was about 250. However, in 1993, this number increased to approximately 5,700 people with the settlement of the northern !Xun in South African Defense Force (SADF) army camps. Since 2011, there has not been any official census; the current population in the Tsumkwe Settlement has only been estimated at between 2000-3000 people. However, the Namibian Household Income and Expenditure Survey (NSA, 2016, [NHIES]-2015/16) stated that Khoisan-spoken households accounted for 1.5% of the population, with an average household size of 5.6 people. Further analysis and estimation regarding the population have been elaborated in the sub-section of the quantitative findings (Chapter 4.3).

### Installation of a Mini-Grid in 2011

Following the independence of Namibia, a solar-diesel hybrid mini-grid (Figure 4.1) was installed in 2011 with hybrid funding sources alongside other community-related international research projects for the development of the settlement.

*Figure 4.1. Mini-grid in Tsumkwe*

*Source: Author*

The South African military supplied the basic power infrastructure for the military base, including a diesel generator that provided electricity to government institutions (Chiguvare, 2019). Following independence in 1990, two generators remained in Tsumkwe, supplying electricity to the connected places for primarily governmental/institutional use, operating 18 hours daily. However, Eisenbach et al. (2009) describe this electricity supply as unstable, insufficient for generating income activities, and inadequate to supply inefficient appliances. Thus, the grid was making electricity generation and consumption unsustainable while creating a monthly operating deficit of 60,000 NAD, excluding maintenance costs (Eisenbach et al., 2009). Under these circumstances, the Tsumkwe Energy Project was launched in 2008 with a total budget of US$ 3.9 million, funded by the European Commission (75%), NamPower (14%), and the Otjzondjupa Regional Council (11%) (Boyd et al., 2012). Additionally, there were a total of 107 households connected to the mini-grid, and over 86 households were planned to be connected in Damara Location, Gauteng Location, ||haras Location, Tsumkwe Central, Sewe Huise, and Tsumkwe North to the electrical grid (Boyd et al., 2012). Furthermore, more households showed interest in connecting to the mini-grid; however, due to the impermanent structure of the houses, the Local Government repeatedly declined their requests (Boyd et al., 2012).

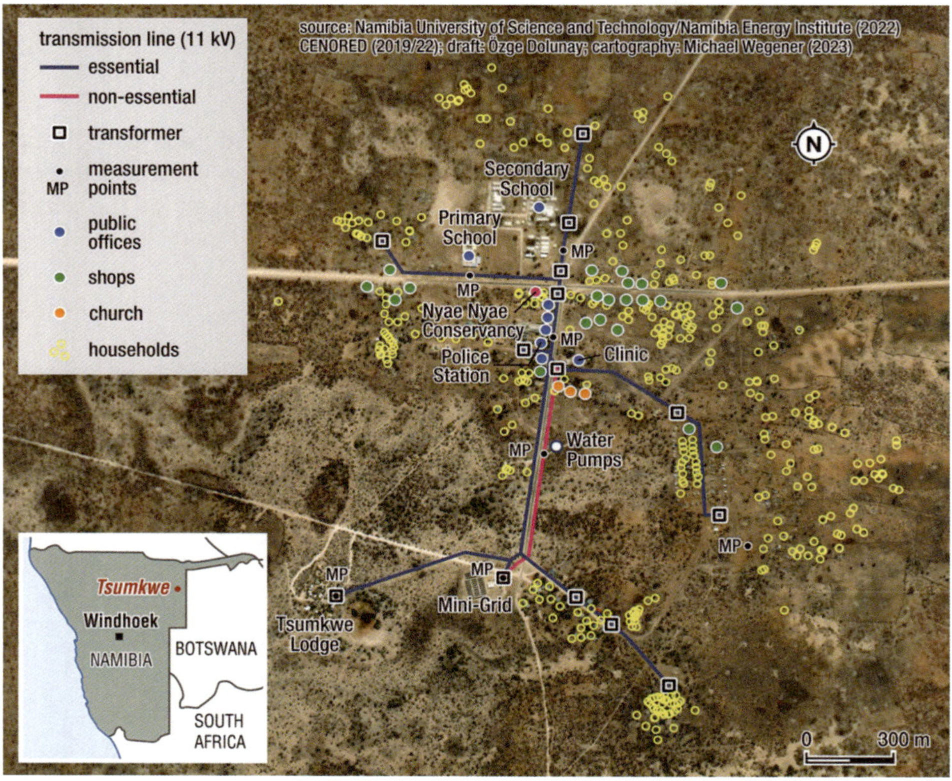


*Figure 4.2. Dwelling Units (2011) in Tsumkwe with the Electricity Distribution Lines*

Figure 4.2 shows the essential and non-essential lines currently installed in Tsumkwe: The essential line is connected to the clinic, schools, police station, and water pumps, whereas the non-essential lines are connected to households, local shops, and public offices. The local dwellings shown in the figure belong to the census in 2011 (it is currently unknown if any new housing has been established in the meantime). The essential and non-essential lines that were provided by CENORED in 2019 and 2021, showing that no new 11 kV transmission lines have been reported. The map also illustrates that the majority of shops and public offices have been established on the Main Road and the Main Street in Tsumkwe. In some locations, the dwellings are scattered around these roads; this makes it costly for those that are situated 80 meters distance from the connection points. In addition to this hybrid infrastructure, which has been maintained until present, in 2022, the PROCEED Project Team inaugurated a 26 kWp solar PV rooftop system. This rooftop system was built on the roof of the Tsumkwe Secondary School and connected, as a decentralized system, to the mini-grid. The PROCEED project also maintained an active research project to investigate the technical performance between the main mini-grid and the decentralized new system.

### 4.1.4 Local Setting and Main Actors in Tsumkwe

Tsumkwe Settlement is situated in the Tsumkwe Constituency within the Otjozondjupa Region in North-Eastern Namibia. Its geographical coordinates are 19°36′0° South, 20°30′0° East, at an altitude of 1000-1200 meters above sea level (Chiguvare, 2019). The majority of the population in Tsumkwe is Ju/'hoansi, surrounded by the San villages around the settlement area. Other ethnic groups, such as Ovambo, Ovaherero, Zambezi, Kavango, and Damara/Nama, also reside in Tsumkwe (Boyd et al., 2012). The settlement consists of households, local shops, offices of several ministries, the office of the Regional Council of Otjozondjupa Region (Tsumkwe Constituency), the Settlement Administration, Traditional Authority Tsumkwe East, primary and secondary schools with their dormitories, police station, Nyae Nyae Conservancy Office, NGOs, Tsumkwe Lodge, office of CENORED, NBC Radio (Figure 4.3).

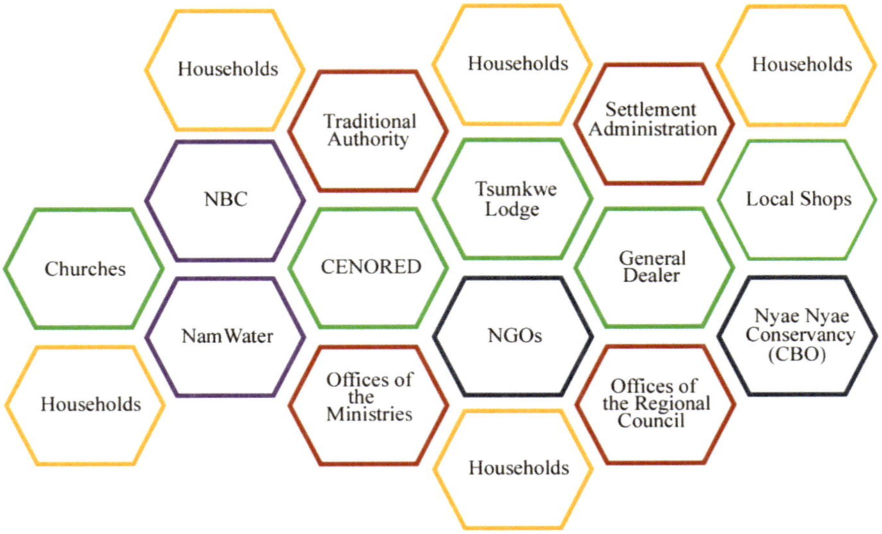


*Figure 4.3. Local Community Actors in Tsumkwe*

*Source: Author*

**Community-Based Organizations**

The Conservancies are legally registered areas that enable conservation and development over the defined large areas, empower rural development, and are run by a constituted body of management by the community (NACSO, 2012). The elected members serve the interests of all members for the development of residents and the sustainable use of the wildlife, whereby its members can add income (NACSO, 2012). Prior to the independence of Namibia, there were few rights for the residents living in the communal areas. After the legislation in the mid-nineties, a new approach to conservancy was adopted.

The Nyae Nyae Conservancy (NNC), established in February 1998, is the first communal and the second largest conservancy in Namibia: It covers an area of 8,992 square kilometers and is among 86 conservancies in Namibia as of 2023 (NACSO, 2012). Approximately 2,300 Ju/ʹhoansi live in the area, whereby they are actively involved in the activities of the Conservancy. The Nyae Nyae Conservancy area neighbors the Botswana border in the East, the Khaudum National Park in the north, the Ondjou Conservancy in the south, and the N≠a-Jaqna Conservancy in the West. The introduction of wildlife in 1999 increased the amount of game in the area and enhanced its value for tourism and trophy hunting (NACSO, 2012). The primary income activities in Nyae Nyae are craft sales, camp and cultural tourism activities, trophy hunting, veld products, and live game sales. Trophy hunting remains the largest source of income for the Conservancy. The Nyae Nyae region is divided into four sections for management purposes. The structure of the NNC consists of a board, representatives of each area, a chairperson, and a conservancy manager (NACSO, 2012). The board is responsible for appointing the management committee and the manager, while focusing on strategic planning. The management committee includes the conservancy manager and a minimum of five members. It is responsible for appointing the staff, making short-term decisions, and supervising performance. The Conservancy employees include 14 rangers, a CBNRM field officer, a craft trainer, a craft shop coordinator, an agriculture coordinator, a devil´s claw coordinator, three water team staff, an office administrator, and a cleaner (NACSO, 2012). In addition, G!hunku Craft Centre also belongs to the Nyae Nyae Conservancy. This management and operation structure in the Conservancy is also representative of the division of labor within the San community´s Conservancy (van der Loo & van Reijen, 1992).

#### *4.1.4.1 Non-Governmental Organizations*

There are three main NGOs present in Tsumkwe, the Nyae Nyae Development Foundation´s office, the Ju/ʹHoansi Traditional Fund, and the Palms for Life.

#### *4.1.4.2 Public Institutions and Parastatals*

There are several public institutions in the social stratification (Table 4.14, Table 4.15). The presence of the offices of the Regional Council, the Settlement

Office from the Ministry of Regional and Local Government and Housing, the Ju/'hoansi Traditional Authority, the Magistrate's Court, the offices of the Ministry of Agriculture, Water and Forestry, Ministry of Environment and Tourism, Ministry of Gender, Equality and Child Welfare, Ministry of Works, Ministry of Urban and Rural Development shows a high representation of the state institutions in the field along with the Primary School, the Secondary School, the Clinic, the police station, the veterinary and the Community Development Centre.

It is important to note the vitality of the electricity for the clinic and the schools. Tsumkwe schools receive their students from Tsumkwe and surrounding villages. Tsumkwe Primary School had a daily attendance of 751 (Figure 4.4), of which 258 boarded in the school's hostel, with 26 teachers, at the time of the research. Tsumkwe Secondary School (Grades 8-12) was attended by 280 students, of which 200 were boarded in the school's hostel and 15 teachers. The sum of students in both schools reached 1031, representing 32% of the Tsumkwe population estimation in 2021. On the other hand, Tsumkwe Clinic, also in the same essential line as the schools and water pumps, had 17 employees, with 40 to 60 patients daily and 54 TB patients in an annex building. The police station has 19 employees, and the Traditional Authority has 11 employees.

*Figure 4.4. Students of Tsumkwe Primary School*

*Source: Author*

Parastatals are the state-owned enterprises belonging to the Ministry of Public Enterprises in Namibia. There are a total of 98 public enterprises categorized as commercial, non-commercial, and extra-budgetary. The National Statistics Agency, the Namibia Water Corporation (Namwater), the universities, the Electricity Control Board, the National Broadcasting Corporation (NBC), and the Meat Board of Namibia are relevant to this research. The NBC in Tsumkwe employs 9 people, very active radio station in communication.

In addition, there are churches that belong to the Dutch Reformed Church, the Seventh Day Adventist Church, the Christ Love Ministries, the Roman Catholic Church, the Evangelist Church, the Pinkster Church, the Prophets Church, and the Zion Council of Churches (ZCC).

#### *4.1.4.3 Private Businesses*

Tsumkwe is home to many local small shops as well as two relatively larger companies. The local shops (Figure 4.5) in Tsumkwe have in general one or two employees, excluding the general dealer with 9 employees and they are operational seven days of the week. Some of the local shops are food shops and supermarket (OK, Savanna, Hanna´s, Karumbu etc.), the fuel station and general dealer, Tsumkwe Hardware, the restaurant, several shebeens (Iparo, Top Life, Rock Star, OK Bottle Store, Hours Inn, Omupanda, Bush Paradise etc.), first and second-hand clothing shops, butcher, hairdresser and barber shops that serve the residents in Tsumkwe and surrounding villages.

*Figure 4.5. Local Businesses*

*Source: Author*

On the other hand, a larger scale private business is the Tsumkwe Country Lodge hosting the travelers visiting the Tsumkwe region or stopping by during the journey from Botswana to Namibia or the other way around. Until 2022, the Lodge belonged to the TUCSIN NGO[8], dedicated to improving the educational standards of young Namibians operated by them with the employment of people from Tsumkwe. In 2022, the ownership of the Lodge was transferred to the Company Pinar. During the time of the field research in 2020, the Lodge employed 22 people in total.

As mentioned earlier, the Central Northern Regional Electricity Distribution is a private company, one of the five Regional Electricity Distribution companies, established in terms of the Namibian Companies Act in 1973, whose shareholders are mainly Local Authorities (LAs), the Regional Councils (RCs) and Nampower, who transfer their assets and customers to REDs in exchange for shares

[8] The University Centre for Studies in Namibia

(Electricity Control Board [ECB], 2007a). CENORED´s local representative office is situated in Tsumkwe for both Gam and Tsumkwe. In 2017, CENORED took over the operation and maintenance of the mini-grid from the Ministry of Works.

### *4.1.4.4 Households*

The Namibian Government Gazette no.4314 (Republic of Namibia, 2009, p. 2) defines households as "one or more persons, related or unrelated, who live together in the same homestead or compound, but not necessarily in the same dwelling unit, who has a community of life with common catering arrangements, is answerable to the same head of the household and occupy dwelling units, which may be a building or a part of a building or a group of buildings regardless of their physical condition". The Namibia Inter-censal Demographic Survey Report (NSA, 2017, [NIDS], 2016, p. 101) specifies the households as private places excluding school hostels, police barracks, hospitals, and prisons; while differentiating them according to their types as a detached house, semi-detached/townhouse, apartment/flat, single quarters, mobile home (caravan, tent) traditional dwelling, improvised housing unit (shack) and other. In Tsumkwe, the households (Figure 4.6) that were observed as traditional dwellings, concrete single quarters, shacks, and wood houses. Further details will be provided in the quantitative section.

*Figure 4.6. Households in Tsumkwe*

*Source: Author*

In summary, past struggles, alongside Namibia´s independence, have influenced Tsumkwe´s pathway to development. The timeline in Figure 4.7 illustrates the engagement of different actors in the Tsumkwe and Nyae Nyae region starting from the 1950s and today. It seeks to understand the present situation for future reflections and potential visions in the provision of electricity and its governance. The government´s activities before and after the independence, the active participation of the community together with the Community-Based Organizations (CBO), the research and NGO activities, the mini-grid electrification through different partnerships, the role of CENORED, and the international projects such as LIFE and PROCEED shall be summarized within this chronology. It also highlights the community´s involvement in the recent history of Tsumkwe. Thus, it ensures the importance, necessity, and success of these activities.

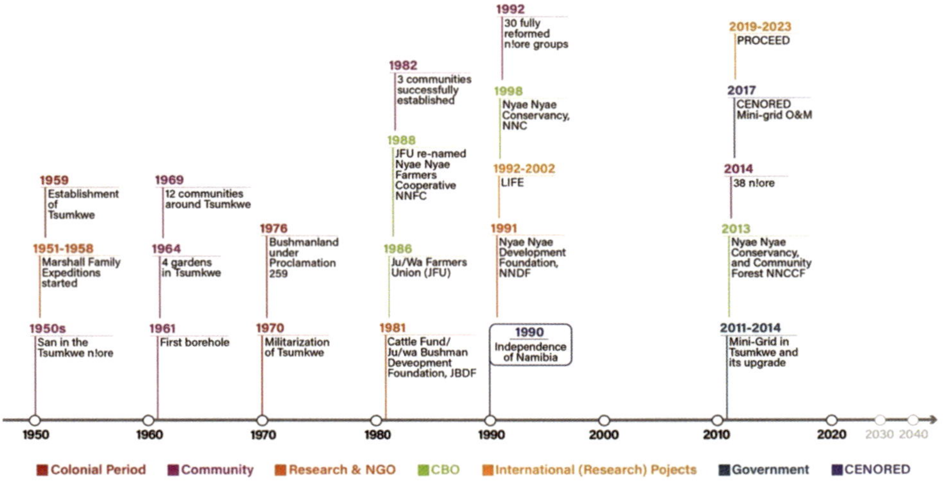


*Figure 4.7. Timeline of Main Events in Tsumkwe (1950 – 2023)*

*Source: Author*

## 4.2 Qualitative Findings

This sub-section describes the findings of the interviews and open-ended survey questions, which were analyzed inductively within the context and explained with the process of modernization. Fourteen interviews were conducted online and in person with the men and women residing in Tsumkwe who accepted interviews for this study. A fifteenth interview took place briefly in the seventh grade of the Tsumkwe Primary School

### 4.2.1 Process of Differentiation

The diversity of the local actors in the Tsumkwe Settlement informed the ministry's work carried out at the local level. In addition to the local shops selling food and other necessities, the NBC Broadcasting Company, NamWater, and Tsumkwe Lodge are among the medium to big-size companies present in the field. This section presents the findings according to Table 4.1.

*Table 4.1. Differentiated Groups and Structures*

| Type | Differentiated Groups and Structures | |
|---|---|---|
| **Place-Based Ethnicity** | San groups | Non-San |
| **Mini-Grid Ownership/O&M** | Before CENORED | After CENORED |
| **Local Actors** | Public | Private / Community |

*Source: Author*

#### *4.2.1.1 Place-Based Ethnicity*

"The past is present in places in a variety of ways" (Massey, 1995, p. 186).

Everyday life takes place! A place is beyond its daily meaning used to imply its location or territory in its physical form. A place is "a territory of significance, distinguished from larger to smaller areas by its name, by its particular environment qualities, by the stories and shared memories connected to it, and by the intensity of meanings of people give to or derive from it" as described by Relph (2009, p. 24), including the past (Massey, 1995). From one side Relph (2009, p. 24) highlighted places "occurring where the webs woven by the human beings touch the earth and connect people to the world"; while Massey (1995) focused on how places stretch through the time. Erfani (2022, p. 452) suggested that multiple forms of interrelations between people and places can bring diversity and complexity in their mutual relations, shaping one another. As a result, he conceptualized it within the interrelationalities of the individual, community, and place, including attachment, identity, satisfaction, and dependence (Erfani, 2022, p. 459).

Relph (2009, p. 31), described the sense of place as a "synaesthetic faculty" that combines the senses, movement, purpose, and anticipation, which is individual and intersubjective as well as linked to the community, personal memory, and self. Furthermore, he derivated or studied further a "pragmatic sense of a place" that combines an experience of the place together with the background of contemporary daily life within "the uniqueness of a locality" that is in touch with the regional and global contexts (Relph, 2009, pp. 30-31). On the other hand, Massey (1991) argued the sense of a place from a time-space perspective, in which movement and progress have been equivalent to "time"; statis and reaction have been equivalented to `space´/´place´. In her alternative interpretation of place from this perspective, she investigated it as intersecting points or a "meeting place" of movements and communications and networks of social relations beyond its history, on a wider scale as local, being an integrated part of the global. She also pointed out that "places are processes" of social interactions in action, which do not have single ´identities´ and also involve conflicts, and which are "continually reproduced with their own uniqueness (Massey, 1991, p. 29). According to this view, each place has its own local social relations, in which they are differentiated in a wider world; globalization of social relations creates a geographically uneven development; and along with the accumulated history, they are unique. She explored in the global sense of place ´an understanding of its character´, which can be constructed by linking that place to places beyond (progressive sense of a place)" (Massey, 1991, p. 29). Thus, from these above-mentioned definitions and descriptions of the "sense of place," it can be deduced that the phenomenological approach outlined by Relph´s pragmatic sense of place shall be the starting point in the research "need for a global sense of the local" (Massey 1991), that would pave the way for a "progressive sense of place" of the accumulated historical time-space concurrences within the "processes of places" (Massey, 1991).

When considering the history of Tsumkwe and the San people in the Nyae Nyae region, it is crucial to understand the multi-layered challenges related to 'ethnicity.' The land issues in the Nyae Nyae region have deep roots, stretching back to the colonial period and pre-independence era, as detailed in the first section of this chapter. These issues are intricately tied to the recognition and citizenship of the San people in Namibia.

The quote below, from interview 14, exemplifies the place-based connection of the San people to the ancestral land within the Nyae Nyae region that is currently managed under the Nyae Nyae Conservancy´s statutory conditions. Having the "San" identity does not give direct access to the communal land. The membership connected to living in the area for a particular timeframe outlines the constraints within the Conservancy membership. In this way, two levels or types of San identity appear in the context of San: those who are originally from the region and those who are not. The data from this interview is insufficient to state whether the "San" mentioned is meant for the Ju/´hoansi only or other San sub-groups. However, it illustrates the differentiation within the San groups in their social life, besides the differentiation among all Tsumkwe´s residents.

> "So far, **the land issues**, we, we, we, because there are a lot of people that want to come and stay now in Conservancy. But we have **rules**, and we have **a constitution**. Only, so, **it´s not something that´s made now**. It was made by the elders. **They only take San people** as members of the Conservancy. But it doesn´t mean, **if you are San**, you can come **from somewhere** and stay. No! **You have to be here**. There is a timeframe where you have to stay. So, you can **be accepted** as a member" (Interview 14, Pos. 30).

The differentiation of the identity/belonging to the statement about the structure of the Constituency´s rules, as well as the existence of a constitution, enables us to see the formalization of the rules of elders under a written constitution, which has evolved from 1986 with the establishment of Ju/wa Farmers´ Union (JFU) until the official registration and recognition of the NNC in 1998. In this level of recognition, a somehow community-based hierarchical structure appears in contradiction with the traditional San living, hunter-gatherer communities living more independently. Furthermore, this also represents one of the transitions from the oral tradition towards the written rules for the case of NNC.

This ethnic differentiation, which is also closely linked with structural differentiation, inevitably introduces some bureaucracy but makes the ethnic representation compatible with the existing structures of the country. The NACSO is the official entity supporting 86 conservancies in thirteen regions and rural associations in improving the conservation through CBNRM and managing their resources for their benefit (NACSO, 2023a). Thus, with the introduction of an officially institutionalized body to carry out their specific scope of work (division of labor), the bureaucratical aspects have also appeared or multiplied, which are also part of the rationalization process (detailed in Chapter 2). From a different perspective, NACSO took its position at the top level of the hierarchical governance structure while supporting the Conservancies at the local levels and some other associations at the regional levels.

Indeed, this process of institutionalization is very closely linked to the birth of NNC as a successor of Ju/wa Farmers´ Union by the leadership of Tsamkxao Oma=[9], also known as Chief Bobo, who saw the only way of survival of Ju/´hoansi in the 1970s, was to come together, connect, support one another other and move out of Tsumkwe to start farming and obtain self-sufficiency, when the people were under severe conditions of poverty, sickness, death, malnutrition, violence and crime. On the other hand, taking into account the young age of the respondent, he also confirms/demonstrates that the efforts of Chief Bobo have been passed on, still accepted and recognized, and enabled the next generation to pursue them, to stay in the unity of the community, to protect their land and develop further.

This success over the history of NNC, with the leadership of Chief Bobo, throughout the challenges of the colonial and post-colonial eras, has been supported by the Marshall family from the United States. They made their first expeditions in the region between 1951 and 1961 and, in later years, contributed to the research, science, San communities, and their recognition. They implemented a type of participatory action research amidst the challenging times of Ju/´hoansi. It established the first Cattle Fund and continued with the Ju/´hoansi Development Foundation in 1981 with John Marshall, Claire Ritchie, several Namibians from Windhoek, and some Ju/´hoansi who knew the father Laurance Marshall (Biesele & Hitchcock, 2011). They established the Integrated Rural Development Programme (IRDP) and assisted with establishing the JFU. As a result, participatory research led by action assisted, supported, and empowered the Ju/´hoan communities and paved the way for future generations.

## Ethnic Differentiation in Education

Beyond the land rights, the differentiation of San continues to exist in education. Upon completing the first grades in the village schools, San students come to Tsumkwe to continue their studies. The quote from interview fourteen below shows how the village schools contribute to the education of San children, but at the same time, they face difficulties when they arrive in the Tsumkwe Settlement, which potentially leads them to be shy compared to "other" ethnic groups in the classroom.

> "In Tsumkwe, we, we, I really do not know how to explain why **we are shy**. But it´s maybe because I would say, most of us, we grow up in vil-

[9] Present Ju/´hoansi Traditional Authority

> lages and whereas, some villages have schools, so. Where that, in those schools, **we only have our tribe**. So, once, maybe those schools may go up to Grade 5. If you are done, then you come here. And then, now, you are used to your tribe. **It´s hard to fit in with others, now"** (Interview 14, Pos. 54).

In line with the "shyness" of the San students, Interviewee 10 confirms it as well. It emphasizes the language of instruction at the schools in Tsumkwe as one of the main reasons San students are shy compared to their classmates. He informs that 225 children in the village schools are taught by eleven teachers from Grade 1 to Grade 3. The class sizes vary from 32 to 51, and the language of instruction is Ju/´hoansi, with the second language being English. While he highlights the addition of a secondary school as a positive outcome over the past years compared with when there were only non-San teachers and staff at the hostel, he claims that there are currently San staff at the hostel, but the San teachers are lacking.

> "**What I see** is now some changes are there because that time it was only Tsumkwe junior secondary school, and the majority of teacher were Non-San people now after that and also the workers at the hostel. But after **that now at least** we are having now two schools. The other one grade 8 to 12 and the primary grade 1, grade 0 to grade 7. But the most challenging things is we don't San teachers at the school, although we have San cleaners at the hostel. We **still want more teachers of San** who can **understand the life and tradition** and the situation of the San people, San kids. Because perhaps kids are having problems then it is difficult to come to a teacher and to explain. Sometimes it's the language barriers which is a problem, but if a kid can see someone whom he is **comfortable with he** can talk and explain his problem" (Interview 10).

From his point of view, having San teachers and San staff in the education who understand the traditions and life of San people would enhance the communication with the San students and enable them to express the issues that they face during their education in Tsumkwe. Very closely connected to the San teachers and the language of instruction, he also claims that Tsumkwe Primary School only offers English and Afrikaans, which creates additional challenges for the adaptation of the San students who come from the village schools who are only taught in Ju/´hoansi and English.

> "Yeah, **what I see is now there** must be some changes that have to take place to maybe in future to have another school. Which has to implement this the medium of instructions to be Ju/´hoansi to be taught, because now currently Tsumkwe primary is just offering **English and Afrikaans**. Now it's a problem to our children, a challenge as they are coming here for, let me say transgression [transition] now they have to start now grade 4 now English and Afrikaans. And **never been** taught

> Afrikaan …It's just English and Ju/'hoansi and I want that there be some changes as which has to happen. To have a new school which offer Ju/'hoansi and English so that our kids as they are coming from villages, they can adapt to the situation" (Interview 10).

This ethnic differentiation in education brings on surface the San identity at the early ages of a child, who is going to become the future San.

### Ethnic Differentiation and Business "Being Non-San" in Tsumkwe

Until now, light has been shed on ethnic differentiation from the perspective of the San people. However, in Tsumkwe Settlement reality, the interviews showed the other side of the medallion. Business is very closely connected to the process of rationalization, as well as the process of differentiation. Interviewee 13 is a businesswoman who has already lived in Tsumkwe for a very long time. In the interview, she talked about her struggles in working with the people from the villages and the impact of the NNC at the beginning while setting up her business. From the village where she started her business to the interruption by the conservancy, there is a movement and action created by her in a different place than where she lives. Subsequently, the voices of the people from the villages gained weight for her to do her business since her work was found helpful for the villages, and that is how she succeeded in growing her business relations to six villages within Nyae Nyae.

> "Ok, first, the way I started with the one village only. So, the other village, when they heard that I am working with them only, they started complaining, they want me to go and work with other villages. Then, I was like, I tried my best **to go there**, but then, the **conservancy, they wanted to stop me**. They don't want me to go and do the job saying that because **I am not a San**. Yeah, that's what they are trying to do. But then, the members, the villages, they said "no", with us, because we **see** that she is **helping** us, **we cannot stop her,** she must come and work with us. So, **from there**, I was working XX[10], XX and XX and then I came here in Nyae Nyae itself. It was Dinui, from Dinui, I went to Neemu iye, then Mahamise, then, they were six village" (Interview 13, Pos. 2).

In light of this quote, which builds on the previous quotes on ethnic differentiation, one can see the dependency of San in the villages on non-San for generating income and the constraints created by the Conservancy. In contrast to this situation, there is community solidarity speaking up against the Conservancy's position in the interruption itself.

[10] Village names that are indecipherable.

#### *4.2.1.2 Differentiation in the Electricity Supply Structure*

The diesel generators installed during the apartheid time in Tsumkwe/Namibia were left behind and continued supplying electricity to government institutions such as schools, police stations, and clinics. When the Tsumkwe Energy Project installed a solar PV system to run in a hybrid format, some households were connected in 2012. In the subsequent years, more households and local shops have been connected to the mini-grid systems. The government, through the Ministry of Works, operated the system and supplied electricity only during the day with many interruptions. The operational challenges faced during those initial years of the mini-grid ended with transferring the mini-grid´s ownership, operation, and maintenance to CENORED. Thus, the division of labor to undertake specific electricity supply services for the Tsumkwe Settlement occurred.

What did this mean from local and national perspectives? CENORED´s presence in the settlement had various benefits for the community at the local level. It was represented by only one person in charge of everything, such as the sales of the prepaid tokens, handling of the new applications, listening to the complaints, responses to the technical issues that occur in the clients´ connection points as well as the mini-grid O&M. The community members had a responsible person that they can talk to, submit their connection applications, and buy their tokens. The applications are not processed from the CENORED´s office at the settlement but from the central office in Otjiwarongo. For the application, the distance of the new connection point from the mini-grid transmission line has been calculated. The regular connection fee is 2,500 NAD up to 80 meters. A new fee for the connection has been calculated if it is a longer distance. Although CENORED has an internal computerized application process, this could not be observed/followed by the client. At this stage, we see the functionality of CENORED in the locality from the differentiation perspective. However, it should also be noted that the differentiation does not happen within the O&M of CENORED in their local office, with only one person in charge of everything.

From a national perspective, although CENORED has been established as a private company, it provides the distribution and supply of electricity unbundled from Namibia's centralized apartheid period. The unbundling has not been privatized but instead changed costume without losing the government's power in governing processes. The author observes this as part of the process of differentiation, which will be elaborated in Chapter 6.

#### *4.2.1.3 Differentiation through the Visions*

What is vision? Kantabutra (2008, p. 130) analyzed the theoretical and empirical literature on vision, whereby he classified the typical characteristics of the theories investigated starting from the 1990s under four main categories. Accordingly, "the vision is about a desirable future," it influences an "action towards a common goal;" it designs a pathway for "organizational members to proceed," and it is considered to be "important for the strategy and planning".

Kantabutra's research (2008, p. 132) suggests the role of vision as a critical point in leadership and management in achieving goals and found a positive impact on performance outcomes in "enhancing and sustaining organizational performance." He also highlighted a shift in the leadership from top-down to bottom-up (Kantabutra, 2008, p. 135). Building upon the definition of Strange and Mumford (2005), he defined the vision as a "cognitive construction or specifically a mental model, a conceptual representation used both to understand system operations as well as guide actions within the system" (Kantabutra, 2008, p. 135).

Whereas Tomaselli (2005, p. 124) informed about the Ju/'hoansi interpretation of time, the past, and the history through a quote by Lorna Marshall (1976, p. 53):

> "Although they look respectfully to the past, they are not history-minded. They make no effort to hold actual past events systematically in mind or teach them to their children – neither events that concern the living people nor those concerning their forebears. They remember what they happen to remember their father and grandfathers telling them."

Looking at visions, or envisioned futures, the qualitative interviews suggested the visions within the process of differentiation under two main categories: expectations and wishes, while they have also contradicted with Lorna Marshall's quote above by reflecting on their past and future.

#### *4.2.1.3.1 Expectations of "San Town in the Future"*

Panitz et al. (2021, p. 1) defined expectations as "probabilistic beliefs about the future that influence our perception, affect, cognition, and behavior in many contexts", which also enables us to navigate the future in the making or yet to come. The quote below from interview 6 highlighted both San people and Tsumkwe as "sort of left behind" but also underlined their "willingness to stay on their own, in the bush" and becoming "a town". Pike et al. (2024) described the "left behind places" as a "leitmotif" term that has increased in use after 2008 crisis, in describing geographical inequalities in the rural and urban geographies of the Global North. The term has initially been used by the United States of America after the war in Vietnam in 1960s as a motto for the army, followed up in the policy use for combatting poverty, later on in education and in the academia in 1970s, also adopted in the EU and international organizations' policies (Pike et al., 2024). The "left behind" people and places experience similar spatial disparities "in accessing economic opportunities and services" (Pike et al., 2024, p. 5). The relationality of places and people have demonstrated its presence in the elaboration of the place-based and people-centered dimensional policies, which are linked to six potential dimensions with "left behind" terminology identified by Pike et al. (2024, p. 6): in economic, social, environmental, institutional and governance, political, cultural, and infrastructural domains. The interviewee 6's opinion about San people and Tsumkwe both being left behind, contributes to the research in the spatial geographies that the people and places are inseparable. Furthermore, the use of this terminology within the context of the policy development in the poverty, education, health or development by having a place-based and people-centered approach, and thus the respondent in this case of Tsumkwe, uses the terminology itself. But there is also an accepted or a chosen situation of staying there, which is linked to their historical past.

Furthermore, moving in the direction of the town of the future, Table 4.2 summarizes the symbols of this envisioned town by the interviewees from a temporal perspective between the present and the following five to ten years. Their descriptions fit primarily into the infrastructural and public services such as a tar road, health centers, training centers, and electricity, as well as financial institutions and shops with a variety.

*Table 4.2. Symbols of the Envisioned Town in Tsumkwe*

| Characteristics | Tsumkwe Settlement (present) | Tsumkwe Town (in the next 5-10 years) Like Grootfontein, Otjiwarongo and Windhoek |
|---|---|---|
| Road | Gravel road starting from the national road | Tar road until Tsumkwe |
| Health Centers | Clinic | Hospital or a second clinic |
| Financial Institutions | An ATM machine at the general dealer | Banks |
| Shops | Local shops such groceries, butcher, food shops etc. | Malls, all different shops, cinema |
| Training Centre | Community Development and Learning Centre | Computer center, including trainings |
| Electricity | Partial electrification in essential and non-essential lines | Electrification for all households and for all services |

*Source: Author*

***"Tsumkwe is left behind, San people are left behind, San people need their own town"***

> "I think Tsumkwe is a place which was a sort of left behind. San people were a **sort of left behind within**, with the different with the other ethnics of the country. So, San people were people who were out of everyone willing to stay on their own, in the bush. So, with that development is also bringing to it that **San people also need their own town like any other ethnic in the country**" (Interview 6, Pos. 219-227).

This quote has been categorized as an expectation by being a "need" that cannot be achieved overnight but a more robust notion than being only a "wish." By electing to have a town representative of San ethnicity, he makes a comparison with other ethnic groups. He underlines a willingness to recognize the country in a higher hierarchy among different types of settlements, creating a link between "left behind" and "development-driven" inspirations. Pike et al. (2024) considered this a critical point of the multi-dimensional and relational approach to spatial geographies that contributes to developing policies for the development potentials of "left behind" places.

## (Ju/´hoansi(San)) Town of the Future

> "The ministers who are responsible for developments are also coming to talk to us" (Interview 9, Pos. 110-122).

Confirming the approach above, interviewee 9 interpreted the "development" as a responsibility of the ministers, implying the role of the government, while appreciating the fact that they came and talked to them. When looking closer at the hierarchical structure of different settlements in Namibia, three main categories are defined by the Local Authorities Act (LAA) of Namibia (1992): municipalities, towns, and villages. The Namibian Economic Policy Research Unit´s paper (Seckelmann, 2001, p. 30) states that the level below the villages are settlements not subject to the LAA but to the Regional Council´s Act by being rural centers. Seckelmann (2001) states that the degradation at the level of 11 villages in the settlements is due to insufficient financial resources to provide primary services such as water and sewerage. The electricity and gas supply has only been undertaken under the responsibilities of the "towns." Kalkfeld provided an example that lost its *village* status to *settlement*, in which a regional representative was quoted by Seckelmann (2001, p. 55):

> "No income was coming into town, especially from the community for services rendered to them. Which means no income except the subsidy from central government. (. .) Since the place is no more generating funds, it is just depending on subsidies from central government. ( . .) So that's why its title was lowered (ET K 19)."

Reflecting to the development expected from the government or governmental actors, the hierarchical levels and related responsibilities fall back to the ability of the residents of that particular settlement to pay its services that fall in the category of settlements. On the other hand, interviewee 1 expressed one´s expectation from the government in bringing the tarred road, the bank, and the shops like there are in all other towns.

> "I could like **to see it**, I could like to see Tsumkwe as a **one of the towns** in Namibia (……). I could like **the government** to put **for us a tarred road** with **the bank** and all the **shops** which are in **all other towns**" (Interview 1, Pos. 81-81, 88-90).

Tsumkwe "town" has been expressed in between the lines of expectations and wishes. They combined some ongoing works or some news with their expectations and wishes: "I see Tsumkwe…Not as a resettlement area, but as a town

(Interview 3, Pos. 74-75)" was the simplest way of expressing their expectation, emphasizing a (infra)structural upgrade, which could also demonstrate from a temporary settlement to a permanent town. According to the Namibian administrative scale, this upgrade would also introduce the infrastructural and public services. Furthermore, some interviewees focused more on its ethnic representation among other towns of Namibia as interviewee 7 specified this expectation of a "little town for Ju/'hoansi people, for not to struggle," in the sense that within this expression, the "township" would also involve protection for San people.

> "I think, **I expect Tsumkwe** to be like **a little town** for the **Ju/'hoansi people** to not struggle. Like I said going somewhere to do your things for example go to the office. To Grootfontein to Otjiwarongo to Windhoek. Yeah, for sure and in ten years Tsumkwe will be, must be changed, will be **changed for sure**" (Interview 7, Pos. 9-15).

Others put more emphasis on the symbols of the towns they mentioned, like Grootfontein, Otjiwarongo, or Windhoek; "no more going shopping to Grootfontein" sounds like advertising to live in Tsumkwe also highlighted the willingness of people living and staying in the settlement interlinked with their desire to supply their necessities from town-shops.

> "Of course, we are **no more going to go for shopping** to Grootfontein. We are going to do our shopping in Tsumkwe" (Interview 3, Pos. 84-88).

"Where people will come together" could also be seen as a unifying point through electricity, signaling a shift in their lifestyles since neither cinema nor computer centers are native to the settlement but exit in their future expectations as a recreational and learning center of interest.

> "Yeah, I think that we will also need ... Electricity will be. There will also be centres. Like uh. If it is not **cinemas, computer centre where people will come together.** People will want to do **computer trainings**. They will also have centres where computers are installed where they **can come and have their courses**" (Interview 12, Pos. 112-114).

On the other hand, in line with the town's service provisions, Interviewee 3's expectations about the malls and hospitals were connected and conditioned to the tar road. In his perspective, without the tar road, the eligibility for these services could not be achieved in Tsumkwe.

> "I hope so because with the **tar road**, everything and those five years that I am **seeing forward from here**, things will just be going smoothly. **Malls and all the different shops**, even the **hospital**, will be built, you know. Everything, from five years onward **as long as there is** a tar road. I believe, yeah" (Interview 3, Pos. 104-111).

The tar road was a focal point of interest in their visions of a town, followed by a hospital, shops, and banks. Whether the gravel road will be upgraded to a tar road and extended to Tsukmwe or not, there is an expectation of interviewee four that they are waiting for it. In this case, there is an expectation from the government to build the road, and the developments are connected to it.

> "What I'm seeing here now, even I think maybe the **tar road will be here** if I'm not mistaken. I think they will extend even maybe **the secondary, the clinic**. A lot of development I see that still comes. **We are waiting for it**" (Interview 4, Pos. 81-89).

In the summary of their expectations of becoming "(San) Town," all interviewees detailed their expectations of a town-like Tsumkwe by visualizing them during the interviews. In this sense, how they shared their expectations was genuine and imaginable, as was their place-based connectedness. Moreover, beyond the meaning of enhancing everyday life activities linked with service provision, on the ethnic place-based connotation, this also implied the recognition of San people and their town-to-be-settlement as in an equal administrational status with other ethnic groups before the authorities, beyond their past struggles.

#### *4.2.1.3.2 Wishes, Hopes, and Desires for the Future*

As the Oxford Learner´s Dictionary (2023) defines, a wish is a hope, a desire, or a longing for something to happen. It does not exist in the present moment but would be desired in the future, regardless of its (im)possibility. The wishes for the future that the respondents mention have a material or calculable form towards the future but also contain some abstractness as well. The findings are presented in Table 4.3. Employment opportunities shall be observed and quantified, but the qualities of everyday life and equality have different indicators to evaluate the degree of their normality and equality.

*Table 4.3. Wishes, Hopes, and Desires for the Future*

| | |
|---|---|
| Wishes, hopes, and desires for the future | More opportunities, more development |
| | Employment Opportunities for Ju/'hoansi |
| | "Normal Life as Brothers and Sisters" |
| | "I want it for Everybody, I want us to be equal!" |

*Source: Author*

Interviewee fourteen's quote contributes to the intersection of expectations with the wishes toward growth and development in Tsumkwe, recognizing "electricity" as one of the "place-changing factors" in enabling this desired "useful growth." Electricity plays a vital role in the envisioned future.

> "I wish to also **to get more opportunities coming**, because **more electricity for us**, the whole villages, and as well as **more development**. I want to **see the place grow**. Because **I know the electricity is one of the changing factors** of **building a place up of, making a useful**

> **growth**. So, that´s what I would like to see in the future" (Interview 14, Pos. 62).

### Employment Opportunities for Ju/'hoansi

> "Actually, if there were **projects for the Ju/'hoansi** that they are capable of then maybe, It could **maybe be gardening and anything to their needs**. But I can not specify for their needs but even things **like a chicken project** would be something that would **change their lives if they take it by heart**" (Interview 11, Pos. 201-207).

Likewise, the wish for development is linked with electricity, it is also linked with employment. Within the ethically differentiated context, interviewee eleven expressed his wish to have projects for Ju/'hoansi and projected some ideas such as gardening or poultry or anything that might accommodate their needs. In this quote, there is an acceptance of the (in)capability of San people from his side to achieve this type of work. Nevertheless, as a wish that might come true, he foresees it as a life-changing possibility if it is worked on and taken seriously. The struggle of the San people with the risk of losing their land in the past, the "place of death" in Tsumkwe, and the realization for themselves that the only way to survive has been through farming and settling in their n!oresi, this interviewee expresses the 21st century struggles for Ju/'hoansi in surviving in Tsumkwe with "projects" and having livelihoods that our semi-dependent for sustaining life. This partial dependency is also related to creating livelihoods under the conditions in Tsumkwe.

### "*Normal Life as Brothers and Sisters*"

On the other side of the coin of ethnic differentiation, interviewee five´s wish is "to have a normal life as brothers and sisters", expressing her desire to be one family, regardless of their ethnic or racial background, and also showing the ethnic differentiation within Tsumkwe among different ethnic groups or even tribes to one extent. According to her, everyday life means a family in unity, in which she explains her desire to bring them together in the second quote below. She also emphasizes that there is distance and discomfort among "black people" and "Bushmen" with the exclusion of some.

> "I mean **normal life** is like, people's normal. People... Let me say this side people are like: No, **he's a black person** and he's **Bushman** and he's **a white**. Just like **a normal person as one family.** Not things like that. No, you cannot live with people because you are the tribe. **I cannot do with this** (….) because **you are not my tribe. I just … a normal life as brothers and sisters**" (Interview 5, Pos. 226-237).

> "It's like… These things of… like you are a Bushman and you are what tribe. They are just not, not getting along man. So, **I want to bring**

> **them together even as one**.... (Interview 5, Pos. 245-250) No. It's just among..., Let me say here in, Tsumkwe it's among Bushmen. Bushmen and blacks, they are not that close. Only some of them" (Interview 5, Pos. 253-256).

### "*I want it for Everybody, I want us to be equal!*" - Equality Among all Residents in Tsumkwe

From a non-San perspective, the wishes belong to the community as well as the whole. Interviewee 13 underlined community while refusing to have individual wishes or wishes for herself. There was a dislike for thinking about individuality, but the entire community perspective was well-adopted and accepted. Different from the previous wish for employment, this one shall take place or not. This quote, "I want us to be equal," contains a stronger desire and more determination for it to happen in the future for both San and non-San who belong to the entire Tsumkwe Settlement Community without any ethnic differentiation. Interviewee 13´s future wish includes everyone in Tsumkwe living in harmony and working together in a quiet, safe, and pleasant land away from illegal things.

> "In the future, it will be a nice place **if** us, the community, **we are working together**, not bringing illegal things in this area. Because it´s a **nice land, very quiet and safe**, not like other places. (...) What I can wish myself in the future? With me, I don´t like, like... how can I say it now? **Like benefiting myself only, the future I want, I want it for everybody in Tsumkwe**. It´s what I want**. I want us to be equal**" (Interview 13).

The quotes above illustrate some ethnically differentiated wish(es) with undifferentiated ones while considering the current issues in Tsumkwe among different ethnic groups. From a broader perspective, Ju/´hoansi wishes, if one might say, depict a vision of "survival" or being "self-sufficient" or "self-sustaining" in their everyday life. In contrast, non-San wishes depict more expansive harmony among people in their daily lives, living together. As a result, although one could conclude that the expectations, wishes, hopes, desires and overall visions are place-based ethnically, there are also place-based visions for the future from non-San groups who envision a future life in Tsumkwe-town-to-be.

### 4.2.2 Process of Rationalization

The definition utilized by van der Loo and van Reijen (1992, p. 31) is the "ordering and systematizing reality in order to make it predictable and controllable, which also implies that our thinking and acting is more and more subject to calculation, justifiability, and control, that happens in different and interrelated levels". These levels include the level of worldview, the collective action, and the individual action levels. From an action perspective, the rational action implies a "well-considered" action, seeking means and methods that are effective and efficient in result, which create the surrounding reality that is controllable (van der Loo & van Reijen, 1992). It was also considered a cultural perspective of differentiation, considering the cultural reality with the perception, values, meanings, norms, ideas, and symbols that give meaning to our actions and have a role in the culture (van der Loo & van Reijen, 1992).

The rationalization in action (within the PROCEED project) situates itself in a supra-national, national, regional, and local level of rational action in order to analyze and provide measures for the mini-grid(s) to be predictable, controllable, and also sustainable. The mini-grid's social, economic, and technological analysis has incorporated predictability and controllability. Since the national actors such as the Ministry of Mines and Energy, the Namibia Energy Institute, the Namibia University of Technology, CENORED are directly involved in the research with the supra-national actors such as the University of Bayreuth, the University of Applied Sciences of Neu Ulm, the Technical University of Applied Sciences of Ingolstadt and industrial partners from both countries; one could describe the rationalization at the national and regional levels in the process of reaching to the local level by different means. Based on the previously explained literature by van der Loo and van Reijen (1992), the change in the worldview, turning into a knowledge-based society, relying on scientific evidence to forecast and control, has contributed to the modern scientific worldview. Weber´s "disenchantment of the world," which signifies the "disappearance of the magical, enchanted garden life towards a rational, controllable and calculable future," makes a twist in the case of electricity in Tsumkwe. As much as the rational actions supported on supranational, national, and regional levels, at the local level, the meaning of the electricity could be best described as the "Re-enchantment of the worldview" or "enlightenment, taken the invisible form of "electricity," which is most likely where the magic is associated with.

#### *4.2.2.1 Re-enchantment of the Worldview*

Landy and Saler (2009, p. 1) suggested that Weber´s disenchantment with the world was "incomplete" and "each time there was a withdrawal of religion in a particular area, the remaining was filled with a secular strategy for the re-enchantment would cheerfully emerge to fill in the void," which was making the strategies themselves enchanting. One could also interpret this situation with electricity: energy/electricity being a modern tool or a secular tool for development, bringing local, regional, national, and international actors of MLG together.

From a broader perspective, the renewably generated electricity in the off-grid regions and peripheries is the "magical," "enchanting" tool in reaching energy access, reducing poverty, and creating new jobs and livelihoods. It is also responsible for enhancing the conditions of the healthcare centers by reducing child mortality and other diseases (through improved medication storage), improving the educational infrastructure (through better materials and conditions), providing power for the use in agriculture and animal husbandry (to grow crops, to harvest, to self-sustain). These are just some of the magical characteristics mentioned in almost every RE-based project or policyt, which are supported by the seventeen Sustainable Development Goals as well as in many visions and development plans of the countries looking forward to having 100% energy access for their residents.

"**I thought electricity was something magical, something big**. There would be big candles everywhere. I heard electricity in physics but never thought that the light to be something so big" so described a young lady in her early twenties in a random walk down the Tsumkwe Main Street. The times when she met with the electricity were her early teenage years; one could only hear the cheerful resonance of this moment blended with the memories of her first thoughts and expectations. However, doesn´t this have a similar connotation with the re-enchantment in the rest of the levels of governance with the implementation of renewable energy-based electricity enriching the imaginations of many?

#### *4.2.2.2 Meaning of Electricity*

The meaning of electricity has been compiled as illustrated in Table 4.4 and the findings have been explained accordingly.

*Table 4.4. Meaning of Electricity*

| | |
|---|---|
| Meaning of electricity | "It means light to me!"<br>From the Optical Visibility to "*Video, ergo sum*" (I see, therefore I am) or "*Scio, ergo sum*" (I know, therefore I am) |
| | "Source of Life" |
| | Reliability and Autonomy |
| | Everyday Life Matters: From Basic Functional Needs to Comfort |

*Source: Author*

### From the Optical Visibility to "*Video, ergo sum*" (I see, therefore I am) or "*Scio, ergo sum*" (I know, therefore I am)

Further meanings also supported this powerfully random description of the "magical" electricity in describing it for the community in Tsumkwe. The respondents often mentioned that "I/we don´t sleep in dark/ness anymore" to answer the question about the changes in their daily lives after their houses were connected to the mini-grid. Descriptions would often also continue with "we have light now!". However, the meaning behind this light was not meant to have only a bulb illuminating a space; this light meant "optical visibility"; this light meant to be "(sub)consciously aware" that when they wake up at some point in the night for whatever reason, they would see around without any trouble; it would be a safe space. Further respondents emphasized it; "we were struggling before, especially sleeping in darkness," "when we wake up in the night, we can turn on the light to see around, " or "I know that there is light, no more darkness. " One could also consider this „knowing" as one of the rational reflections. "When we wake up in the night, we can turn on the light to see around". As Relph (1970, p. 194) suggested from the phenomenological approach in looking at the life-worlds, the experiences of humans and the relationality between the world and human beings provide meaning(s) for "our desires, our evaluations and for the landscape we see". The relationship between the eyesight in the night with its immediate environment and being sure without any doubt that the person is alive represents a (inter-)connectedness to life. Thus, the "darkness" representing the former life has transformed into the enlightenment in modern life. Therefore, an inspiration from Descartes´s "I think; therefore, I am" statement could be formulated as "I see; therefore, I am" and even furthermore, "I know, therefore I am," which may support further research fields and endeavors.

In this way, the community in Tsumkwe has added to the substantial meaning of "optical visibility" as a practical, calculable, controllable condition for their everyday life, beyond a simple bulb quantified in most research. Having heard this resonance in stressing the word "light" caused the author to look and think of it differently. Thus, long after the first research and the interviews, the author had the experience of being exposed to the blinding and dazzling Christmas light decorations in one of the big airports during travel under many restrictions. Indeed, it was a moment of recognition of the re-enchanting characteristic of light in every geography through this research.

## "Source of Life"

From the re-enchanting magical meaning of light or electricity, the physical notion of visibility was also very closely connected to the reliability of the electricity throughout the day, throughout the week (24/7), and being a vital source of life. This view was mentioned in several interviews. Interviewee 3 attempted to compare its vitality with water, which is the closest approximation and is considered "one of the most important things in a person´s life." Through this statement, the relationship between the individual and one´s surroundings has gained prominence.

> "It is **one of the main sources of life**, you know. Some… It's… Let me **not compare it with water**, but it **is almost**. Let me bring it nearby, but it is **one of the most important things in a person´s life**" (Interview 3).

In the following quote from the interview 14, the physical characteristics of energy as a source was emphasized, with a comparison with "sun". One could make an analogy here and say "light, as the sun of the night."

> "But as I understand electricity, it is **a source of light**, **source or energy**, which we get. Most of us, **we use it**, that´s **where we get light bulbs**, light bulbs **in houses**. We have **the opportunity to switch on** lights and **switch off**. So, electricity for me, **it´s more of light**. **It´s more like the sun**" (Interview 14, Pos. 10).

Beyond the source of "life" and source of "light," interviewee 13 related light's functionality in her everyday life experiences and the multiplicity of the experiences that she can achieve with it. "We´ll be stranded here" is closely connected with the "left behind" statement highlighted in the previous sub-section. However, in contradiction, the interviewees claim that they are beyond that point and can do anything they want to do with electricity. In its absence, her analogy is "being like blind" and not "seeing anything going on the other side," which brings us back to the optical light and, to one extent, beyond the optics, considering her life-world experiences. However, what is "the other side"? In this case, one of the meanings could be "seeing afar," seeing distant points nearby, and the immediate surroundings become optically visible in the dark hours of the night, which makes night hours busy with whether leisure, security of supply and safety or any other activity the person seems interested in doing.

> "For me, **it means a lot**! Because there **is more experience**, and we **need electricity to use it**. Without electricity, we **cannot do anything**. We´ll **be stranded here** and we´ll be **like we are blind**; we **cannot see** what is going on other side. But now, **it´s very fast for us to use anything which we want to use**" (Interview 13, Pos. 2).

**Reliability and Autonomy**

*Table 4.5. Reliability and Autonomy*

| | Past | Present |
|---|---|---|
| Reliability of electricity | "We **didn't have light before."**<br>**"Before** it was **not good**, switch on-off form morning to lunch. From lunch to afternoon was not good." | "We have **permanent light**."<br>"It´s **always available."** |
| Reliability of water | | "The disadvantage is water because **water is connected to electricity**. Sometimes the electricity goes off." |

*Source: Author*

The meanings of electricity as "light" or source of life have also been completed with additional meanings (Table 4.5.). Another respondent would also touch the base with its permanency and its vitality, especially with the connection of water: "We have **permanent** light. We didn't have light **before**. The disadvantage is water because **water is connected to electricity**. Sometimes the electricity goes off, we don't know the reason. For 2-3 days…" The essential electricity lines also connected to the water pumps supplying water to the settlement are the reason for this water shortage. When the power goes off, it immediately affects not only the "light," permanency, or other electric appliances but also cuts the water off. Some would also recall when the electricity was supplied with intervals or interruptions: "Before it was not good, switch on-off from morning to lunch. From lunch to afternoon was not good." The interruptions have not been appreciated. Another reason for this misappreciation, besides water and electricity shortages, is dependency. When they do not have it at home, they become dependent on other people and become limited in their daily activities.

The reliability of electricity was also linked in a way to a sense of freedom, independence, or autonomy "I can do **whatever I want** with light **at night,**" which was also widens its doors to self-esteem: "It´s **always available**. **No more** asking other people for charging phones...**please**…**please**…". The multiplicity of the home activities they can do on their own and the ability to do them without the need of other people brings autonomy through the reliability of the electricity supply (Table 4.6).

*Table 4.6. Dependency and Autonomy*

| Dependency | Autonomy |
|---|---|
| "No more **asking other people for charging phones...**please…please." | "I can do **whatever "I" want** with light **at night."** |
| | "It´s **always available."** |
| | "It makes it **easy for me** to charge my phone and not going to **the houses of the neighbors."** |
| | "I don't go to **people´s house** to watch TV." |

*Source: Author*

Another respondent would support it "I am having light. I don't go to people´s houses to watch TV." These quotes from the respondents also reveal the individual action level described by van der Loo and van Reijen (1992). The practicality and usefulness also contribute an added comfort or easiness into the life: "It brings light, and it **is useful**. It makes it **easy for me** to charge my phone and not going to **the houses of the neighbors**". Electricity makes life easy for people, and for some, it makes them "happy."

> "I work with electricity; it **makes me happy**. Now I have light at home, no more **sleeping in darkness**" (Quotes from open-ended questions [QOQ]).

### Everyday Life Matters: From Basic Functional Needs to Comfort

The threads of energy/electricity (Pasqualetti, 2021) had already been interwoven within the residents' daily lives, with what Weber had described as a rationality of purpose, too. In this sense, technique/technology could be interpreted as the "embodiment of the rationality of purpose," which has provided "autonomy and expansion in the modernization" (van der Loo & van Reijen, 1992). It is essential to note the nuance between independence and autonomy here. Morris (1994) suggested that independence is the ability to do something without assistance, whereas autonomy brings the ability to exercise control on doing something. Electricity use makes people independent from other people while also bringing the autonomy of doing what they would like to whenever they want to do. However, this also creates a dependency on electricity and technology.

A respondent would describe it: "Now I am able to work at night. We have electricity 24/7. I can iron my clothes!" or "I can buy it anytime; it tells you what and how much you are using." This quote informs us about the purposeful use of electricity and the rationality and self-control behind it. Another example would indicate:

> "It was something very new because before we didn't know about electricity and with electricity there are lots of thing that you can do, and you can do plug-ins and you can charge your cellphones and that is all the thing that we didn't have in the past. It is completely a new thing" (Interview 12).

The following examples are multiple quotes in exemplifying the meaning of the electricity interwoven in the daily life:

> "We **have power, electricity** at **home**. I **can do my studies in the light** not with a candle. I **can warm up the food in microwave**. The **fridge is working**, I can **iron my clothes**. Those are the basics for electricity" (QOQ).

Electricity provides an ease, continuity, and a zone of comfort in the daily struggles of the residents: "I am using it **for everything in the house**. It is not giving me any problem. **Everything changed**! Fridge, microwave, TV, **no stress anymore**" (QOQ).

> "I am happy because it is **giving light** and I can **use electric appliances**. I have **permanent light inside the house now**" (QOQ).

On another level, this permanency, this reliability, brings the comfort of security on an individual and household level, as is the case for the country´s desires and visions of having an energy-secure future for self-consumption and self-autonomy. As van der Loo and van Reijen (1992) suggested, the rationality of purpose had provided and also enabled autonomy and expansion in modernization. This (self-)autonomy has also led to pluralization/specialization, which is linked to the division of labor.

*Box 1. "We have Light, now!", What is Light?*

In light of the meanings mentioned here and will be elaborated further about electricity, the "light" steps forth with its multi-disciplinary or multi-dimensional meaning. The literature suggests that *light* has been subject to research from Greek Antiquity to ancient Egypt, the Islamic period, the scientific revolution, and the light in the $20^{th}$ century in optics (Zubairy, 2016; Al-Amri et al., 2016). As the interpretation of Zubairy (2016, p. 5) suggested, the vision was the way/sense/action our eyes stretched out to "touch" or "feel" something at a distance." The different theories of vision that were investigated in antiquity attempted to explain how objects were perceived by humans from the point of distance, their size, their color, and their shape and prepared the foregrounds for the subsequent research and development in the studies of vision, physics, namely, geometrical optics, physical optics, and quantum optics (Zubairy, 2016).

*Figure 4.8. Reflection and Refraction of Light*
*Source: AI-generated image using DALL-E (OpenAI, 2024)*

McDonough (2010) discussed Descartes´ contribution to optics, *Dioptrics,* published in 1637, in three main subjects under ten *discourses*: (1) the nature of light and the laws of optics, including the law of reflection and the law of refraction, (2) human vision with its anatomical description of the eye and the trajectory of light through the pupil, the refraction and reception of the image in the retina, (3) the improvement of the human vision. Ribe (1997, p. 42) claimed Descartes´s Dioptrics as a "practical philosophy" in rendering human beings "masters and possessors of nature," which also refers back to the process of differentiation and the process of domestication within the steps of modernization in the $17^{th}$ century.

*Box 1. "We have Light, now!" What is Light? - Continued*

Based on the quote of Descartes cited by Ribe (1997), Descartes introduced his practical philosophy:

> "By means of this philosophy we could know the force and actions of fire, water, air, the stars, the heavens, and all the other bodies that surround us, as distinctly as we know the various of our artisans; and we could employ them in the same way for all the uses to which suitable, and thus render ourselves as it were masters and possessors of Nature. This is not only for the invention of innumerable devices which would allow us to enjoy with the fruits of the earth and all the comforts to be found there, but also principally preservation of health, which is without doubt the chief good and the foundation of in this life (Ribe, 1997, p. 43)."

In his practical approach, Descartes, being a mathematician and philosopher, used philosophy in the quest from "light to vision," whereby he first investigated the nature of light and worked on its physical characteristics. This was followed by human vision, how it was interpreted and perceived by the brain through the use of the eye and retina as an organ of interpretation, and finally, concluded in the ways of improvement of human vision. As mentioned earlier, this light, reflecting and refracting, has been bringing size, shapes, colors, and distance for a healthy human (eye). In a way, this light enables human beings to look and see the world surrounding them in many colors, sizes, and shapes to make sense of this visual information that they receive firsthand. In its absence, when there is no light, there is nothing to see in the distance nor in the horizon for the human eye. Hence, Descartes, with his rational philosophy, has also contributed to the Enlightenment era of philosophy and science.

In this multitude of different studies not only over the past century but from the antiquity, light has been described as rays, particle, waves, and both particle and waves to understand the vision of human (eye), which is an essential organ to have a phenomenological perspective on our life-world. Zubairy (2016) claimed that "What is light" is still a hot topic although many advanced studies in mathematical theory have been achieved. However, the phenomenological journey from the human vision towards the future vision of electricity/energy could potentially bring many insights to its multi-level governance.

#### *4.2.2.3 Interwoven Processes of Differentiation and Rationalization*

**"Electricity was essential for everyone but not everyone has electricity!"**

> "Like this one, this one that is currently on, **electricity is part of life, everywhere**. It was actually implemented last, **2018**. There was already **a sort of**, form of electricity which was being used. But it was not convenient to supply the whole area or with everyone. Even the current one is **still not convenient** to supply everyone in the area. So, **not everyone** is having electricity. So, it is **for those who can afford** and to the **designated area**. Where you situate yourself is where the electricity is, but most of the **San people** most of them are **unemployed**. Ninety-nine percent of the San people are unemployed. So, they are only relying on the tourism sectors. But they **cannot get that fund to install electricity for everyone**. Since you, whereby you install electricity to your place you need, there is a certain amount of **money that you have to pay**, is when they will connect you electricity in your house, but if you are unemployed, you'll **not be able** to get electricity. So, the **benefit of electricity** is only the **side of government sectors, business areas employed people**. Is **only** where you can find where people are having electricity. So, electricity was essential for everyone but not everyone has electricity" (Interview 6, Pos. 319-350).

**Affordability and Partial Electrification**

"Electricity is part of life," and "everywhere" is the acceptance and the recognition of electricity within life. However, only "designated areas" have electricity, confirming the partial electrification in Tsumkwe; not every household or every shop is electrified. The quote above emphasizes the differentiation and rationalization within the energy/electricity sector from its affordability perspective. A connection fee must be paid to get a house or shop connected to the mini-grid. Interviewee six highlights the advantages of people employed in the government sectors, in the business who have an income to pay this fee. In contrast, the San people claimed to be primarily unemployed or work in the tourism sector. Therefore, according to interviewee six, San people are disadvantaged since they are unemployed and cannot "all" get connected to the mini-grid. Thus, he underlines that although "electricity is essential for everyone, not everyone has it."

The meaning behind this statement is that it highlights the irrationality with which they live, with a rational comparison. During field research, (un-)employment availability has been a significant challenge to residents. In particular, the San people mentioned that they only buy food from the local shops when they earn or receive some money. Electricity is essential for life; it should be for everyone. However, the interviewee also recalled the period before 2018, when the electricity was supplied in intervals or poorly to the entire area, and considered the current system similar to earlier since it is not reaching everyone. The interviewee exemplified that the possibility of San people´s houses connecting to the mini-grid is low due to the lack of income for connection; only employed people can connect where the electricity is.

#### *4.2.2.4 Future Visions*

The visions of the future occur where people are located and where they live their everyday lives. Willow (2022, p. 56) claims that in advancing sustainability research, environmental and social sciences have emphasized more in the study of the past and present while seldom touching the future, where future visions influence and inspire action. The visions are community-constituted, and every community has different logic, wishes, and expectations. With the introduction of energy/electricity in daily life, the respondents mentioned that "everything changed" in their lives by enabling them to have "a normal life." The bodily connections they experience through everyday life have made people more connected to a view of everyday life while, at the same time, becoming dependent on technological appliances and electricity. However, electricity's practicality, usefulness, and comfort are more advantageous. Electricity has become indispensable in the process of modernization. It does not only become indispensable but also irrevocable. The interviews in the Tsumkwe Settlement indicated an overall, community-wise accepted, and constituted vision that Tsumkwe is the *capital town* of the San people.

These visions in Table 4.7 are categorized under expectations and wish to show different social dimensions in the community within the process of rationalization.

*Table 4.7. Visions in Tsumkwe*

| **Visions** | |
|---|---|
| **Expectations** | Differentiated expectations for growth and Non-growth |
| **Wishes** | Individual/self to community – material to abstract/immaterial |
| | Place-based infrastructure - Town |
| | Employment – Vocational Training Center |
| | "Live a Better Life" out of Tsumkwe |

*Source: Author*

##### *4.2.2.4.1 Expectations*

**Differentiated Expectations for Growth and Non-Growth**

Similar to the expectations of the process of differentiation, employment also has a significant place both in the expectations and in the wishes. Following both quotes, employment was found to be a key and changing factor in life. In the quote from interview 11, there is a conditioned thinking that "if more employment opportunities could come the Ju/´hoansi or San people´s way, then, I will say that there are some changes." He interprets employment as a changing factor, compared to electricity. Since he does not see employment opportunities for Ju/'hoansi, he thinks "life for Ju/'hoansi going backwards". In this sense, one could also say that there used to be some opportunities earlier, but not anymore. Income generation through employment is vital for their life.

> "I think it would be better if **more employment opportunities** could come the **Ju/'hoansi** or **San people's way**, then, I will say that there are some changes. What **I'm seeing** is like the **life of the Ju/'hoansi is going backwards**" (Interview 11, Pos. 160-165).

On the other hand, interviewee 14 is more optimistic about employment. He stated that he knew that "the electricity is one of the changing factors of building a place up, making useful growth." In this quote, one can see "electricity" as a condition (one of the conditions) for the growth of a place, the progress of a place, and its development. Electricity can also bring opportunities, not only for the settlement but also for the surrounding villages. This quote also underlines development positively and usefully. Although the quote starts with "I wish," it has been categorized within the expectations due to its conditionality as well as the stronger desire for it to happen "I want to see...", beyond a dreamlike wish to have.

> "I wish to also to get **more opportunities coming**, because **more electricity for us**, the whole villages, and as well as more development. I want to see the place grow. Because I know the electricity is one of the changing factors of **building a place up of, making a useful growth**. So, that´s what I would like to see in the future" (Interview 14, Pos. 62).

#### *4.2.2.4.2 Future Wishes "to live a better life"*

From the rational expectations of having employment opportunities for San people in order for them to improve their lives and electricity that brings growth and development to the place, the wishes depict individualized wishes as well as communal, in material or immaterial characteristics (Table 4.8). The rationality of purpose becomes prominent as an overarching theme within the process of rationalization for four main categories: wishes (1) for life, (2) for business and employment, (3) for a Vocational Training Center, (4) for Development that is illustrated in Table 4.9. The interviewees have been clear about being self-sustaining or self-sufficient to live a "better life" or a "normal life." In their wishes, they demonstrated their own business or employment as a form of income generation to live their lives and simultaneously serve the community with their work. An imagined Vocational Training Center providing training and work opportunities has been highlighted as a "need" for Tsumkwe, along with having banks, offices, hospitals, shops, and malls within the development in the settlement (Table 4.8 and 4.9).

*Table 4.8. Wishes under Four Sub-Categories*

| **Wishes (Sub-category/Type)** | **Individual** | **Community** |
|---|---|---|
| **For Life** "Live a Better Life" | "I wish to have **my own place** whereby I **will be staying myself** and then have **everything that I want** for **myself… my own car**..." | |
| | "I would **humbly** wish to have things like **cattle and goats**, because **if** you have these**, then** your **future life will be better."** | |
| | "For the future I would wish **to have a village**, but **not to stay in Tsumkwe**. So that I could think of doing things like gardening and things that would **bring some kind of development**. But now I'm just sitting here in Tsumkwe **asking to have a borehole** and that is **not so easy** to get in order **to live a better life**." | |
| | "**Nice future**, **normal future**. **Live with people**, **laugh with them**…. **Just as one"** (Interviews 5, Pos. 39-40). | |
| **For a Vocational Training Center** | "... But for me, I just want each of the youth or any other person with legal document or something from school to get employment". | "Just one thing, it would have been nice, one of the **biggest need** in Tsumkwe would be a **Vocational Training Centr**e, a Vocational Training Centre. That would **need electricity**. Because in that Vocational Training Centre would have, maybe electricity, they can learn about how to fix electricity, fit antenna or even auto mechanic (…)." |

*Source: Author*

*Table 4.9. Wishes under Four Sub-Categories -2*

| Wishes (Sub-Category/Type) | Individual | Community |
|---|---|---|
| **For Development** | | "For example, it will be like, as I said; **banking systems, the offices and the hospital** which **must be much better. (…..)**. And then **other offices and also where you can get money**. Sometimes you'll need to borrow money, you can go to…" |
| | | "(..) there will be lots of developments. (…)I hopefully believe there will be **lots of shops and malls** by and from five years on." |
| **For Business & Employment** | "I will wish I have my **own business and take care of my kids**. That's the only wish I have and live **a normal life**. The kind of business, the business that I'm planning of doing is like a **food business**. So, like now in Tsumkwe people are like struggling a lot because they don't have a where can people buy food. They are just taking, there are only small shops I can say. So, I wish to have a **big shop** to sell food supply the family around here." | |
| | "I would like to **be just a businessperson**. Yeah, not a person to work **under Government** or in other job, but if that job can be able to **fulfill my dreams**, I can able do that. But my dream is to, **job opportunity just to work or to continue** with my business whereby I can **eradicate the unemployment**, which is on in Tsumkwe." | |

*Source: Author*

## For Life - "Live a Better Life"

> "**Nice future, normal future. Live with people, laugh** with them.... **Just as one**" (Interview 5, Pos. 39-40).

The interviewees expressed their wishes for the future in a materialistic way, providing their conditionalities to have a "normal" future or "better" life for themselves. In this regard, interviewee two wishes to have his own place, car, and everything he wants. However, he also stated that having a gravel road challenges this. In this way, he immediately thought of the "real-world" conditions while expressing his future wishes. Thus, while this quote also partially belongs to the process of individualization, at least in the wishes, it does present logical roots to the present life, which is calculable and controllable.

> "I wish to have **my own place** whereby I will **be staying myself** and then have **everything that I want for myself** (Interview 2, Pos. 49-51) (Laughter....). My own car – Laughter" (Interview 2, Pos. 56-57).
>
> "For example, yeah. But the challenge now we don't have a tarred road. We still have the gravel road" (Figure 4.9) (Interview 2, Pos. 59-61).

*Figure 4.9. Main Road and Main Street in Tsumkwe*

*Source: Author*

On the other hand, interviewee 11 "humbly" wishes to have cattle and goats" for a "future life that will be better." The sentence has been expressed conditionally. One cannot talk about a better future life if he does not have the cattle and goats. This condition goes back to the earlier times of the San people in Tsumkwe when they struggled with food and water, and they learned that self-sustainability is crucial if they would like to live in Tsumkwe and survive in Nyae Nyae. "Humbly" shows modesty in wishing something for oneself, which might conflict with the community or its values. However, reflecting on a future unknown for that moment has not been an issue for him.

> "I would **humbly wish** to have things **like cattle and goats** because **if** you have these, then your **future life will be better**" (Interview 11, Pos. 223-225).

Furthermore, the wish of "not" staying in Tsumkwe was also mentioned by interviewee eight, who is looking forward to living in a village and gardening;

this would bring him a sort of development. This wish also contributes to the self-efficacy of San people's lives in their core understanding of living in Tsumkwe and Nyae Nyae. The borehole has also been part of the wish and the issue since he is waiting for it. Otherwise, he will be unable to "live a better life."

> "For the future, I would wish to have a village but **not to stay** in Tsumkwe. So that I could think of doing things **like gardening** and things that would bring some kind of development. But now I'm just sitting here in Tsumkwe asking to **have a borehole** and that is not so easy to get in order **to live a better life**. Only then will I be sorted. That is the only thing that I am wishing for" (Interview 8, Pos. 60-69).

Previously mentioned wishes described a self-sustaining and individualized lifestyle with material objects. In contrast to this depiction of life, interviewee five underlined a nice and normal future from a unified community perspective, living and laughing together. This quote also relates to the differentiation in Tsumkwe with the conflicts between the San people and blacks (or Non-San). The idea behind this is to bring them together, be equal, and live without conflicts, contributing to the well-being of people living in peace.

**Business & Employment**

Intertwined characteristics of business and employment for livelihood with the individualistic and communal linkages also appeared in the interviews. The interviewees have been wishing to have their "own business" in the service of the community. The individual intention of owning a food shop (the quote from interview five) or not working for the government but having "my job" or "a job that will fulfill my dreams" (the quote from interview six) shall be seen from this perspective. Not working under the government has also been perceived as not suitable for the character of interviewee six. However, both wish to do a job to serve people. Interviewee five wants to make it happen via the "big food shops" where food is supplied to families. The underlying aspect is that "big shops" have been associated with the "town" ideal or wish, where everything is available for them, unlike the small local shops. They will not need to go elsewhere or "out" of Tsumkwe to supply food for themselves, their families, and other people in the settlement. Therefore, the "normal life" becomes a life whereby one has a business and takes care of one´s family. On the other hand, interviewee six wishes "to eradicate unemployment" through this business opportunity.

> "I will wish I **have my own business** and **take care of my kids** That's the only wish I have and live **a normal life**. The kind of business, the business that I'm planning of doing is like a **food business**. So, like now in Tsumkwe people are like struggling a lot because they don't have a where can people buy food. They are just taking, there are only small shops I can say. So, I wish to have **a big shop to sell food supply**

> the family around here. The kind of food I mean, are like sugar, tea, maize meal, and cooking oil, salt… Yeah those are the main one. Like extra, fast food" (Interviews 5, Pos. 196-217).

> "I would like to **be just a businessperson**. Yeah, **not a person** to work **under Government** or in other job, but if that job can be able to **fulfill my dreams**, I can able do that. But my dream is to, **job opportunity just to work or to continue** with my business whereby I can **eradicate the unemployment**, which is on in Tsumkwe" (Interview 6, Pos. 272-281).

### Vocational Training Center

Following up on the employment wishes, an institutionalized differentiation with a bureaucratical output has been shaped and expressed in interview 14. The interviewee´s wish is to have youth or any other person certified to get employment. This differentiation through certification shall be linked with the low employment opportunities in the settlement and the wish to encourage the youth to get an education. Furthermore, he focused on one of the needs in Tsumkwe, a Vocational Training Center whereby people can learn different skills, use electricity, fix electricity, exercise in auto mechanics, and do similar activities. Thus, they could generate income once they graduate and get a certification.

> "Regarding employment, I don´t what other places are... But for me, I just want each of the youth or any other person **with legal document** or something from school to **get employment**" (Interview 14, Pos. 36).

> "Just one thing, it **would have been nice**, **one of the biggest need in** Tsumkwe would be a **Vocational Training Centre**, a Vocational Training Centre. That would need electricity. Because in that Vocational Training Centre would have, maybe electricity, they can learn about how to fix electricity, fit antenna or even auto mechanic. So, lots of those stuff can be placed in that building, in that training center" (Interview 14, Pos. 46-48).

### For Development (to become a town)

The accumulated wishes of "a normal life" with employment create a future with a more "developed" settlement that underlines the characteristics of a town with many shops, banks, offices, and a hospital. In the quote from interview three, one can also realize a detachment from the community by borrowing money from a bank instead of an acquaintance from family. This shows an ongoing process of individualization as well as differentiation (with institutionalized settings) in order to live a "normal life" that is set based on rationality of purpose.

> "For example, it will be like, as I said; **banking systems**, the **offices and the hospital** which must be much better.(…)And then **other offices and also where you can get money**. Sometimes **you'll need to borrow money**, you can go to" (Interview 7, Pos. 18-27).

> "Because I believe from now onwards I think with this process there will be **lots of developments.** There will be… I **hopefully believe** there will be **lots of shops and malls** by and from **five years on**. Is how I see it" (Interview 3, Pos. 77-82).

The findings presented within this sub-section belong to the envisioned futures of the residents in Tsumkwe, with electricity being one of the primary facilitators (besides the tar road) of the services in everyday life.

### 4.2.3 Domestication of "Energy Practices" of Everyday Life

Dependence on human beings also arose in pre-modern times when people depended on nature. Some technical support helped them adapt to the natural environment, enabling them to escape from the limitations of their environment and body. Van der Loo and van Reijen's (1992) description also suggests it is the "taming the of the natural and biological forces and thus, the optimization of the possibilities, which nature and body can offer". Herewith, one can draw attention to the "optimization," which is closely linked to the process of rationalization explained in the previous section, demonstrating how modernization processes are closely intertwined. With increasing domestication in modern times, people have become less dependent on the natural environment and the fluctuations of natural circumstances. The progressive improvement of techniques/technologies in life expectancy, nutrition, hygiene, sewerage, water purification, and medicine have also been developed (van der Loo & van Reijen, 1992) and are still developing. Hence, many things related to daily activities have also started changing through electric appliances, such as cooking, heating, cooling, working, water provision, and commuting to work or elsewhere. In the case of Tsumkwe, the changes in daily life after the connection to the mini-grid took place in terms of cooking, cooling, heating, and using electric appliances for leisure, work, and study.

These qualitative findings of the life-worlds have been conceptualized under the local energy practices of everyday life by considering "energy" as an "ingredient of dynamic social practices" (Shove & Walker, 2014, p. 51). Social practices have been considered as the intersections, bundles, and constellations based on three elements: materials, meanings, and competencies (Shove et al., 2012, p. 14). Accordingly, `materials` include "material arrangements," which might be energy sources, technologies, and infrastructures—that point out the practices and changes in the practices (Schatzki, 2011). `Competences` include practices that cognitively experience skills, know-how, and knowledge into a routine. `Meanings` encompasses symbols, ideas, and aspirations that might be socially observed (Usenyuk et al., 2016). Framing energy as a social practice enables us to understand how social practices develop, change, or intersect to understand energy demand and supply (Shove & Walker, 2014, p. 47). Thus, it would be possible to contribute to energy governance and policy strategies from a broader perspective than is often only addressed as a part of the techno-economical-(political) debates.

#### *4.2.3.1 Overarching Meaning(s) of Energy Practices*

In the previous sub-section, the overarching meaning of electricity/energy had been defined as “magic,” a “magical tool” that is a source of light and a source of life, which provides them (in)dependence and autonomy and changes their everyday lives. During the field surveys in the settlement area, the open-ended question regarding the changes in people's daily lives was asked to understand their daily practices and the meaning structures around the energy, including electricity. Bringing phenomenological insights toward the unquestioned life-worlds that originated on the ground has enabled the following local energy practices to be grouped under household practices, household (house-shop) and small business practices, business practices, and office practices.

As highlighted earlier at the beginning of this chapter, the Tsumkwe settlement, already had a functional generator during the era of the South African military post. This generator's use continued after Namibia's independence and extended an essential line of use primarily for public/government use. When the mini-grid was inaugurated in 2011, more households were connected in the non-essential line. They used to use the electricity for a limited period, during the morning and the afternoon/evening. When the respondents were asked about the changes, their comparison was with the time they had electricity with interruption and later when the CENORED took over the ownership in 2017. The reliability of the electricity, having it 24 hours a day, seven days a week uninterruptedly, has had a significant impact on the daily life of the residents:

> “Before it was not good, switch on/off from morning to lunch. From lunch to afternoon was not good” (QOQ).
>
> “We have electricity 24/7” (QOQ).

These quotes were repeatedly mentioned during the surveys. Thus, through the 24/7 uninterrupted supply, one of the main meanings is (in)dependence and (self-)autonomy. Thus, through these interrelated meanings, materials and competence bundles of energy practices have been studied and were illustrated in Figure 4.10. The energy practices below include household, commercial, and public sectors as well. The quantitative data regarding the appliances will be provided in the *quantitative findings* section.

*Figure 4.10. Energy Practices in Tsumkwe*

*Source: Author*

### *4.2.3.2 Energy Practices*

**Lighting**

The lighting of the house or the shop is first and foremost of the energy practices. For this purpose, the materials utilized are light bulbs (60W), energy-saving light bulbs (11W), torches, fluorescent lights (for schools), solar bulbs to illuminate indoor places, and streetlights for the lighting of the streets. Furthermore, additional materials for this practice have been hybrid solar PV mini-grid systems, transformers, poles, cabling, and prepaid/conventional meters that form a bundle to supply and transmit electricity to households and shops.

Reliable and uninterrupted electricity brings self-efficacy with the use of electric appliances throughout the day and night. The practicality and usefulness of house appliances are generally attached to practicality/comfort and ease. Life becomes easy with electricity; the struggles are reduced or disappear, and it brings safety. When it is used for business purposes, it becomes an income-generating activity. For the schools, it extends the students' studying opportunities.

> "Electricity is something **useful, helping people, giving people lights**,"
>
> "**Make things easy**, so you **don´t have to struggle**,"
>
> "I **know** that **there is light**, **no more darkness**" (QOQ).

As a result, the "light" as a point of lighting makes the surface into a bundle of energy practices to start with in people's everyday lives. Then, independence and, to some extent, autonomy have been experienced through these practices, although they become dependent on energy/electricity as well as technology on the other hand.

The lighting available throughout the day and night has enabled students to study at night, others to work at night, or do anything difficult or impossible without electricity.

> "I can **do whatever I want with light, at night**" (QOQ).

There are lights to keep shops open during dark hours, and shebeens (local bars) remain open and can generate income. Schools benefit from the lights in the early or late hours of the day when necessary. The police or teacher housing also provides lighting at their residences. Without any doubt, all these lighting practices are enacted through the visual sense.

> "When we wake up in the night, we can turn on the light to see around"
>
> "I am happy, definitely yes. We don't have any problem. Better than those using electricity which are not using solar. We were struggling before, especially sleeping in darkness. When we wake up in the night, we can turn on the light to see around. Studying at night became easy" (QOQ).

When a respondent recalled the time before the use of electricity at her/his place, the danger of fire appeared due to the use of candles indoors. Now, s/he is safer.

> "Before we used to have **candles**, they **would burn things** and it was **not bright.** Now, **I am able to work at night**. We have electricity 24/7. I **can iron my clothes**" (QOQ).

### Communication

The second item used by the people was the "mobile phone" for communication and financial purposes. The Mobile Telecommunications Limited (MTC) in Namibia provides network coverage for the use of mobile phones through the satellite systems in Tsumkwe. A stand-alone solar PV mobile charging booth (Figure 4.11) was also observed during the field research. Herewith, the author suggests that "internet" shall also be part of this communication material. Although it was not directly mentioned during the interviews, smart phones have been used in the settlement.

*Figure 4.11. Stand-alone Solar PV Phone Charging Booth*

*Source: Author*

> "It is completely **a new thing**. It **was something very new** because before we didn't know about electricity and **with electricity there are lots of thing that you can do**, and you can **do plug-ins** and you can **charge your cellphones** and that is all the thing that **we didn't have in the past**" (Interview 12, Pos. 87-89).

> "It was very important. Because like now, as we are now talking on the phone without electricity it was, it could have been impossible to make such a call. You are there**. Over there where I cannot see you**, but now I can sit here. I **can see your face** on my phone and **can chat**. I can **hear your voice**, and this **is all about electricity**" (Interview 12, Pos. 94-95).

The first competence is, without any doubt, communication among people. The quote above emphasizes the importance of electricity in communication practice, including digital calls, chats, and video calls. The interviewee is aware of the multitude of activities that one can do with electricity. These activities are fully sensed with hearing, speaking, writing, and seeing. Besides the communication practice(s) connecting people from different places, backgrounds, and geographies, the interviewee experiences these practices via digital devices and the internet, which introduces the digitalization component of the suggested process of domestication in the following sub-section. The time interval between the first analog mobile phone he used and the digital phone he has been using is relatively short, meaning that there are intertwining and overlapping processes regarding the appliances used for communication practices.

Two additional vital functions have been identified: Mobile money and mobile electricity payment systems. The mobile money is a money transfer system operated by the MTC.

The operation runs through mobile phone numbers to send and receive money from several service/vending points, from number to number, from person to person. Any mobile phone shall be used for this purpose. Mobile electricity payments shall be done only by smartphone holders and paid for by internet banking systems. The main banks, FNB´s e-wallet, Standard Bank´s blue wallet, and Bank Windhoek´s easy wallet, have been used for online banking. Figure 4.12 illustrates one of the dissemination posters of the FNB Cash Plus service point.

*Figure 4.12. FNB Poster*
*Source: Author*

From a commercial perspective, the parastatal NBC has also been working in Tsumkwe and broadcasting media coverage (Figure 4.13) for a long time. Before connecting to the mini-grid, they ran their own genset.

They were buying five drums of diesel per month, which was quite costly. With the connection to the mini-grid, it has become cheaper and easier for them now. They pay 42,000 NAD monthly, which is equivalent to 16,000 units. They have been having issues when there is rain and lightning because their electrical equipment gets damaged from this unstable electricity load. The equipment is running 24/7. They have a central control unit, two radio transmitters, and broadcasting equipment with their former genset that they are not using anymore.

*Figure 4.13. Broadcasting Equipment*
*Source: Author*

As a result, from a commercial perspective, it creates savings for the business, making it easier for them to manage the energy provision from the mini-grid. They said they did not need to worry about buying the fuel and running the generator. Now, it has become easier for them. They pay the necessary amount, and they receive it.

> "It makes it **easy for me** to charge my phone and not going to **the houses of the neighbors"** (QOQ).
>
> "I can buy it **anytime**; it **tells you** what and how much you are u**sing**" (QOQ).

The respondents expressed their satisfaction with the practicality and ease of communication practices, with the advantage of practicing them any time they would like. This way, they have become independent without needing to go to their neighbors´ houses to charge their phones. The second quote underlines the traits of a conscious consumer who is aware of his/her own consumption, besides the meanings of independence and autonomy.

Furthermore, Figure 4.14 shows online prepaid token purchase messages informing about the transaction such as corresponding units purchased, ECB and NEF levies paid, and the type and number of consumers. In addition, a customer service number is provided for queries. On the other hand, the use of such online systems also demonstrates digital literacy among those users.

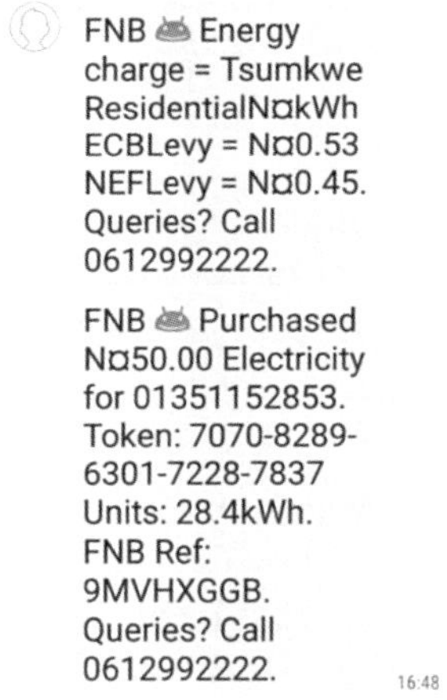


*Figure 4.14. Mobile Prepaid Electricity Token*

*Source: Author*

## Cooking

Cooking is a practice that belongs to the bundle or constellation of household practices; it also includes boiling water in the kettle or using the microwave to warm food. The respondents mentioned that they made their lives easy. On the other hand, it can also be interpreted that with all the home appliances, they do not feel deprived of anything.

The surveys demonstrated that firewood, electricity, and gas are used as energy for cooking. The appliances used are hot plates and gas cookers, in single or combination, depending on their availability. The respondents informed that firewood had been collected from the bush; hot plates are used for electric stoves; and gas bottles are necessary for gas cookers/stoves to cook food. On the other hand, the survey findings identified the use of microwaves and kettles for warming up the food and boiling the water.

Some respondents mentioned that they stopped using gas bottles due to the (un)availability at the general dealer. The inconvenience caused them to return to firewood use. Some others mentioned that they mainly use gas cookers when they have gas bottles; when it is over, they switch to hot plates. The users of hot plates claimed that they could pay for the electricity; it was there, and they did not need to look for gas. Moreover, the firewood users in the open fireplace do not need to look for wood.

> "**No more looking for woods** in the bush. We **can watch TV**, **listen to music, store milk and meat in the fridge**. Wow" (QOQ).

On the other hand, many NRAs involved in the Tsumkwe Energy Project (TEP), as well as CENORED, have been informed about the high energy consumption of these devices, which are not adequate for the off-grid systems due to their instant power consumption, especially in the peak hours of the day. During a workshop in 2023, the community members mentioned the "demand side management measure" organized by CENORED with the involvement of the community, in which they have been informing the users of high-consumption appliances for not using them in order to have more energy available on the peak times.

The appliances such as kettles and hot plates are not considered within the electric appliances in use in an off-grid electrification system; only (LED) lights, mobile phones, refrigerators, radios, televisions are taken into consideration (Hirmer & Guthrie, 2017), and in addition, fans (Sayani et al., 2022).

> "I am happy because the first solar was not good, it was always on-off. I am not bored at home. I have TV 24 hours. I can iron in the light, **boil water in the kettle**, **very easy**" (QOQ).

## Cooling and Heating

In Tsumkwe, household fridges and freezers, commercial refrigerators and freezers, electrical fans, and air conditioning have been used as cooling devices at the household and commercial levels.

In addition to the bundle of energy practices at the household level, cooling practices are also used. The respondents with fridges use them to store milk, meat, or drinking water. It makes life easy; it makes "things" easy. In this way, the food that is indispensable for life has not been wasted but stored. Further, freezers are used to keep the meat for more extended periods. The following four quotes refer to this life-changing practice, perceived, confirmed, and internalized with the senses of touch, taste, and, to one extent, even the smell of fresh and unspoiled milk or meat.

> "We are happy because it was not so good in the generator times, only. Now it is better. A lot of things changed! **We connect the fridge; we can freeze the meat**. Children can study at nights" (QOQ).
>
> "No more looking for woods in the bush. We can watch TV, listen to music**, store milk and meat in the fridge**. Wow" (QOQ).
>
> "It **changed my whole normal life**. Now I am **drinking cold water** and **ironing at my own** house" (QOQ).
>
> "Electricity is important for **everything**, **makes things easy: fridge, washing**" (QOQ).

A respondent also used the cooling practice to make ice blocks to sell and generate income for herself, in addition to the other bundle of practices exercised at the household level (the quoted interview three). On the other hand, the local shops have also stated the use of refrigeration and freezers to store drinks or meat to keep them cold and fresh. It is an essential element and practice in their everyday lives as well.

> "It has really changed my life because I could now, **my main aim with electricity is now**, I **can watch television and sleeping in the light**, you know. I can use, I **can iron my clothes**. I can do everything. **Yeah, I can make ice blocks and sell**. Yeah. That's, yeah. I can say" (Interview 3).

Besides the cooling practices related to taste and touch (food and drink), there are also cooling practices for the space with the use of air conditioners and fans, which are not typically used in an off-grid system. The respondents reported that the dry season might get quite warm, and that is when it becomes necessary for them to use fans.

Based on the above-mentioned cooling practices enabled with different appliances, the respondents also experience these practices through their multiple senses in their everyday lives. Drinking a glass of cold water, cold beer, or fresh milk, keeping meat for longer days, having a breeze on warm days, and generating income all contribute to a better life for the respondents.

The surveys did not identify any extensive heating practices, except the use of air conditioning in one place.

### Leisure/Entertainment/Enjoyment

Among this bundle of practices, television, radio, music system, and hi-fi shall be considered the electric appliances mentioned by the respondents as belonging to watching TV and listening to music or radio as a practice at the household level. The *shebeens* have also been utilizing music practice at several locations throughout the settlement.

The quote below summarizes the energy practices at the household level, also underlying their access to portable solar PV sets when they needed electricity.

In this case, they had the electric appliances already before they were connected to the mini-grid.

> "**We are happy** because **it helps us so much**. Sometimes we don't have firewood, and **we use the kettle**. We **can use light in the room**. Everything **electrical makes life easy**. **My wife has a sewing machine**, she uses it. It **was so difficult before**. I **had a TV**, but I couldn't watch it. I had **borrowed a generator from friends** (solar PV and battery used). We used (to) **have candles when they were finished it became dark.**"

Another respondent (quote below) expresses her satisfaction with having electricity so she can have lights and TV, enabling her children to stay in the house and not go to the streets or visit other people´s houses. The following respondent mentioned charging their phone and watching a movie or music in the neighbors´ houses. The following respondent emphasized its reliability for 24 hours during the day. Watching TV is a practice that involves audio-visual senses and occupies time in everyday life. However, as a sense-inclusive energy practice, it can also be combined with taste by eating and drinking.

> "I am **happy** because **I have lights**! **I have a TV**. Kids don't run around, in the street or visit **other** people's houses" (QOQ).

> "We don't go to the neighbors' houses anymore to charge the phone, to watch a movie, music, no candles needed" (QOQ).

> "I have TV 24 hours" (QOQ).

The quote below demonstrates a comparison from earlier times after electrification. The respondent complained about how boring it was, the inability to watch TV in those days, not even drinking cold water, and the darkness overall, expressing the changes they went through with their senses. Furthermore, there is also a comparison with other places that have electricity.

> "I mean very bad, because **I couldn't do, anything that I wanted to do**. Watching television, you know. **It was just bored.** You could **not even drink cold water if is hot**. You know people who were having, they were enjoying but us, **we were not having** and **our location** was really dark at all" (QOQ).

Shebeens depicts a different competency line as a place for people to gather, drink, dance, and socialize with music that satisfies the audio and other senses, reaching all central points in Tsumkwe. During the second field research, it was observed whether the number of the shebeens or the number of music systems had increased. After a few talks regarding this with the locals, they mentioned that they did not notice it, and one of the neighbors stated that he was not disturbed.

Leisure practices such as watching TV and listening to music or radio did not only mean independence and autonomy for people because they could practice them in their own place but also exercise them on their own during leisure times and having them all around the day provided this independence. For shebeens, having the music system is an asset in attracting more clients and certainly generating more money for this purpose.

### Sewing, Ironing, Washing

> "Electricity **is very important for everything**. It **makes things easy**. I can **do washing or use fridge**" (QOQ).

Sewing, ironing, and washing are also other mentioned practices in the settlement and have the power to change the entire everyday life. Iron is one of the most energy consuming appliances and is also not included within the off-grid systems as an electric appliance. Thus, CENORED has included it from the demand side management measures that they have implemented since 2022 (Figure 4.15). Moreover, it has the power to change the entire normal life "It **changed my whole normal life**. Now, I am drinking **cold** water and ironing at **my own** house.

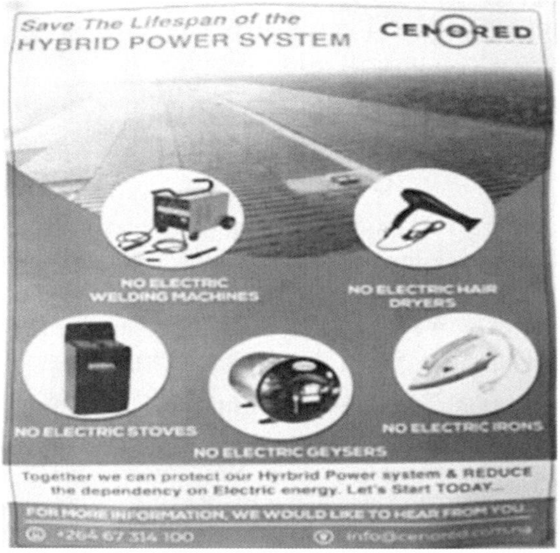


*Figure 4.15. CENORED's Public Announcement for Demand Side Management*
*Source: PROCEED Workshop*

As a result, energy practices of everyday life such as washing clothes, ironing, charging phones, cooking with an electric stove instead of collecting firewood or looking for a gas bottle, watching TV, studying at night, having light, listening to radio/music, not going neighbor's houses for those services, using a kettle, sewing clothes, freezing meat, having cold drinks, using a microwave, having water are all part of the everyday life activities. The overarching meanings of independence, autonomy, self-efficacy and doing business from these practices, in its case, provide people to live a "better" life compared to their previous years, and they experience them by using all their bodily senses 24/7, all day around without interruption. Thus, the expression "electricity is life" or "electricity is a source of life" comes from living these bodily senses throughout the day and night all year round.

#### *4.2.3.3 Contribution of Digitization and Digitalization to the Domestication of Energy Practices*

One can see how the residents' daily lives have been equipped with the electric appliances that facilitate their everyday lives; they bring comfort and a better life without suffering. Moreover, besides the (progressive) domestication identified within this study, digitization and digitalization have also been observed in the field. Digitization converts analog data to digital data, an integral part of digitalization (Fusaro, 2023). It is mainly used in libraries, archives, the music industry, the arts, and the film industry, but is not limited to these. An essential distinction between digitization and digitalization is that what has been digitized is the information and not the process. An example would be the digitization of paper-based documents into digital formats.

However, before this process of digitization, San in Nyae Nyae had gone through a transcription process for education purposes. NNFC had observed some struggles of San students at the Tsumkwe School and had taken this issue seriously for their future, to have an education for their future employment, especially as teachers for San people (Biesele & Hitchcock, 2011). In pursuit of this, right after the independence, the Namibian government adopted an educational policy that integrated minority-language education in the first three years of schooling, in which Ju/ʹhoansi was included, and pilot projects were started in Namibia. Nyae Nyae Village School Project (VSP) was conducted with USAID/Namibia and Nyae Nyae community members, which created a "self-literate" population and enabled the Ju/ʹhoansi language orthography to be taught in the schools (Biesele & Hitchcock, 2011). In 2002, Ju/ʹhoan Transcription Group was formed for language development and cultural heritage preservation projects in a digital format. The audio-textual recordings of the "legacy collection" collected from the 1970s until 2011 included dreams, narratives from political meetings, oral history to healing, interviews, and conversations that have been transcribed and translated by the VSP computer-literate native speakers (Biesele & Hitchcock, 2011) not only for the archives but for the Ju/ʹhoansi as well. In 2008, the Programme was extended to the Youth Transcription Project, and the future with the "spirit of ʹeach one teaches oneʹ" to train especially young people, young women, for transcription and translation (Biesele & Hitchcock, 2011, p. 241). As a result, the trainees have been digitally recording folk tales, stories, medicine songs, and chronicles from the elders, video recording them and keeping them in the Nyae Nyae Tape Archive for their future.

Following the digitization period of San cultural heritage with solid support from international research participation in 2012, the mini-grid system provided electricity to a broader community than the generators were put in operation; this has enabled the use of mobile phones and, at a later stage, smartphones when they were available. However, this has not only been taking place in Tsumkwe; it has also been taking place in Namibia, in Africa, and all around the world, with short time differences between implementations in different geographic locations. Smartphones have only been one object using electricity and the internet to carry information for digital processes. In this sense, digitalization refers to the "process of using the digitized information and data and new technologies to create new digital ecosystems, automated process, and greater efficiency in the processing, storage, and transmission of this data" (Fusaro, 2023). This has been used in business, in automating many processes, entertainment, teaching, and media, to name a few (Fusaro, 2023). Communication and collaboration among people have also been enhanced and simplified through digital tools. Banking and finance have been among the sectors that have adopted digitalization in their everyday business. Hence, mobile money in Namibia, used to purchase prepaid electricity tokens, has been one of the examples that has reached remote areas like Tsumkwe Constituency.

In line with this development in education for San people in Nyae Nyae, starting from the 1990s, the computer developments that had started in the 1940s and 1950s made their way to the internet and digital communications in the 1980s. Rapid digital sector growth regarding equipment and internet development has occurred. Mobile phones have been in use; smartphones have been introduced, and the steps to the digitalization process have started progressively. In 2020 and its subsequent years, the COVID-19 outbreak has caused a boom in digital services, expanding its use in several areas of our lives and intensifying its presence in our everyday lives.

It is important to note here that the birth of this digitization in Nyae Nyae was due to the struggles of the San students in the end-apartheid period in Tsumkwe and the necessity of education for the future of San children and the San community. The independence of Namibia set the required basis for this development in education from a minority perspective and integrated it into the education curriculum. In 2020, Namibian schools adopted digital learning tools embedded in their educational system. The rapid growth in the use of digital appliances has also taken place in Tsumkwe, albeit at a different pace, but it is there. The residents of Tsumkwe and their acquaintances in the villages have been experiencing the process of domestication both from an electrification side, at their households and shops. However, they have also been experiencing it from the digitalization side.

This particular point was outside the scope of this study but surfaced during the analysis of the collected data, which included observation and earlier studies. The available data does not provide any further interpretation for its future rather than confirming the beginning of these processes that are evolving in parallel and intertwined with each other.

With the past years' rapid and progressive technological developments, the world and human beings have found themselves at the beginning of the digital age or the age of IT. Supporting van der Loo and van Reijen's (1992) point, people, society, and technology are inseparably linked. Although electrification and energy/electricity access have been debated in government strategies, plans, and policies, digitization and the (process) digitalization have yet to be noticed. Since both of them are directly connected to the presence of energy and the internet, to some extent, for digitalization especially, the author strongly suggests that they can be incorporated within the process of domestication under the communication practices to start with.

#### *4.2.3.4 Visions about the Future Energy Practices*

The visions about the local energy practices within the domestication process will be quantified and presented in detail in the quantitative section, since the interviewees´ visions contributed to more overarching visions such as Tsumkwe becoming a town or having own business. Only one respondent stated his wish to have a car for

> "I´d like to use electric appliances, **get a computer, iPad, washing machine**" (QOQ).

Interviewee 12 envisioned cinemas and computer centers where people would come together and do training. Although one could also classify computers as individual appliances, there is a vision for collective use and benefit, for an upgraded communication practice with digital appliances and the internet—the next generation of Tsumkwe, her children like to play games on the phone.

> "I think that we will also need … Electricity will be. There will also be centers. Like uh. If it is not cinemas, **computer center, where people will come together.** People will want to do computer trainings. They will also have centres where **computers are installed where they can come and have their courses**. (…) My children, **my first son is very fond of cellphones and machinery, and he likes playing games on the phone**" (Interview 12, Pos. 112-116).

On the other hand, NBC stated that they dream of having another central control unit that will link Tsumkwe to Windhoek or to the world. Then, they will be able to broadcast live from Tsumkwe to the world. This dream is to broadcast to the capital and the world, be independent and autonomous, and connect to the whole in real time. This overarching vision directly contributes to the "global sense of local."

### 4.2.4 Process of Individualization

The process of individualization defined by van der Loo and van Reijen (1992) suggests that the individual who breaks away from one´s collectivity of immediate environment has growing importance of the individual, whereby the development of the individual identity, as well as individual behaviors towards groups, society, and state, occur. The individuals start dealing with social units different from their collectivity, thus reaching a certain level of independence. In the case of Tsumkwe, both the interviews and the surveys suggested that a certain level of individualization occurs in place while also keeping community ties. This section presents the findings of the process of individualization within the following framework in Table 4.10.

*Table 4.10. Process of Individualization*

| | |
|---|---|
| Process of Individualization | Living a "Western Life" |
| | Independence and autonomy |
| | Esteem (respect, self-confidence) |
| | Ownership |

*Source: Author*

#### *4.2.4.1 Living a "Western Life"*

> „I was born and raised here in Tsumkwe. First, I stayed with my parents and that time it was not like today, we are now having the western life. But we were **more like traditional** and I **used to move around with my parents** the times that we moving around **gathering or doing hunting**. So, then thereafter I went to stay with my father, uh my sister. My first sister. My older sister as I was growing up I went to school in Tsumkwe and after I finished here at the junior, primary phase, junior secondary phase I continued to Grootfontein High School. And it's there, **where I ended**, and then, I got employed by government. So, I started, (Indecipherable) I was about twenties and then I started making family and so… When I started with my family, it **was always in Tsumkwe**. (……) First I started with the government ministries, I was working for the Ministry of Health and Social Services as a laboratory assistant and as **I saw that the work didn't suit, I resigned**. Then, I applied for a position in the Ministry of Agriculture, where I was working as a clerical assistant for seven years, **then, I resigned**" (Interview 12, Pos.31-38).

The quote above belongs to an interviewee in their mid-forties, who considered themselves living a "Western life." Unlike most San people with difficulties in education, he successfully completed his primary school degree in Tsumkwe and his secondary school degree in Grootfontein. He used to live with his parents, "moving around," hunting, and gathering, which is part of the traditional San living in Nyae Nyae. He stated they have been living a "Western life" compared to his youth. He settled in Tsumkwe and started his family there. The government also employed him at the time. However, when he realized that the work did not suit him, "he resigned" as an independent and autonomous individual who could make personal choices and act accordingly.

#### *4.2.4.2 Dependency and Autonomy*

The Table 4.11, highlights the matrix of dependency versus independency with/out autonomy of the process of domestication and demonstrates mutual ties with the process of individualization. The use of mobile phones and having their own televisions at home have marked an important place in the everyday life of people in Tsumkwe.

*Table 4.11. Dependency and Autonomy*

| Dependency | Autonomy |
|---|---|
| "No more **asking other people for charging phones...**please...please." | "I can do **whatever "I" want** with light **at night."** |
| | "It´s **always available."** |
| | "It makes it **easy for me** to charge my phone and not going to **the houses of the neighbors."** |
| | "I don't go to **people´s house** to watch TV." |

*Source: Author*

The respondents often mentioned:

> "I **don't visit the neighbors´ house** for charging my phone anymore,"
>
> "It´s **always available**. No more asking other people for charging phones... **please... please**" (QOQ).

On the one hand, there is reliability and availability of electricity all around the day, 24/7, from a technical perspective; on the other hand, this availability represents independence. Performing the activities of charging the phone, watching TV, cooking on an electric stove without facing any struggle in collecting wood, and freezing meat with the designated electric appliances for each energy practice have led them to do things for themselves without depending on another person (Toledano-Gonzales et al., 2019). This independence has opened ways to autonomy, that is, "the ability to act and make personal choices" by taking control over the events happening in their daily activities (Toledano-Gonzales et al., 2019, p. 277). Furthermore, through this independence with/without autonomy reached their premises, a notion of comfort, ease, and happiness also emerges, expressed as:

> "I am **happy to have it**. Now I can charge **my phone at my house**" and "it brings light, and it **is useful**. It makes **it easy** for me to charge my phone and **not going to the houses** of the neighbors" (QOQ).

Research conducted by Pinto et al. (2017, p. 11) concluded that comfort and well-being are terms used interchangeably, referring to the concepts related to quality of life; however, they are not surrogates but share common attributes. On the other hand, Toledano-Gonzales et al. (2019) defined *quality* of life as an "optimal state of well-being in the dimensions of health (physical, mental, social, emotional, spiritual)" while the multidimensional concept of well-being is the "result of an active process aimed at improving individual lifestyle." In this respect, Maslow´s hierarchy of needs (McLeod, 2024) provides an approach to the needs of individuals based on hierarchy, in which the physiological needs such as food, water, shelter, and clothing take the first row within this pyramid, that is linked with the attributes of the quality of life and/or well-being. According to the self-actualization process that an individual pursues in life, an individual can focus on a higher level in the pyramid once the needs of the lower order have been met. Thus, with the independence acquired through energy practices, the respondents accomplish their physiological needs for food, water, shelter, and clothing.

#### *4.2.4.3 Esteem*

*Esteem* is a broad term that covers several aspects: The fourth order of Maslow´s hierarchy of needs, under the "Esteem needs," whereby is classified as "esteem for oneself" and "esteem of others" in general (Maslow, 1980). A desire "for achievement, for strength, for confidence, for independence, and for freedom" is included within the self-esteem (Maslow, 1980, p. 13; Maslow, 1943). In the esteem of others, there is the desire for reputation, recognition, respect, and appreciation from others. Self-esteem and earning respect are closely linked with autonomy (Marmot, 2013), and there is a set of characteristics interconnected for different purposes in human life, such as respect. It is also essential to differentiate self-respect from self-esteem. "Self-respect" is the value one gives to oneself, whereas "self-esteem" is the value one gives to one's abilities and skills.

**Respect**

In the following quote from interview 13, the businessperson narrated their thoughts on increasing sales and marketing their products. While talking about the potential places, she mentioned a place with a "well-known" supplier, compared to herself. She considers herself not well-known. She respects them and supports the San people like herself. Therefore, she would not intend to go there to sell her products; she would instead go elsewhere for this purpose. Thus, by showing respect in two ways, she also demonstrates her own (business) ethics based on respect for others.

> "I don´t have any connection **there**. There are already people who are selling the **same** product **with me** and **with them**, **they** are **well known**, **not like me**. So, they took that market now. Because them, **they are al-**

> **so supporting the San**, here. So, **I cannot go and disturb, again, I have to look for somewhere else"** (Interview 13, Pos. 2).

Meshanko (2013, p. 11) defined respect as "an active process of nonjudgmentally engaging people from all backgrounds. It is practiced increasing our awareness and effectiveness and demonstrated in a manner that esteems both us and those with whom we interact." Respect or esteem is linked with leadership (DeLellis, 2000) or organization (Meshanko, 2013) and situates itself as the core of the relationships that drive the esteem in others and us. Thus, leaderships have been constructed through this relationality of low or high-esteem-driven organizations.

#### *4.2.4.4 Land Ownership*

The notion of ownership has diverse uses depending on its context. However, ownership refers to the juristic notion whereby an entity, a group, or an individual has legal rights over a property, an asset, or an item. It enables the owner the right to use, manage, transfer, or dispose (Sárközy, 1982). It can be in different forms, such as individual, joint, corporate, government, and collective ownership. The different ownerships have been found in legal, economic, sociological, and socio-psychological aspects and may evolve based on societal, technological, and economic changes. Jami et al. (2021) suggest a positive link between self-esteem and psychological ownership, self-esteem, and prosocial behavior, which will be discussed further. Thus, although Maslow´s pyramid is not directly linked to ownership, individualistic characteristics and the notion of ownership will be considered from individual and community perspectives based on the interviews.

The quote from interview seven underlines the importance of the land for the San people, the right of the land as a sense of the place, through which independence and self-sufficiency emerge. One could own a business on a land to which one has the right. It also highlights that the "land" shall not be owned but ′have the right′ to use it since it is ancestral and originally belonged to the San people. One could argue that this quote should also be interpreted as a bureaucratized procedure, which should also be acknowledged within the process of rationalization. Once an authority has granted the right, one shall exercise farming, own business, cultivate, and sell food. This quote also emphasizes the "accepted" official authority to give the right to land within the established hierarchies, such as the Traditional Authority. In this sense, there is not a complete detachment from the community ties; however, an individual is willing to be self-sufficient by providing one´s own needs.

> "You **have the right to your land**, and you can, **you could do your own businesses**. At **your farm you can**… It's a really, really good thing" (Interview 7).

#### *4.2.4.5 Future Visions for Self-Development, Learning, Business, Solidarity and Respect*

The visions classified below under co-management for the desired training center, self-esteem and independence, self-development and lifelong learning, love and belonging, and well-being are all wishes (Table 4.12).

*Table 4.12. Future Visions for Self-development, Learning, Business, Solidarity and Respect*

| | |
|---|---|
| Future Visions for Self-development,<br>Learning, Solidarity, and Respect | Co-Management of the Desired Training Center |
| | Wishes for Individual Business Ownership and Autonomy |
| | Self-Development, Communication, and (Life Long) Learning |
| | Solidarity and Respect |

*Source: Author*

**Co-Management of the Desired Training Center**

Referring to ownership and its link to esteem, management, and organizations, the quote below from interview 14 presented a co-management wish and future opportunity regarding the Vocational Training Center they wish for. The quote itself does not inform any "ownership" *per se*. However, it clearly denotes a co-management and co-ownership wish to operate this future training center. Their potential idea would be to co-run the center between the Government and the Conservancy. One could also interpret this co-management as a collective-based co-ownership opportunity. Since the interviewee mentioned that normally, they do not want only the government. The details of such ownership will still be open to discussion. On the other hand, the quote informed about the acceptance of San people´s shyness and low interaction. To one extent, one might claim low self-esteem in general for San people in contrast with non-San or tribes other than their tribe in the village. From this basis onwards, the co-management of the training center has been based on the rational grounds that there is a need for the government to run it together in order to bridge the gap caused by shyness or low self-esteem.

> "We **normally** do **not want a government only to run** it. As from my point of view, I would like it like, it´s the government and then, the Conservancy, it´s linked. As long as it can be linked together. So, **we can try and run the center**" (Interview 14, Pos. 49-50).
>
> "That **would help**, because we can now…I will not lie, **our tribe** we have a **high number of shy people**. They **don´t like to interact with other communities** or **other people**. So, this is the opportunity. If we **are linked to the government** regarding the **training center**, we can

> **try and identify the kids with potential** and get them there" (Interview 14, Pos. 52).
>
> "In Tsumkwe, we, we, I really do not know how to explain why we are shy. But, it´s maybe because I would say, most of us, we grow up in villages and whereas, some villages have schools, so. Where that, in those schools, we only have our tribe. So, once, maybe those schools may go up to Grade 5. **If you are done, then you come here**. And then, now, you are **used to your tribe**. It´s **hard to fit in with others, now**" (Interview 14, Pos. 54).
>
> "We would be. If it´s regarding **our Conservancy and our villages**, then, it **would be nice for us to be part of such projects"** (Interview 14, Pos. 60).

Moreover, the underlying reason for such a training center is the educational and vocational development of San kids coming from the villages after they complete their curriculum over there in order to introduce and integrate them into the society in Tsumkwe as well as into life in Tsumkwe. In this way, the interviewee thinks that these types of projects could enhance the participation of San people, but more precisely, youth, not only for Tsumkwe but also for the villages. Thus, such an operation and participation in co-management would create empowerment within the San communities, which could enhance the participation and the completion of higher educational degrees.

### Wishes for Individual Ownership and Autonomy

The quote below was mentioned in the process of rationalization, from a rational point of view, regarding owning his business and not being employed by the government. From the individual ownership point of view, "own business," "myself," and "personally" demonstrate the individualistic standpoint of the person. Attaching this wish to the "just a businessperson" but *nothing else identity*, ownership of this business identity appears as a part of the individual ownership ideals, which are expected to fulfil dreams. Moreover, this businessperson's identity would also bring a higher self-esteem, given his lower education. As a result, a pro-community expectation follows the interviewee: By owning his business and creating employment, he could "eradicate the unemployment."

> „Oh. **Myself, personally, I'm planning to have a big shop**. **I want to** have a shop **in my future**. Because I, actually, I've never been at the tertiary schools. I just studied at a lower level whereby I completed my matrix, **so I just know the basics that I can help with myself**. **So, my dream for the future. I would like to be just a businessperson**. I would like to be just a businessperson. Yeah, **not a person to work under Government or in other job**, but if that job can be able to fulfill **my dreams, I can able do that**. But my dream is to, job opportunity just to work or to continue with my business whereby **I can eradicate**

> **the unemployment**, which is on **in Tsumkwe**" (Interview 6, Pos. 384-401).

## Self-Development, Communication, and (Life Long) Learning

Interviewee nine depicted a very different training wish than the training center imagined for the kids leaving the villages or learning skills. His thoughts presented a center for adults since he was relatively at an older age. "Such a lower-level education" expressed, in a way, a contribution to lower self-esteem, but this did not hinder him from wishing for further education. However, the average education age has long passed by. "He would humbly" take part if there were any courses offered. He confirmed his literacy and wished to learn further in literature and computer skills to stay in touch with his friends on the computer.

> "The only thing that I can think of is about **education**. If there was an institution that offers some kind of **training maybe for adults** then I would like to enroll this institution to further my education, since I have **such a lower level of education**, **despite my age**. I am thinking of things **like literature or computer skills,** since I **am capable in literacy** and if there are such trainings, I **would humbly** like to take part. I would like to be **computer literate** to do **correspondence to friends on the computer**" (Interview 9).

Studies suggest that studying literature enhances skills such as cognitive development, reflexive thinking, empathy, cultural understanding, communication, creativity and imagination, personal growth, and becoming more aware of our world. Curtis (2015, p. 2) shared a more profound meaning by stating that "Literature is about the human experience". From a phenomenological viewpoint, by shedding "light" on learning literature, one could claim that the person recognized the value of the literature (of some writers) that can make the mind transcend the time and space (boundaries) of the actual physical location while fostering curiosity and imagination. Thus, the individual enables oneself for self-development through literature, the *written* literature.

In Western culture, literature is perceived in a written form. However, there are many traditions whereby storytelling exists as a type of literature performed and transferred to the next generations orally; the San culture belongs to this group. As stated earlier, the San transcription project, digitizing the cultural heritage, has been a way of recording them. Therefore, interviewee nine's wish also illustrates an interest in learning more from the written literature, despite his older age, and contributes to the process of individualization with the self-development component, but it is not limited. Using computer skills to stay in touch with friends is an additional way of socializing in this context using digital communication tools, which contributes to the process of domestication at the same time.

## Solidarity and Respect

These quotes from Interview Five and Interview 13 focus on the social connectedness, solidarity, and esteem one needs based on the fourth order of Maslow´s hierarchy of needs. They both envision a future in harmony with the community in Tsumkwe, not responding to the question from an individualistic perspective but rather from the community perspective, *for* the community and with the community. In this respect, interviewee 13´s future, where people and community work together in a tranquil and safe place like Tsumkwe, defines the notion of solidarity in a way. Individual or social (community) solidarity can strengthen the cooperation, unity, and support among individuals or groups to promote harmony, justice, and equality. However, she also delivers her concerns about illegal things. She claims that only then it could be an excellent place for all, equal for all.

> "What I can wish myself in the future? With me, I don´t like, like... how can I say it now? Like benefiting myself only, the future I want, **I want it for everybody in Tsumkwe**. It´s what I want. **I want us to be equal**. With me.. laughters. If I ... Let me say it... How can I say it now? In the future, **it will be a nice place if us, the community, we are working together,** not bringing illegal things in this area. Because it´s a nice land, very quiet and safe, not like other places" (Interview 13, Pos. 2).

Interviewee five wishes to have a nice and normal future, with people laughing, living together, helping each other, and sharing. She describes Tsumkwe´s people as helpful already compared to the place she lived before. In this way, she foresees a better future than now.

> "**Nice** future, **normal** future. **Live with people, laugh with them…. Just as one.** It should getting better like living with the people. They are more helpful. When you don't need anything they, they will you, not like the way, where I was. **Because here when you need something, you need help from others they will help you with what you need**. That's how it's becoming better" (Interview 5).

Both of these quotes belong to non-San people residing in Tsumkwe. They express the need for recognition and extend a call for solidarity for all in unity. Originated from Latin-solidare, meaning "to combine parts to form a strong whole, to make firm" (Komter, 2010, p. 1460), it also denotes, in the process of differentiation, when different specialized units or groups or individuals come together in solidarity. Therefore, to one extent, even this vision demonstrates different groups, in this case, based on different ethnicities, who envision a future together. Furthermore, this recognition involves respect, love, and (social) esteem (Mark, 2016; Driessens & Nærland, 2022), which connect it to Maslow´s pyramid of needs on the fourth order. (Self-)Esteem, (self-)love, and (self-)respect are all interlinked and intertwined concepts that have a significant and direct influence on social recognition. Their individual and social absence might cause struggles and conflicts (Honneth, 1996). As a result, the process of individualization sheds light on different notions of individual characteristics for a person who lives in a society.

### 4.2.5 Paradoxes of Modernization

As indicated in the Chapter 2, modernization process is a process accompanied with its paradoxes for each of four processes that have intertwined each other. Thus, similarly the paradoxes also follow a similar pattern.

#### Paradoxes of the Process of Differentiation

The paradox of the differentiation has been interpreted with the scale reduction and the scale enlargement happening on either side of the same coin (van der Loo & van Reijen, 1992). As explained earlier, the division of labor in a society starts making it organizationally complex, bringing about different forms of cooperation, groups, and associations. Specifically, in Tsumkwe, the establishment of Nyae Nyae Conservancy or its earlier version, Nyae Nyae Farmers′ Cooperative, the NNDF, several NGOs show the differentiation at the community level. In contrast, several public institutions demonstrate the same at the public level; the growing number of local shops and the establishment of the CENORED Office are key in indicating the multiplication of specified activities, the division of labor, and the social stratification. Under this circumstance, solidarity has also been adapted to this complexity for Ju/′hoansi only, which is happening with NGOs supporting the Ju/′hoansi community in Tsumkwe and collaborating with national and international institutions. A formation of different groups has not been observed or heard for non-San groups in Tsumkwe.

Furthermore, as a result of this process, the boundaries of the original group become blurred, while many new groups might emerge from the original group. In the case of Tsumkwe, the emergence of new groups, such as NNC under the CBNRM, NNFU for the farmers, and NNDF NGO, have emerged during the colonial period in Tsumkwe (as well as in Namibia) as solidarity in action in defense of land and culture. At this stage, this situation could be interpreted as the protection of "we-identity" for San people, which did not immediately result in a progressive individualization linked to social changes due to social differentiation (van der Loo & van Reijen, 1992). Therefore, even today, the protection of the "San" identity continues while incorporating some individualizing links to "I-identity," which introduces the other side of the "we-identity" coin. Furthermore, the unemployment issue, which contributes to unequal income levels, creates the first strata between employed and unemployed and then a second strata between different public and private employment types.

#### Paradoxes of Rationalization

From the cultural lens of the process of modernization, an increasing self-control/mastery takes place in the levels of the worldview, collective action, and individual action with increasing systematization and ordering of the social and natural reality in order to make it predictable and calculable (van der Loo & van Reijen, 1992). At this point, the "narrowing of vision" to the purpose of independent entities, organizations, or institutions achieve their own goals, which are functional by their character and have specific values and norms; that is also connected to the rationality of purpose and rationality of action (van der Loo & van Reijen, 1992). Without any doubt, especially when the institutions and organizations are concerned, the bureaucracy also appears as a functional tool in realizing these specific goals. On the other hand, the re-enchantment of the worldview through the use of electricity and the meaning of electricity, as well as its extended meaning of "light" and "source of life," have a more "widening of vision" in contrast to the narrowing of vision.

Furthermore, due to the interdependent processes of differentiation and rationalization for electricity provision, the residents and local business owners in Tsumkwe have become aware that without a prepaid electricity token, it is impossible to have power in households and businesses. Thus, the awareness created about electricity consumption has been met by rational thinking about where the consumption will take place and how much it will be. However, there is an interesting paradox in the electricity provision prior to receiving the electricity. The goal-oriented and functional characteristics of the rationalization process incorporate an increasing level of bureaucracy to make it predictable and controllable. In this sense, while there is a specialization of service for mini-grid connection, a set of bureaucratic steps must be completed to achieve this goal.

In the case of Tsumkwe, the applicants must fill in certain forms to apply for a mini-grid connection within the mini-grid connection. After the screening procedures by the CENORED, whereby the duration is not standardized, the connection points less than 80 meters from the poles are commonly accepted with an application fee of 2500 NAD. This fee is standard up to 80 meters; additional installation fees are requested from the applicant depending on the technical connection requirements for longer distances. During the field research, the author heard connection fee requests that went up to 8,000 NAD from some applicants complaining about its unaffordability. Some other informants had accepted this 2,500 NAD as affordable, albeit expensive, in return for the great benefit they would receive from electricity.

In this sense, regardless of the number of specialized institutions, businesses, or households that have become pluralized by "narrowing down the vision" (van der Loo & van Reijen, 1992); each of them who is willing to connect to the mini-grid electricity, goes through the same procedure, which creates waiting application folders at the CENORED premises and usually the connection point. As a result, this situation also creates a "generalization" or "widening of view" (van der Loo & van Reijen, 1992). The differentiated functional units blend in a general, overarching category for mini-grid electricity connection. Under the current tariff structure, the tariffs set for households, prepaid businesses, and institutions depict the generalization, although there might be more functional independent units in the future.

**Paradoxes of Domestication**

The paradox of becoming dependent on technology or technical appliances while liberating themselves from the natural or physical barriers occurs in the process of domestication in Tsumkwe as well. The more people would like to connect to the mini-grid, the more they become dependent on the electricity and the electric appliances. The more people are connected to the limited mini-grid capacity, the more limited they become in using it. Although they see electricity as a tool or a means to a better, easy, or more comfortable life, the interrupted connections create "struggles" for them and cause dissatisfaction. This paradoxical situation also requires more systematic self-control for people in their behaviors with each other. As indicated earlier, technology is not only a "thing," but also creates its social network.

Elias (1995, p. 7) described the technization as "the process by which, as it progresses, people learn to exploit lifeless materials to an increasingly greater extent for the use of humankind, by treating and processing them, in war and in peace, mostly in the expectation of a better life." He also specifies the better life referring to "a social process in the development of which living conditions do not become good in an absolute sense but become better in reference to an earlier phase" (Elias, 1995, p. 8). Similar to the process that Tsumkwe residents have been going through with incorporating electric appliances. Thus, collecting firewood for cooking or heating becomes redundant. The kettle, the gas bottle or the "hot plate" serve best the needs of those moments. Furthermore, Elias (1995, p. 7) interprets and defines the process of technization as well as civilization as the "unplanned long-term processes, without having long-term aims but moving in a noticeable direction." In this way, he also explains the "unplanned," which is "arising from the interweaving, the conjunction, the cooperation, and confrontation of many planned activities" (1995, p. 7).

Although Elias had based his research on processes spread over centuries, the rise of digitization and digitalization might have brought a faster pace in the twenty-first century for the Tsumkwe case. The communities have been experiencing the processes of digitization and digitalization simultaneously, with the ongoing process of domestication linked with electrification. The far is nearer with the use of mobile phones and much nearer with the use of smartphones. The internet is the following tool or means after electricity, which is strictly dependent on electricity and enables many digital processes to reach remote areas, including Tsumkwe. As a result of the accumulated processes, an "unplanned" situation takes place in Tsumkwe as a result of many planned activities, such as electrification, domestication, and digitalization that resulted in an unplanned consumption scenario limiting the residents in using the electricity and reaching the limits of the mini-grid capacity. In addition, on the other side of the digitalization coin is the risk of cyber-attacks and data leaks (Fusaro, 2023).

> "Electricity is **very important**. It has **advantages and disadvantages**. If you chock, you're **dead**. You **musn´t play** with it. The advantage is that you can **have everything in your** house" (QOQ).

Lastly, the quote above underlines the recognized associated advantages and disadvantages of electricity, intersecting lines of life and death at home: the possibility of having everything in the house with the possibility of dying by electrocution in a poorly installed case. However, this also informs how much the residents get acquainted with electricity and its use.

**Paradoxes of Individualization**

The challenge of pluralism appears in a modernizing culture involving many values, norms, and meanings that complement and conflict, as discussed by van der Loo and van Reijen (1992). The multiplicity of choices provides freedom and a detachment from the traditional collectivity that people are born into and on which they are dependent. In this modernized societal structure, the individual comes first, whereas, in traditional society, the individual is subordinate to the collective (van der Loo & van Reijen, 1992). In the case of Tsumkwe, the individualization of having their housing, mobile phones, and television, as well as their willingness and happiness to charge them or watch them in their places, demonstrate the beginning of the process of individualization with its paradoxical stance. The loss of dependency on other people and the abstractedness of their connection to non-place-based contacts via mobile phones and television or, very shortly, via the internet allows more people to follow each other in this path of modernization.

Domestication and individualization intertwine to provide the multiplicity of different electric appliances at a household and individual level, interrupting traditional collectivity and bringing forward more individualized household settings. However, the collective "San-identity" remains in the settlement, even though there has been an individualization in process.

## 4.3 Quantitative Findings

The quantitative findings in this section are composed of surveys and other collected data from institutions and archives, describing the process of modernization from a quantitative point of view.

### 4.3.1 Quantified Local Setting

#### *4.3.1.1 Population in Tsumkwe*

Tsumkwe Settlement is a location that appears relatively more in the literature due to its ethnic background. However, the household and population data remain challenging to research and analyze due to missing and insufficient information. Apart from the growing population, housing, and local shops, the establishment of the offices of public institutions and branches of national companies have also taken place, in addition to the Nyae Nyae Conservancy, Craft Centre, and some churches. CENORED is the only company that is related to electricity and electricity provision.

Within the context of insufficient data, during the research period, nearly to the end, after the travel restrictions have been lifted due to the pandemic, the author could reach the National Statistics Agency (NSA) for their support regarding the official statistics. As a result, the Table 4.13 demonstrates the official statistics for Tsumkwe Settlement in the 2001 census (884 inhabitants) and 2011 (1404 inhabitants), with a total housing of 290 households. Even though the planned census could not take place in 2021, an estimation of the population (based on the population growth data recorded in the previous years) for 2021 and 2031 has been calculated by the NSA only for this research. Accordingly, the estimated population in Tsumkwe is 3215 inhabitants (in 2021) and 5026 inhabitants (in 2031). On the other hand, to better analyze the survey data for Tsumkwe, the average number of people per household was calculated based on the census of 2011 (4.84 people/household). In this way, it has been possible to estimate 664.07 total housing (in 2021) and 1038.31 total housing (in 2031) to use in the further and future analysis of the findings of this research (Table 4.13).

*Table 4.13.Tsumkwe Population and Household Estimation based on the NSA Data*

| Years | **2001** | **2011** | **2021*** | **2031*** |
|---|---|---|---|---|
| Population | 884 | 1404 | 3215* | 5026* |
| Total Housing | N/A | 290 | 664.07** | 1038.13** |

*Estimated data provided by NSA in 2022
**Average inhabitant per household has been calculated at 4.84 (2011).

*Source: Author*

The population and the total housing data are crucial for calculating and designing infrastructure-related activities. Therefore, concerning the population of Tsumkwe Settlement, based on the estimations of NSA and the data collected from the literature, the population, and total housing growth have been illustrated in Table 4.13. In the 1970s-2001, the population has been fluctuating due to the San people's land struggles and Namibia's independence in 1990, as explained in section 4.1. From 2001 until 2011, the population's change rate was 58.82%. On the other hand, the change rate from 2011 to 2021 has been calculated at 128.99%. The estimation until 2031 is 56.33% growth in the population. Given the population of San in Namibia 34,000 (Biesele & Hitchcock, 2011), the Tsumkwe Settlement´s population reaches nearly 9.5% of the overall San population in Namibia, and Tsumkwe Constituency´s (Tsumkwe East and West) nearly 26.5% of the overall San population in Namibia.

#### *4.3.1.2 Social Stratification and Division of Labor within the Local Actors in Tsumkwe*

The social stratification and the division of labor to one extent that has been occurring in Tsumkwe since the independence of Namibia shall be observed by the diversity and the quantity of the public, private, and community actors. Accordingly, division of labor is splitting activities within a production process or society by specialization. At the same time, social stratification refers to the hierarchical structure within individuals or groups in a society. The government of Namibia has twenty ministries, whereby the nation-state has been governed; the responsibilities and activities have been divided to serve the public better at all levels. Thus, the national and regional state representative offices are present in Tsumkwe Settlement (Table 4.14), along with the community representations, national and local private companies, and the residents in general.

At the national level, eight ministries out of twenty have been present in Tsumkwe, along with two parastatals, NBC and NamPower, and seven public institutions. From a regional level, the Otjozondjupa Regional Council´s Office for Tsumkwe Constituency operates in Tsumkwe. The Ju/'hoansi Traditional Authority is the official representation of the San people of the Tsumkwe East, according to the Traditional Authorities Act no. 25 of 2,000. The NNC is a San community representation organization, and two other NGOs are running locally. There are eight active churches. Apart from CENORED and Tsumkwe Lodge, there are no other national/regional private companies; the remaining are locally owned and operated small shops (Table 4.14 and Table 4.15). Given the fact that Tsumkwe´s population is approximately 3,215 inhabitants, and Tsumkwe Constituency is nearly 9,000 inhabitants, a strong state representation is present in the settlement, which creates a social stratification from employment and education classifications, which will be addressed in the household surveys, in detail.

*Table 4.14. Local Actors in Tsumkwe*

<table>
<tr><td rowspan="3">National and Regional State Representa-tions</td><td>Public Institutions</td><td>Tsumkwe Primary School (Pre-primary and kinder garden included),<br>Tsumkwe Secondary School,<br>Clinic,<br>Police Station,<br>Magistrate´s Court (Civic Affairs and Administration),<br>Captain Kxao Kxami Community Learning and Development Centre (CLDC), (Ministry of Education)<br>Veterinary</td></tr>
<tr><td>Ministerial Representations</td><td>Ministry of Home Affairs,<br>Ministry of Gender Equality and Child Welfare,<br>Ministry of Agriculture, Water, and Land Reform,<br>Ministry of Environment, Forestry, and Tourism,<br>Ministry of Urban and Rural Develop-ment</td></tr>
<tr><td>Regional and Lo-cal Representa-tions (all belong to Ministry of Urban and Rural Devel-opment)</td><td>Tsumkwe Settlement Office (Regional Council)<br>Office of the Regional Council (Tsu-mkwe Constituency)<br>The Ju/'Hoansi Traditional Authority</td></tr>
<tr><td colspan="2">Churches</td><td>Dutch Formed Church<br>Seventh Day Adventist Church<br>Christ Love Ministries<br>Roman Catholic Church<br>Prophets Church<br>Zion Council of Churches (ZCC)<br>Evangelist Church<br>Pinkster Church</td></tr>
<tr><td colspan="2">Parastatals</td><td>NBC Radio<br>NamWater<br>NamPost</td></tr>
<tr><td rowspan="2">Community Representations</td><td>Community-Based Organization</td><td>Nyae Nyae Conservancy in Tsumkwe East</td></tr>
<tr><td>NGO</td><td>Nyae Nyae Development Foundation<br>Ju/'Hoansi Traditional Fund<br>Palms for Life</td></tr>
</table>

*Source: Author*

*Table 4.15. Local Actors in Tsumkwe (continued)*

| | |
|---|---|
| **Private Companies** | MTC Tower - No official representation<br>CENORED<br>Tsumkwe Lodge and Chief Bobo Training Center |
| **(Local) Private Shops** | Fuel Station / General Dealer<br>Food Shops and supermarkets<br>Grocery, Butchers, Bar / Shebeens<br>Hairdresser, Restaurant, Second-hand clothing etc. |
| **Households** | 664, (estimation based on the NSA data). |

*Source: Author*

#### *4.3.1.3 Electrified Households and Businesses*

The Central Northern Regional Electricity Distribution is a regionally operating national company with public shareholders and a board of management. The Regional Council of the Otjozondjupa, Kunene, and Oshikoto Regions, the Local Authorities within this area, and NamPower are CENORED´s shareholders. More detailed information will be provided within the national and regional renewable energy governance chapter. The headquarters are in Otjiwarango, and the local representative office is situated in Tsumkwe for both Gam and Tsumkwe. In 2017, CENORED took over the mini-grid's operation and maintenance (O&M) from the Ministry of Works in order to enhance the O&M of the mini-grid as a specialized regional power distribution company. Therefore, this type of company-driven action demonstrates how the state has foreseen the power sector and divided the country into five different service regions to provide electricity, increase productivity, ease management, and better control.

*Figure 4.16. Households and Local Shop with Electricity*

*Source: Author*

Table 4.16 below illustrates the number of connections or clients in Tsumkwe (Figure 4.16) under three main categories: residential (prepaid), business (prepaid), and business (institutional), which also establishes a link from the process of differentiation that has already started in Tsumkwe and continued with the tariff setting of differentiated units. The institutional category is not prepaid but based on conventional metering. Thus, it is leading to further tariff differentiation than households and businesses, which contributes to the social differentiation within the electricity tariffs. Accordingly, there is an 8 to 10% increase in the rate of categorized electrification based on this data from 2020 to 2022 and a 34% increase in the total connections from 2019 to 2022. On the other hand, from a residential perspective, with 330 household connections in 2022 and the estimation of NSA for Tsumkwe with 664 total housing units, the mini-grid electrification rate has been calculated at 49.7%. Taking into account Namibia´s overall electricity access was 55.2% in 2021 (WB, 2023) or the national data that is approximately 50%, Tsumkwe remains shortly below the national electrification rate but above the 33.2 % rural electrification rate, placing Tsumkwe in a relatively better situation (WB, 2023). In addition, counting the unelectrified households realized by the community members in Tsumkwe resulted in 186 households and nine businesses and shops at the end of 2023.

*Table 4.16. Electrification Increase from 2020 to 2022, and its Distribution*

<table>
<tr><th>Years</th><th>2019 (Boamah, 2020)</th><th>2020*</th><th>2022*</th><th>Increase in Electrification</th><th>2023**</th></tr>
<tr><td>Total</td><td rowspan="5">260</td><td>363</td><td>393</td><td>8%</td><td rowspan="5">N/A</td></tr>
<tr><td>Residential (Prepaid)</td><td>305</td><td>330</td><td>8%</td></tr>
<tr><td>Business (Prepaid)</td><td>49</td><td>54</td><td>10%</td></tr>
<tr><td>Business (Institutional)</td><td>9</td><td>9</td><td>0</td></tr>
<tr><td>Awaiting Applications</td><td>69 (67 prepaid, 2 institutional)</td><td>57</td><td></td></tr>
<tr><td colspan="5">Unelectrified Households and Shops</td><td>186 Households and 9 Shops</td></tr>
</table>

*Data received from CENORED in 2020 and 2022.
**Counting realized by the community members for the PROCEED Conference in September 2023.

*Source: Author*

In 2012, at the beginning of the Tsumkwe Energy Project, 107 initial connections were made and upgraded, with 86 additional connections in the same year, summing up to 195 households. On the other hand, 290 households at the time of the last census (NSA, 2011) demonstrated that approximately 50% of future increases have been considered in the design of TEP. Two hundred sixty total connections have been met by 2019, which is nearly the total housing reported by the NSA, describing that the demand forecast made in 2011 was suitable for 2019 or it was when all households were connected. Therefore, it is possible to state that the mini-grid was designed to cover and supply electricity for the existing population in Tsumkwe at the time with some margin for growth, but not within the entirety of considering potential future demands, desires, wishes, and aspirations of a growing society towards future. Thus, a combined future projection is necessary to calculate and foresee future demands for a growing place by all its meanings.

## Survey Data – Sample Size

The recent data regarding the population, the number of households, and the electrified households were unavailable during the first field research. Therefore, 68 mini-grid-connected and eight unelectrified households and businesses were surveyed, corresponding to 26.15 % of the total connections in 2019. It took a very long time to receive the client list from CENORED corresponding to the period of the field research. The percentages in the survey column (Table 4.17) represent the percentages that have been calculated with the figures for 2020 because there has been no other detailed information about the various categories of clients. During the surveys and interviews, some residents mentioned their applications waiting to be processed by CENORED; in 2020, these applications counted as 69, including two institutional and 67 prepaid applications. Two years after, after another lengthy process of data reception from CENORED, the new figures showed 57 waiting applications, an increase of 25 connections in the households with 330 connections, and an increase of 5 in prepaid businesses, resulting in 54 connections in total. The conventional institutional remained the same, with only nine connections.

*Table 4.17. Survey Data in Detail*

| Years | Surveys | Connections in 2020* | % of the Survey with total connection |
|---|---|---|---|
| Total Connection | 68 | 363 | 19% |
| Residential Prepaid | 51 | 305 | 17% |
| Business Prepaid | 11 | 49 | 22% |
| Business Institutional | 6 | 9 | 67% |
| Awaiting Applications | | 69 (67 prepaid, 2 institutional) | |

*Source: Author*

These extraordinary *Force Majeure* circumstances, with the limitations caused by COVID-19 right there during the field research and not having accurate recent data, changed many plans in the research flow, as mentioned earlier. Under these conditions, the sample size of 51 prepaid households, which is more than 28 samples, provides a confidence level of 90% with a 15% confidence interval calculated with the MaCorr calculator. Ideally, a 95% confidence level with a 5% interval would have been presented under normal conditions. However, although the analyzed confidence level and interval have limitations, they still provide a comprehensive overview of the settlement to take into account within the scope of this research. Furthermore, since the sample sizes of 11 prepaid businesses and six institutional businesses are very small, the confidence levels and intervals have not been calculated, but some key data have been shared in the quantitative section.

#### *4.3.1.4 Local Businesses*

Of the local businesses, including prepaid and institutional (n=17) surveyed, 58.8% corresponded to private and 41.2% to the public. The number of employees per business varied between one to 26 people, depending on the type of institution. Of the overall respondents, 52.9% of the businesses responded that they had 1-5 employees. Moreover, a public and private sector differentiation has appeared, whereby 77.8% of the private businesses answered the question by employing with 1-5 people, only.

#### *4.3.1.5 Household Demography, Socio-Economics, and Education*

In this section, the quantitative data regarding household demography, education, employment, and income distribution will be discussed for all mini-grid-connected households (n=51) that have been surveyed.

**Household Demography**

Based on the results of the descriptive statistics (Table 4.18), the mean of residents per household is 6.08 people, with a minimum of one person and a maximum of 22. The average in Namibia corresponds to 3.9 people, in the rural 4.6 people, and in the Otjozondjupa region, 3.9 people per household (NIDS, 2016, p. 89). Tsumkwe remains above all the averages. The dwellings have a minimum of 1 room up to 4 rooms; the mean value is 2.24, with a median and mode both 2 (rooms). Out of these households, only two utilize the place for commercial activities. However, this was not explicitly mentioned during the survey. The question about age was initially asked. However, most ages were unknown; therefore, it was taken out of the survey during the field survey.

*Table 4.18. Descriptive Statistics for the Household Demography*

| | N | Min | Max | Sum | Mean | |
|---|---|---|---|---|---|---|
| | Statistic | Statistic | Statistic | Statistic | Statistic | Std. Error |
| Total Residents | 51 | 1 | 22 | 310 | 6.08 | .593 |
| Females | 51 | 0 | 14 | 154 | 3.02 | .349 |
| Males | 51 | 0 | 10 | 156 | 3.06 | .307 |
| Children | 51 | 0 | 13 | 161 | 3.16 | .385 |
| Adults | 51 | 1 | 12 | 149 | 2.92 | .291 |
| Rooms | 51 | 1 | 4 | 114 | 2.24 | .142 |
| Valid N | 51 | | | | | |

*Source: Author*

Furthermore, the surveyed household demography resulted in the 51% female and 49% male ratio, with 53% children and 47% adults. These ratios are equivalent to Namibia, with 51.4% female and 48.6% male, and 51.6% female and 48.4% male in rural gender distribution (NIDS, 2016, p. 46). However, they differ from the Otjozondjupa regional distribution, with 48.5% female and 51.5% male (NIDS, 2016, p. 46).

**Educational Background**

The education system in Namibia has been composed of four main educational levels, which consist of pre-primary (from 3 years old until 5-6 years old), primary (Grade 1 - Grade 7), secondary (Grade 8 – Grade 12), and tertiary levels of education. Based on the 51 surveys conducted in Tsumkwe households, 31.4 % reported having no education or have quit primary school (21.4 %); 45.1% reported to have completed primary school; 9.8% reported to have completed secondary school and 13.7 % tertiary school (Figure 4.17).

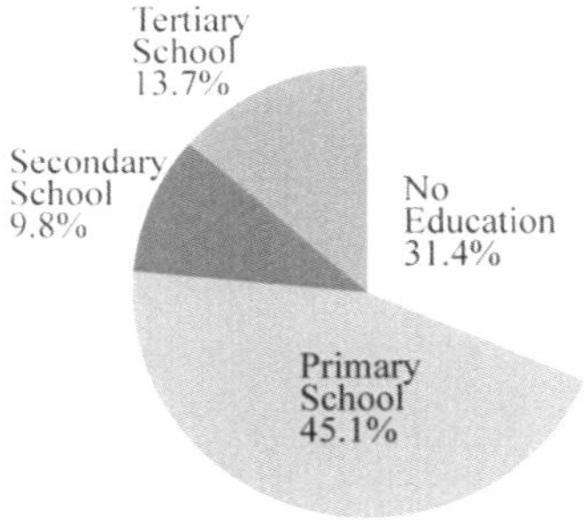


*Figure 4.17. Highest School Degree*
*Source: Author*

On the other hand, according to national statistics, NIDS (2016, p. 75) stated that 0.5% have not completed formal education, 18.4% have incomplete primary, 49.7% completed primary, 22.6% completed secondary education, and 8% completed tertiary.

In this case, Tsumkwe has been over the average of Namibia with incomplete primary education, below the average completed primary education; below the average completed secondary education, and above the average in tertiary education. Although Namibian statistics demonstrate slight increases from 2011 until 2016, Tsumkwe´s survey data describes a substantial ratio of no formal education and incomplete primary.

**Socio-economic Background Household Occupation**

The process of differentiation outlined above indicates that the division of labor and social stratification have started to influence the socio-economic conditions of the households in the Tsumkwe Settlement. The self-employed 7.8% refers to the jobs undertaken within tourism or research-related activities (Figure 4.18). The state employees refer to the respondents who work in the public sector. The unemployment rate among the respondents is 35.3 %, which is higher than Namibia´s unemployment rate of 21.3% (WB, 2023).

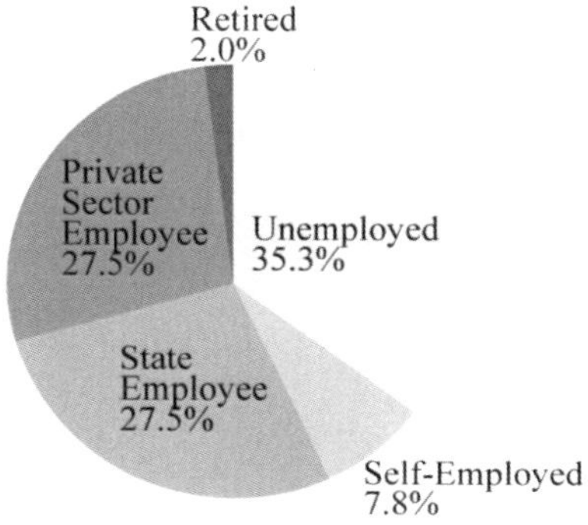


*Figure 4.18. Occupational Distribution in Tsumkwe*
*Source: Author*

This rate has decreased compared to 27.4% in 2012 in Namibia (NSA, 2013). This rate was 26.2% in rural Namibia and 25.3% in the Otjozondjupa Region. Moreover, a total of 55% of state and private sector employees (Figure 4.18) demonstrated similar income levels in comparison with the 53.6% ´salaries and wages´ within the results of the annual consumption by the primary source of income in Namibia (NHIES 2015/16, p. 98). On the other hand, during the field research, some respondents stated that they were unemployed when they did not have a permanent job or retired.

**Occupation and Education**

Furthermore, very closely related to the process of differentiation, the following Figure 4.19 demonstrates the distribution of occupational categories with different educational backgrounds. 56.3 % of the unemployed people do not have any education or quit primary school, and 39.1% have a primary school degree or quit secondary school. Whereby 71.4% of the state employees have a tertiary education degree. Moreover, 60% of secondary school graduates are in the private sector, followed by 34.8% of primary school graduates and 28.6% of tertiary school graduates. In this regard, there is more unemployment with no formal or less education. The state employees have mostly tertiary education, and the private sector employees tend towards any educational background. As a result, the data shows a close connection between education and (un)employment.

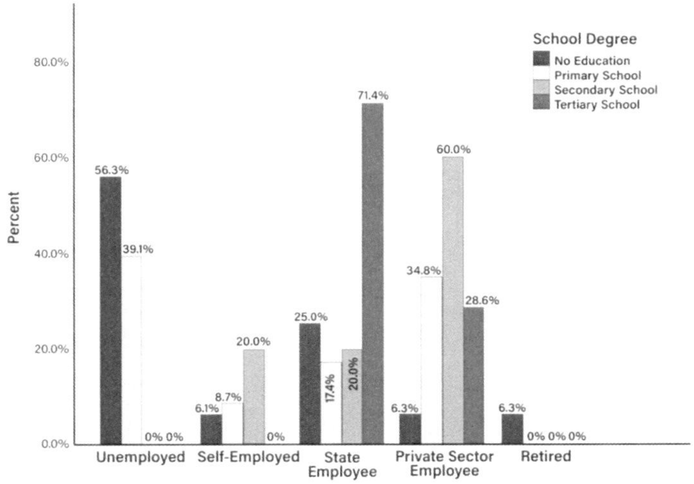


*Figure 4.19. Distribution of Highest School Degree by Occupation*
*Source: Author*

## Income

The income distribution per household has also been investigated in the differentiated occupational categories. The survey on the average income categories was based on the minimum wage of 1,300 NAD/month when the first field research was held in 2020. "N/A" represented the 13.7% of the households who did not want to answer the question; 33.3% stated that it was more than 3,900 NAD per month, and 25.5% mentioned up to 1,300 NAD per month (Figure 4.20). The descriptive statistics in Table 4.19 indicate that the mean average is 1.8039, equivalent to the average range of 1,301-3,900 NAD per month. As a total annual household income, these figures would be equivalent to 0 – 15,600 NAD, 15,603 NAD – 46,800 NAD, and 46,803 NAD and more. The recent literature is quite limited in accessing up-to-date household income ranges. However, the preliminary results of the Namibian Household Income and Expenditure Survey (NHIES) conducted in 2003/04 and 2015/16 provide some insights into this matter.

Accordingly, the NHIES in 2003/2004 (National Planning Commission [NPC], 2004, p. 36) informed that the annual average household income in Otjozondjupa was 34,897 NAD (adjusted[11] 9,457 NAD/capita/year), whereby this figure increased to 43,520 NAD (adjusted 10,357 NAD/capita/year) overall in Namibia. The urban average was noted as 66,625 NAD (adjusted 17,899

[11] "The adjusted per capita income is higher due to the fact that children under 16 years count less than one adult household member, as an adult equivalent scale has been applied to adjust the per capita income (NHIES, p. 36)."

NAD/capita/year), with a very high disparity with the rural average of 27,792 NAD (adjusted 6,137 NAD/capita/year). The lowest income was recorded in Ohangwena, with 22,166 NAD (adjusted 4,294 NAD/capita/year). The survey results concluded that urban households had an average annual income that was three times higher than rural households, and the Gini coefficient for Namibia was calculated at 0.6, which indicated a highly unequal income distribution.

*Table 4.19. Average Income Statistics*

| | | |
|---|---|---|
| N | Valid | 51 |
| | Missing | 0 |
| Mean | | 1,8039 |
| Median | | 2,0000 |
| Mode | | 3,00 |

*Source: Author*

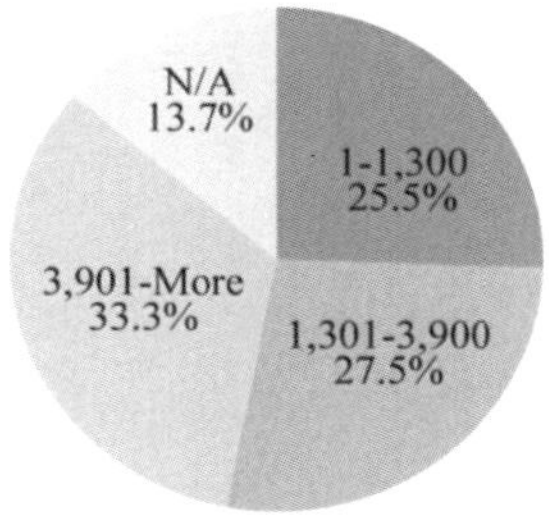


*Figure 4.20. Distribution of Average Income (NAD/month)*
*Source: Author*

Furthermore, the most recent NHIES (2015/16, p. 14) indicated that the Gini coefficient was reduced to 0.56; however, it still keeps its position among the highly unequal income distribution. Conversely, the average annual household consumption was calculated per language spoken in the households. Accordingly, the Khoisan languages spoken in the household had a consumption of 39,606 NAD (7,088 NAD per capita). In contrast, Otjiherero-spoken households had 101,231 NAD (24,827 NAD per capita) and Afrikaans-spoken households had 258,406 NAD (72,017 NAD per capita) (NHIES 2015/16, p. 97). Furthermore, Namibia's annual average consumption was noted to be 119,065 NAD (28,434 NAD per capita). Zambezi region was the lowest on this scale with 55,112 NAD (12,446 NAD per capita), and Otjozondjupa region with 99,805 NAD (19,352 NAD per capita).

*Table 4.20. Namibia's Poverty Lines (current NAD/adult/month), 2003/04-2015/16*

| Poverty lines / (NAD) | 2003/2004) | 2009/2010 | 2015/2016 |
|---|---|---|---|
| FPL (Food poverty line) | 127.15 | 204.05 | 291.1 |
| LBPL (Lower bound poverty line) | 184.56 | 277.54 | 389.3 |
| UBPL (Upper bound poverty line) | 262.45 | 377.96 | 520.8 |

*Source: NHIES (2015/16, p. 104)*

The following indicator figures (Table 4.20) have been the food poverty line (FPL), lower bound poverty line (LBPL), and upper bound poverty line (UBPL) which were calculated as 293.1 NAD/per month per adult, 389.3 NAD/per month per adult, and 520.8 NAD/per month per adult, respectively, which is equivalent to 3,517.2 NAD, 4,671.6 NAD, and 6,249.6 NAD per adult per annum. According to the findings of the average income range per household in Tsumkwe between 1,301 NAD/month and 3,900 NAD/month (equivalent to 445.55 NAD/adult and 1335.62 NAD/adult calculated with a mean of 2.92 adult/household), Tsumkwe has been situated at the border of the UPL. As per the yearly consumption data in Namibia, the results in Tsumkwe are lower than the averages in Namibian, Otjozondjupa region, and Otjiherero spoken household consumption. However, it is in the range of the Khoisan-spoken household average. Moreover, the range between 0 and 15,603 NAD per household income per annum lies below the range of the FPL.

**Income and Occupation**

The primary sources of income outlined in the NHIES 2015/16 (p. 103) in the annual consumption assessments have been identified as salaries and wages (53.6% of the households), pension (11%), subsistence farming (10.6%), business income (9.6%), remittances/grants (9.6%), drought/in-kind receipts (2.7%), commercial farming (0.3%) and others (3%). In comparison with the Namibian main income source breakdown, Figure 4.21 below provides insights that the state employees' monthly income is more than 3901 NAD, which is followed by the average income of 1301 – 3900 NAD. The private employee sector incorporates all income categories, with 1301- 3900 NAD being the highest category (35.7%). The respondents who mentioned to be unemployed also possess monthly incomes of between 0-1300 NAD (53.8%), 1301-3900 NAD (35.7%), 3901-more (11.8%) and N/A (57.1%). As a result, the households have the ability to generate income by other activities that they do not interpret as a job.

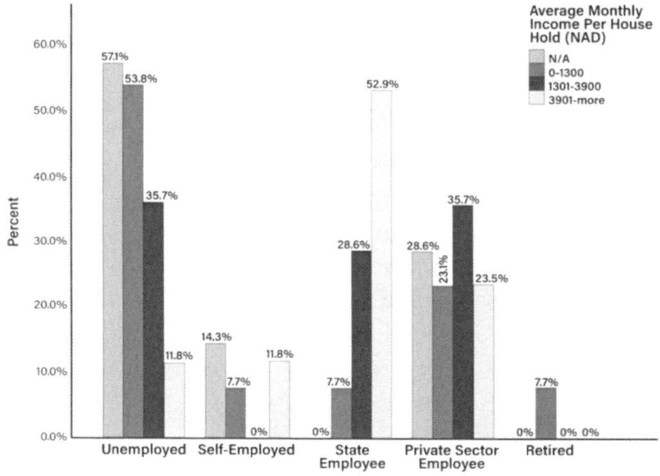


*Figure 4.21. Distribution of Average Income by Occupational Categories*

*Source: Author*

## Ownership of Land, Livestock, Plot, and Car

The ownership open-ended question "Do you have any ownership of livestock, land, plot, car, etc.?" has been answered "yes" by 54.90% of the respondents, of which 9.8% stated both land and livestock ownership and "no" by the 45.10% of the respondents. Moreover, of the 35.30% who stated to have livestock, 23.5% belonged to cattle only (Figure 4.22). The remaining 11.8% owned chickens, chickens and goats, and goats and cattle in small percentages. The livestock has been primarily used for meat and milk and sold when needed. In this sense, it is a cattle bank, whereby they have "live" assets. On the other hand, 31.4% of the respondents stated to have land ownership, 21.6% had ownership of land in the village, 5.9% had a plot, and 3.9 percent had a garden (Figure 4.23). The land in the village has been cultivated mostly with maize, beans, carrots, melon, watermelon, vegetables, and mahangu.

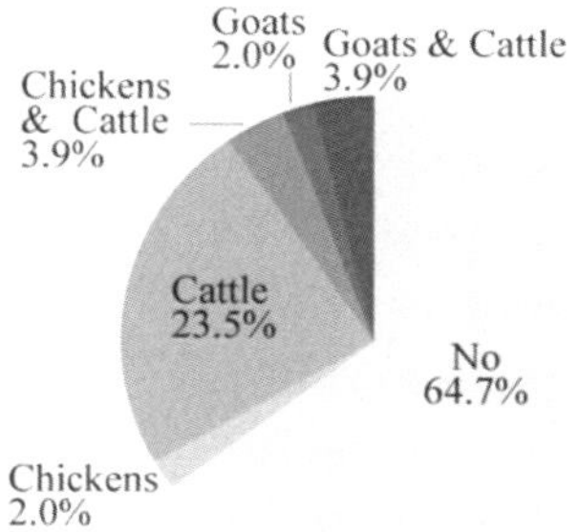


*Figure 4.22. Livestock Ownership*

*Source: Author*

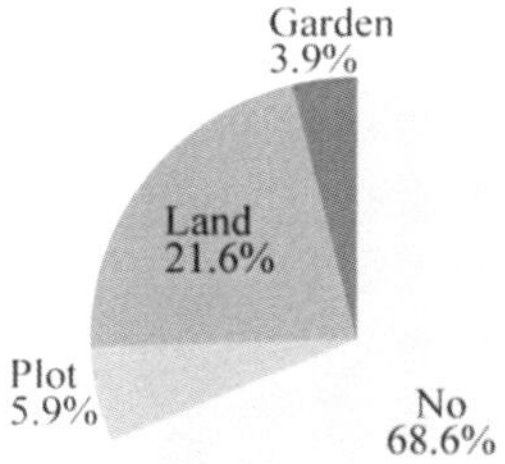


*Figure 4.23. Land Ownership*

*Source: Author*

Although the analysis has been focused on the individual ownerships that have been shaping the everyday life of the residents in Tsumkwe, Figure 4.22 and Figure 4.23 demonstrate how the differentiation has been taking place for live-stock and land ownership. In this sense, of the 54.9% who stated to have a type of ownership, 25.5% have no education or incomplete primary education, and 23.5% have primary school degree or incomplete secondary school, which corresponds in total to 89.25% of the overall respondents with ownership (Table 4.21). Thus, it is possible to interpret that the residents with lower education qualifications have an ownership of either livestock or land.

*Table 4.21. Distribution of Ownership by Highest School Degree*

| | | | Ownership of livestock, land, plot, car | | Total |
|---|---|---|---|---|---|
| | | | Yes | No | |
| Highest School Degree | No Education | Count | 13 | 3 | 16 |
| | | % within Ownership | 46.4% | 13.0% | |
| | | % of Total | 25.5% | 5.9% | 31.4% |
| | Primary School | Count | 12 | 11 | 23 |
| | | % within Ownership | 42.9% | 47.8% | |
| | | % of Total | 23.5% | 21.6% | 45.1% |
| | Secondary School | Count | 0 | 5 | 5 |
| | | % within Ownership | 0.0% | 21.7% | |
| | | % of Total | 0.0% | 9.8% | 9.8% |
| | Tertiary School | Count | 3 | 4 | 7 |
| | | % within Ownership | 10.7% | 17.4% | |
| | | % of Total | 5.9% | 7.8% | 13.7% |
| Total | | Count | 28 | 23 | 51 |
| | | % of Total | 54.9% | 45.1% | 100.0% |

*Source: Author*

On the other hand, the differentiation that has been illustrated in Table 4.22 indicates that 35.7% of the overall ownership belongs to the respondents with a monthly income between 1-1300 NAD, 28.6% with an income between 1301-3900 NAD, and 21.4% with more than 3901 NAD monthly income. The respondents who did not want to answer the question accounted for 14.3%. This table also supports the findings in the previous section on employment; although respondents claimed to be unemployed, the data suggests that there are other sources of income generation other than monthly salaries.

*Table 4.22. Distribution of Ownership by Average Monthly Income*

| | | | Ownership of livestock, land, plot, car | | Total |
|---|---|---|---|---|---|
| | | | Yes | No | |
| Average Monthly Income per Household | N/A | Count | 4 | 3 | 7 |
| | | % within Ownership | 14.3% | 13.0% | |
| | | % of Total | 7.8% | 5.9% | 13.7% |
| | 1-1300 | Count | 10 | 3 | 13 |
| | | % within Ownership | 35.7% | 13.0% | |
| | | % of Total | 19.6% | 5.9% | 25.5% |
| | 1301-3900 | Count | 8 | 6 | 14 |
| | | % within Ownership | 28.6% | 26.1% | |
| | | % of Total | 15.7% | 11.8% | 27.5% |
| | 3901-more | Count | 6 | 11 | 17 |
| | | % within Ownership | 21.4% | 47.8% | |
| | | % of Total | 11.8% | 21.6% | 33.3% |
| Total | | Count | 28 | 23 | 51 |
| | | % of Total | 54.9% | 45.1% | 100.0% |

*Source: Author*

Furthermore, the classification according to the type of employment enhances this argument. According to Table 4.23, 42.9% of the respondents who have ownership are unemployed, 32.1% are state employees, and 17.9% are private sector employees. Whereas 26.1% of the respondents without any ownership are unemployed, and 39.1% are private sector employees.

*Table 4.23. Distribution of Ownership by Occupation*

| | | | Ownership of livestock, land, plot, car | | Total |
|---|---|---|---|---|---|
| | | | Yes | No | |
| Occupation | Unemployed | Count | 12 | 6 | 18 |
| | | % within Ownership | 42.9% | 26.1% | |
| | | % of Total | 23.5% | 11.8% | 35.3% |
| | Self-Employed | Count | 1 | 3 | 4 |
| | | % within Ownership | 3.6% | 13.0% | |
| | | % of Total | 2.0% | 5.9% | 7.8% |
| | State Employee | Count | 9 | 5 | 14 |
| | | % within Ownership | 32.1% | 21.7% | |
| | | % of Total | 17.6% | 9.8% | 27.5% |
| | Private Sector Employee | Count | 5 | 9 | 14 |
| | | % within Ownership | 17.9% | 39.1% | |
| | | % of Total | 9.8% | 17.6% | 27.5% |
| | Retired | Count | 1 | 0 | 1 |
| | | % within Ownership | 3.6% | 0.0% | |
| | | % of Total | 2.0% | 0.0% | 2.0% |
| Total | | Count | 28 | 23 | 51 |
| | | % of Total | 54.9% | 45.1% | 100.0% |

Regarding monthly electricity expenses, Table 4.24 shows that 71.4% of the respondents who have a type of ownership spent monthly 0-150 NAD on electricity and 17.9% spent 151-300 NAD monthly. In contrast, the respondents who don´t have any ownership, 39.1% spent 0-150 NAD, 30.4% spent 151-300 NAD, and 21.7% 301-450 NAD monthly on electricity, suggesting that ownership alone is not linked to electricity consumption.

*Table 4.24. Distribution of Ownership by Electricity Expense*

| | | | Ownership of livestock, land, plot, car | | Total |
|---|---|---|---|---|---|
| | | | Yes | No | |
| Monthly Electricity Expense | N/A | Count | 1 | 0 | 1 |
| | | % within Ownership | 3.6% | 0.0% | |
| | | % of Total | 2.0% | 0.0% | 2.0% |
| | 0-150 | Count | 20 | 9 | 29 |
| | | % within Ownership | 71.4% | 39.1% | |
| | | % of Total | 39.2% | 17.6% | 56.9% |
| | 151-300 | Count | 5 | 7 | 12 |
| | | % within Ownership | 17.9% | 30.4% | |
| | | % of Total | 9.8% | 13.7% | 23.5% |
| | 301-450 | Count | 0 | 5 | 5 |
| | | % within Ownership | 0.0% | 21.7% | |
| | | % of Total | 0.0% | 9.8% | 9.8% |
| | 451-700 | Count | 2 | 2 | 4 |
| | | % within Ownership | 7.1% | 8.7% | |
| | | % of Total | 3.9% | 3.9% | 7.8% |
| Total | | Count | 28 | 23 | 51 |
| | | % of Total | 54.9% | 45.1% | 100.0% |

*Source: Author*

**Financial Aid**

The need for financial aid is also closely connected with the socio-economic section of this chapter. However, since it was asked as an individualized question, it was considered in this sub-section. 47.1 % of the respondents stated that they do not need any financial aid, whereas 52.9% were affirmative. In this sense, the respondents who were in need, received support from financial institutions (28%), from families and friends (48%), and from the government (24%), providing differentiated sources of support other than community ties only.

#### *4.3.1.6 Continuity of Education and Employment in the Future*

The respondents also stated their vision for their children to continue their studies, where it is possible. In this sense, according to the survey findings, out of the 161 children/youth mentioned, 91 of them is attending to school, which corresponds to the 56.52% of the children. Moreover, 71.5% of the respondents claimed that they would like them to continue their studies if there are sufficient funds (5.9%), if the children are successful (7.8%) and, depending on the circumstances (5.9%) (Figure 4.24).

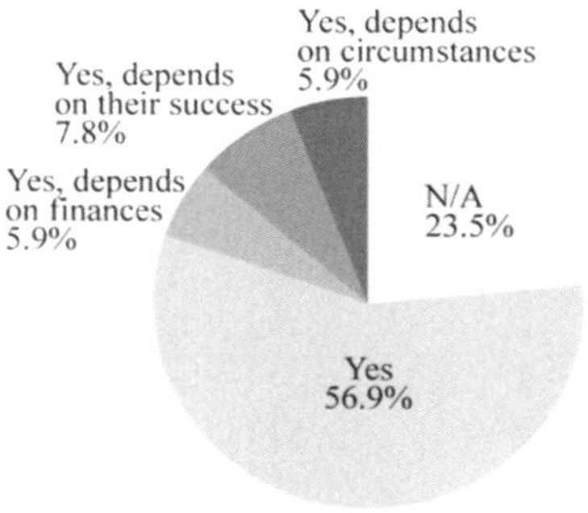


*Figure 4.24. Continuity of Studies*
*Source: Author*

A respondent was determined that her son would study law and her daughters would become doctors and teachers. Another respondent stated that the children "need to be pushed forward" to study. On the other hand, the respondents who stated that they were unemployed stated their willingness to do any work, which corresponds to 54.5 % of the respondents; 22.7% stated jobs such as business, hair design, early child development, government work, while 22.7% mentioned that they did not want to work.

### 4.3.2 Electricity Tariffs, Consumption, Affordability of Tariffs, and Supply Satisfaction

#### *4.3.2.1 Households*

The definition utilized by van der Loo and van Reijen (1992, p. 31) is the "ordering and systematizing reality in order to make it predictable and controllable, which also implies that our thinking and acting is more and more subject to calculation, justifiability, and control, that happens in different and interrelated levels." These levels include the level of worldview, the collective action, and the individual action levels. From an action perspective, the rational action implies a "well-considered" action, seeking means and methods that are effective and efficient in result, which create a surrounding reality that is controllable (van der Loo & van Reijen 1992). It was also considered a cultural perspective of differentiation, considering the cultural reality with the perception, values, meanings, norms, ideas, and symbols that give meaning to our actions and have a role in the culture (van der Loo & van Reijen, 1992).

The close connection indicated in the previous subsection between the division of labor and the social stratification among different employment types with different income levels, and between households and businesses demonstrates itself in the electricity tariffs. In this respect, there is a differentiation between business and household prepaid tariffs, and there is a differentiation between commercial prepaid and institutional conventional metering tariffs. At this point, in the process of utilizing this electricity for daily life, the people/residents have often been mentioned as "clients/customers" for economic interpretations or "consumers/end-users" for technical user interpretations. Accordingly, the households/residents or prepaid clients have a tariff of 2.22 NAD/kWh (2.08 plus fixed charges) in 2020, prepaid business clients pay 4.07 NAD/kWh (3.48 plus fixed charge), and the business institutional customers 2.60 NAD/kWh. The tariff has not changed in the next two years. The fixed charges for the prepaid business clients are ECB[12] levy, NEF[13] levy, LA[14] surcharge, and VAT. The prepaid clients do not pay any VAT. The difference between residential and businesses is almost double. The business institutional customers´ tariff is very close to the residential tariff. Moreover, the government subsidized these tariffs, but detailed and reliable information could not be obtained about the amount of this subsidy.

**Affordability of Tariffs**

The close relationship between social stratification and the tariff system has also mirrored the perception of the affordability of electricity tariffs. When the respondents answered the question, "Is electricity affordable, expensive, or cheap for you?" they did not only consider whether they had enough funds to buy the prepaid token, but they also assessed what they use and how they use it in their everyday lives. Based on this preliminary assessment, they answered the following question: The rationalization of individual lifestyles has started becoming visible in the meaning of electricity within their life-worlds with electricity and its affordability.

Therefore, no correlation has been found between the category of expenses and electricity affordability. However, Figure 4.25 illustrates the responses of the respondents, whereby 5.9% of the respondents found the electricity tariff costly, 25.5% expensive, 47.1% affordable, and 17.6% cheap.

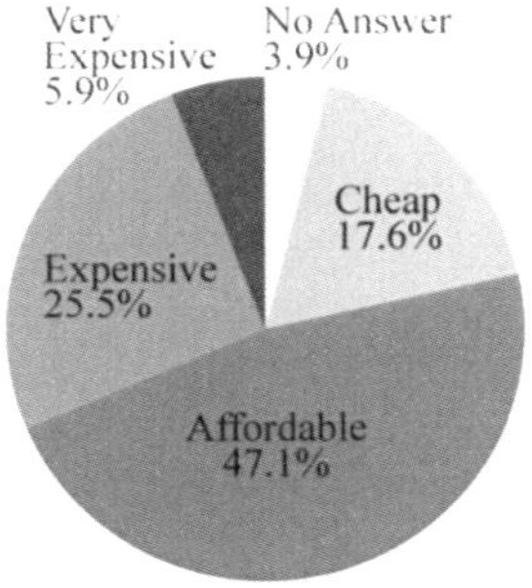


*Figure 4.25. Affordability of Tariffs*
*Source: Author*

[12] Electricity Control Board
[13] National Energy Fund
[14] Local Authority

About the employment categories mentioned earlier, the state and private sector employees corresponded to 55% and 25.5% of the households were below the FPL. Therefore, it can be assumed that the affordability perceived by households with regular incomes with salaries and wages has been distributed among the cheap, affordable, and expensive. For those living below the FPL, the responses would be spread in very expensive, expensive, and affordable.

**Awareness of the Electricity Bill**

Furthermore, Figure 4.26 illustrates what the respondents think about what they pay on their electricity bill. Accordingly, 49.0% of the respondents stated that they do not know, and 45.1% mentioned that they pay for the "electricity" they use. 2% mentioned "maintenance," and 3.9% "electricity and maintenance." A small total of 5.9% recognized electricity and its included services. In Tsumkwe, many community dissemination activities were held before and after establishing the mini-grid in the settlement.

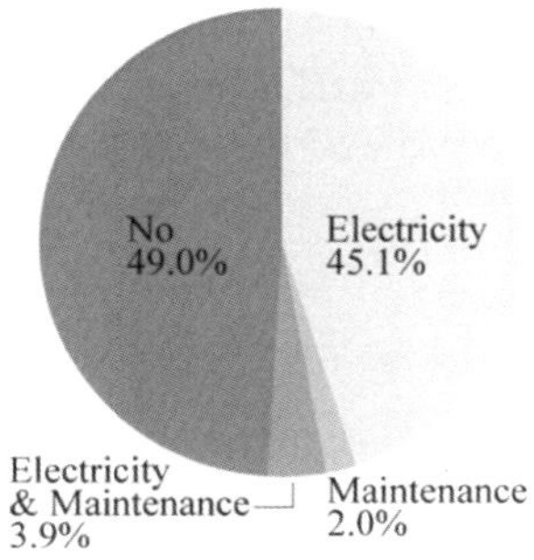


*Figure 4.26. Awareness of the Bill*
*Source: Author*

However, Figure 4.26 also provides a standing point in increasing awareness (raising) activities not only about electricity and energy-efficient uses but also about what they pay and what the electricity price covers. Reinhardt et al. (2012, p. 27) suggest that there might be problems for the individual with too little or too much awareness, but the "key in creating an added value with awareness would be to find an optimal balance". This would monitor energy consumption in a limited-generation scenario with a growing population.

**Awareness of the Electricity Consumption in Units and Expenses**

In addition to the previous awareness, the question that was asked on the awareness of the monthly electricity consumption in units that they have been prepaying resulted in 54.9% of the respondents answering "yes" and adding how many units it would be equivalent to; whereas 41.2% stated that they do not know. This also supports the findings of the Figure 4.26. Accordingly, Figure 4.27 illustrates the monthly electricity expenses, mainly between 0-150 NAD/month with 56.9%. Moreover, 23.5% of the respondents spend 151-300 NAD/month, 9.8% between 301-450 NAD/month, and 7.8% 451-700 NAD/month. The awareness in units, which is equal to kWh, has been known by the majority of the respondents. Moreover, everyone answered the question almost precisely; ironically, it was better known than their age. However, it is how the electricity bill was connected to the reality of their everyday life, which is more than their age. These expense categories will be further evaluated during the process of domestication.

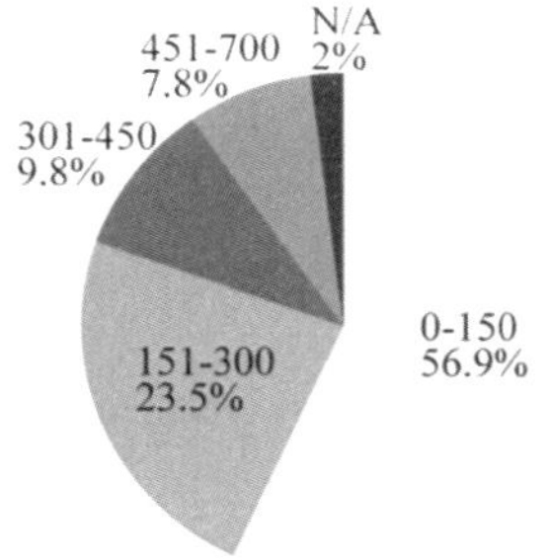


*Figure 4.27. Monthly Electricity Expenses (NAD/month)*
*Source: Author*

*Table 4.25. Descriptive Statistics of Electricity Expenses*

| | | |
|---|---|---|
| N | Valid | 51 |
| | Missing | 0 |
| Mean | | 178.55 |
| Minimum | | 10.00 |
| Maximum | | 700.00 |
| Std. Deviation | | 163.30616 |

*Source: Author*

On the other hand, based on the descriptive statistics about the monthly electricity expenses, the mean value has been calculated as 178.55 NAD/month, with a minimum value of 10 NAD/month and a maximum value of 700 NAD/month (Table 4.25). Moreover, the median is 110 NAD with a mode of 50 NAD, showing a wide spectrum of expenses over the month. The mode 50 NAD was spent by 18% of the respondents, 200 NAD was spent by 12%, and 100 NAD was spent by 10% of the respondents.

Furthermore, the electricity consumption concerning CENORED region in 2018/19 was reported as 2,000 kWh/customer/year and 6,250 kWh/customer/year in Namibia (von Oertzen, 2019, p. 106), which is equivalent to 166,67 kWh/customer/month and 520,83 kWh/customer/month, respectively. Moreover, von Oertzen (2019, p. 106) stated that 50% of CENORED´s domestic customers consume less than 1,200 kWh annually, corresponding to 100 kWh/month. With the existing residential electricity tariff (2,08 NAD/kWh plus surcharges) for Tsumkwe, these figures would be 208 NAD/month, 346,67 NAD/month for average CENORED region, and 1,083.33 NAD/month plus surcharges for average Namibia consumption. As a result, the mean of 178.55 NAD/monthly presents an electricity expense lower than Namibian and CENORED averages.

From the outlined monthly electricity expenses, the distribution based on education as the highest degree obtained illustrates that 31.4% of the expenses have been realized by the respondents without education or incomplete primary school degree, 45.1% by the primary school graduates, 9.8% by the secondary school graduates, and 13.7% by the tertiary school graduates (Table 4.26). 75%

of the respondents were without education, and 60.9% of the primary school graduates spent 0-150 NAD/month in the same categories. 60% of the secondary school graduates and 42.9% of the tertiary school graduates spend 151-300 NAD/month. From Table 4.26, there is a tendency for higher electricity expenses with the higher degrees obtained, but not limited. On the other hand, the lower education level spends the lower category of monthly expenses.

*Table 4.26. Distribution of Electricity Expense by Education*

| | | | Education | | | | |
|---|---|---|---|---|---|---|---|
| | | | No Education | Primary School | Secondary School | Tertiary School | Total |
| Monthly Electricity Expense | N/A | Count | 0 | 1 | 0 | 0 | 1 |
| | | % within Edu. | 0.0% | 4.3% | 0.0% | 0.0% | 2.0% |
| | | % of Total | 0.0% | 2.0% | 0.0% | 0.0% | 2.0% |
| | 0-150 | Count | 12 | 14 | 1 | 2 | 29 |
| | | % within Edu. | 75.0% | 60.9% | 20.0% | 28.6% | 56.9% |
| | | % of Total | 23.5% | 27.5% | 2.0% | 3.9% | 56.9% |
| | 151-300 | Count | 2 | 4 | 3 | 3 | 12 |
| | | % within Edu. | 12.5% | 17.4% | 60.0% | 42.9% | 23.5% |
| | | % of Total | 3.9% | 7.8% | 5.9% | 5.9% | 23.5% |
| | 301-450 | Count | 0 | 4 | 0 | 1 | 5 |
| | | % within Edu. | 0.0% | 17.4% | 0.0% | 14.3% | 9.8% |
| | | % of Total | 0.0% | 7.8% | 0.0% | 2.0% | 9.8% |
| | 451-700 | Count | 2 | 0 | 1 | 1 | 4 |
| | | % within Edu. | 12.5% | 0.0% | 20.0% | 14.3% | 7.8% |
| | | % of Total | 3.9% | 0.0% | 2.0% | 2.0% | 7.8% |
| Total | | Count | 16 | 23 | 5 | 7 | 51 |
| | | % within Edu. | 100.0% | 100.0% | 100.0% | 100.0% | 100.0% |
| | | % of Total | 31.4% | 45.1% | 9.8% | 13.7% | 100.0% |

*Source: Author*

Furthermore, the electricity expenses from different income levels do not demonstrate a clear picture, whereby 56.9%, which represents 0-150 NAD monthly expenses, have been divided into four different income categories, with the income range 1301-3900 being the highest ratio of 19.6%. This also corresponds to the 71.4% within the ratio of the same income range. The second largest category is between 151 and 300 NAD/month, with 23.5%, of which 11.8% belongs to the income group of 3901 NAD and more. Therefore, most of the electricity expenses within the two main expense groups (0-150 and 151-300) have been paid by all income categories (Table 4.27).

In continuation with the same expense categories, the respondents who stated they are unemployed have the highest ratio, with 23.5% of 56.9%, equivalent to 66.7% of the unemployed category. It is followed by the state employees of the same expense group at 13.7% and private sector employees at 11.8% out of 56.9% (Table C.19). Furthermore, although there is a connection between the number of rooms and electricity expenses, some households can still consume more electricity by having 2 or 3 rooms (Table C.20).

*Table 4.27. Distribution of Electricity Expenses by Average Income*

| | | | Average Income (NAD/Month/HH) | | | | Total |
|---|---|---|---|---|---|---|---|
| | | | N/A | 0-1300 | 1301-3900 | 3901-more | |
| Monthly Electricity Expense | N/A | Count | 1 | 0 | 0 | 0 | 1 |
| | | % Income | 14,3% | 0,0% | 0,0% | 0,0% | 2,0% |
| | | % of Total | 2,0% | 0,0% | 0,0% | 0,0% | 2,0% |
| | 0-150 | Count | 4 | 8 | 10 | 7 | 29 |
| | | % Income | 57,1% | 61,5% | 71,4% | 41,2% | 56,9% |
| | | % of Total | 7,8% | 15,7% | 19,6% | 13,7% | 56,9% |
| | 151-300 | Count | 1 | 3 | 2 | 6 | 12 |
| | | % Income | 14,3% | 23,1% | 14,3% | 35,3% | 23,5% |
| | | % of Total | 2,0% | 5,9% | 3,9% | 11,8% | 23,5% |
| | 301-450 | Count | 1 | 0 | 1 | 3 | 5 |
| | | % Income | 14,3% | 0,0% | 7,1% | 17,6% | 9,8% |
| | | % of Total | 2,0% | 0,0% | 2,0% | 5,9% | 9,8% |
| | 451-700 | Count | 0 | 2 | 1 | 1 | 4 |
| | | % Income | 0,0% | 15,4% | 7,1% | 5,9% | 7,8% |
| | | % of Total | 0,0% | 3,9% | 2,0% | 2,0% | 7,8% |
| Total | | Count | 7 | 13 | 14 | 17 | 51 |
| | | % Income | 100% | 100% | 100% | 100% | 100% |
| | | % of Total | 13,7% | 25,5% | 27,5% | 33,3% | 100% |

*Source: Author*

## Electricity Supply Satisfaction and Dissatisfaction

The respondents´ satisfaction has been detailed in Table 4.28 for the question, "Are you happy with the current supply?". 72.5% of the respondents were affirmative, out of which 41.2% were happy for having light and using the appliances, 25.5% for "having it 24/7 and always available", and for not having the danger of fire as in the case of the use of candles. On the other hand, the respondents who were not happy, stated the expensiveness and weather interruptions as their primary reasons.

*Table 4.28. Reasons for Satisfaction and Dissatisfaction*

| | | Frequency | Percent | Valid Percent | Cumulative Percent |
|---|---|---|---|---|---|
| Valid | N/A | 4 | 7.8 | 7.8 | 7.8 |
| | Power and Expensive | 6 | 11.8 | 11.8 | 19.6 |
| | Appliances/Power | 21 | 41.2 | 41.2 | 60.8 |
| | 24/7 & always available | 13 | 25.5 | 25.5 | 86.3 |
| | Expensive | 1 | 2.0 | 2.0 | 88.2 |
| | Weather/Interruptions | 3 | 5.9 | 5.9 | 94.1 |
| | Power and Interruptions | 3 | 5.9 | 5.9 | 100.0 |
| | Total | 51 | 100.0 | 100.0 | |

*Source: Author*

#### *4.3.2.2 Local Businesses*

**Awareness of the Electricity Price**

As outlined in the previous subsection, local businesses, according to the CENORED classification, pay different tariffs depending on the type of metering, whether prepaid or conventional. These local businesses were food shops, restaurants, textile shops, hardware shops, bars/shebeens, police station, state offices, schools, and clinic. It was also found that some public clients do not pay locally; their bills were paid from regional offices of the respective institutions, such as schools or clinics.

It was also found that some public clients do not pay locally; their bills were paid from regional offices of the respective institutions, such as schools or clinics. Looking at the data of the local businesses in Tsumkwe, the category "very expensive" with 11.8% appears in the findings, followed by 35.3% "expensive," 23.5% affordable, and 17.6% cheap (Figure 4.28).

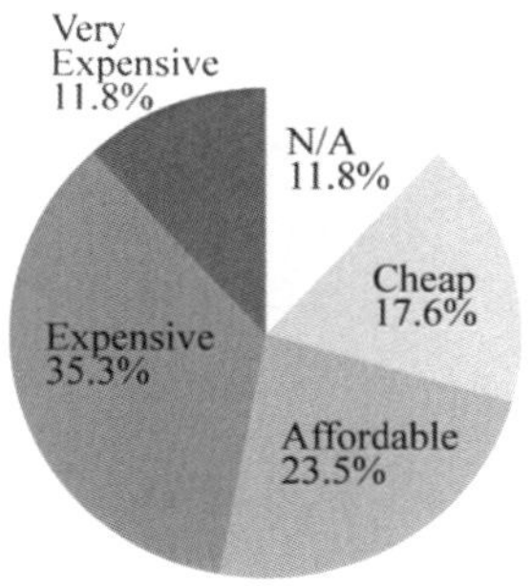


*Figure 4.28. Awareness on the Bill*
*Source: Author*

Considering the tariff for local prepaid businesses, 47.1% of the local businesses find the electricity expensive and/or very expensive compared to what they use. Some shops only use a few lights and a refrigerator; in this case, they stated that it is expensive for the essential items they are using. Some others use more than multiple refrigerators, which increase their consumption. The appliances will be detailed in the process of domestication sub-section.

**Differentiation on the Awareness of the Electricity Price**

Moreover, the differentiation in the awareness of the electricity price has also been investigated based on the public and private businesses. Figure 4.29 illustrates that 50% of the private businesses find the electricity expensive or very expensive compared to public businesses. However, the findings did not show any meaningful/explanatory results. Although the surveys explicitly mentioned that the question addressed the "business" perspective, the researcher observed that the business respondents tended to express their overall opinion regarding the electricity prices without differentiating the business from themselves. Another approach to 28.6% of public businesses finding the prices "cheap" shall only be explained by the fact that they are not the people who are paying. Therefore, they find it cheap (Figure 4.29).

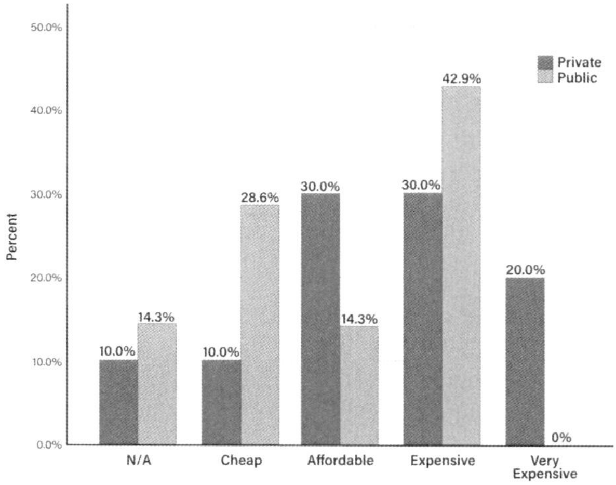


*Figure 4.29. Differentiation in the Awareness of the Electricity Bill*
*Source: Author*

**Awareness of the Electricity Consumption in Units and Expenses**

In connection with the differentiation between public businesses/institutions and private businesses, besides the difference in their associated tariff, the person responsible for the payment also differs. The government pays the public institutions´ electricity bills directly through the regional branches, whereas the private local businesses pay their electricity bills directly in Tsumkwe by themselves. This situation resulted in 58.8% of the respondents who did "not" knowing the monthly consumption, in contrast to 41.2% who knew the consumption. The differentiation in the types of businesses has also led to very different expenses. Since not all the respondents knew their electricity expenses, the findings of this question did not result in meaning for a comparative figure. Nonetheless, the three major classifications shall be made: (1) monthly expenses between 100 NAD – 1,000 NAD for small businesses, (2) 1,000 NAD -5,000 NAD/month for medium size, (3) the relatively big businesses more than 5,000 NAD; in Tsumkwe case was 42,000 NAD/month the highest expense. As a result, it is suggested that the local businesses should also be classified within themselves depending on the energy consumption/practices they perform to generate income. One single business tariff-fits-all approach would not best describe the suitable differentiation in tariff setting, which is currently sided for public institutions. Additional tariffs according to their consumption patterns could be considered.

### 4.3.3 Energy Appliances for Energy Practices of Everyday Life

This mini-grid application and connection for the Tsumkwe settlement also present another paradox. The 303 $kW_p$ mini-grid capacity, designed and put into operation in 2011-12, has specific limits with particular design and operation parameters, such as amperage. In the beginning, five amps were set to limit household electricity use, but this caused many complaints. Hence, the limit has been reset to 10 amp to keep the load on a specified level. As a result, certain electric appliances do not run or cannot or must not run simultaneously. This paradox of being connected to the mini-grid but not using it as one sees it necessary creates a similar paradox of being independent and dependent simultaneously. A limited autonomy has been created with a mini-grid designed with a lower capacity. Furthermore, when more and more people are willing to connect to a limited capacity, the solar generation, and the batteries become insufficient, causing a high level of diesel consumption with the use of the generators in the evening.

#### *4.3.3.1 Households*

Through the use of these appliances, domestication, being one of the dimensions of the process of modernization, as explained in Chapter 3, brings "the mastery of the biological and natural process." This not only improves the quality of life but also creates "a dependence on the means by which mastery is achieved" (van der Loo & van Reijen, 1992). In the process of modernization, in the case of Tsumkwe, the use of electric appliances brings this mastery forth, contributes to the technization (Elias, 1995) and the domestication with the digitization and digitalization processes in the 21$^{st}$ century.

**Number of Different Types of Appliances per Household**

The quantification of the different electric appliances used in households was realized through the survey, which also provides valuable data for technical calculations. According to the findings, nineteen different appliances are found in the sum of all appliances. Of these, only twelve different electric appliances have been used, the maximum in a household; the average number of items is 6.33, with a minimum of one appliance and a maximum of 12. The median is seven, and the mode is eight appliances within this category, corresponding to 17.6% of the households. Two, five, and ten appliances account for 13.7% each as well (Table C.21).

When they are categorized into four categories, with each category having three different electric appliances, the majority of the households, with 33.3%, have seven to nine different items. This majority has been followed by very close ratios of 25.5% for households with four to six different items and 21.6% with a maximum of three appliances (Figure 4.30). 19.6% of the households have ten to twelve different electric appliances.

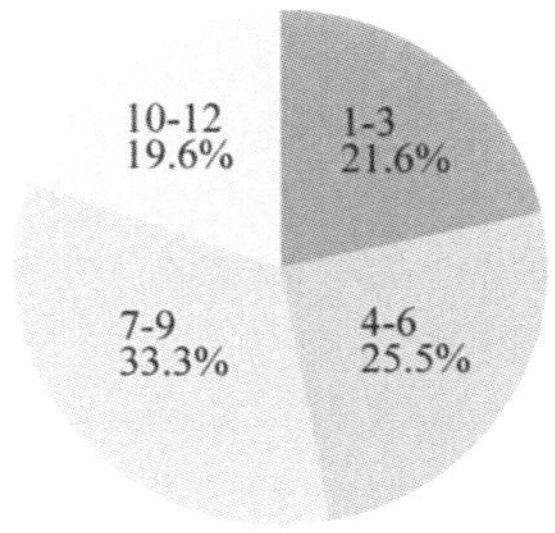


*Figure 4.30. Number of Different Types of Appliances in Groups*
*Source: Author*

Accordingly, Table 4.29 confirms that there is no single tendency for households of different sizes. It is possible to find 7-9 appliances in a single-room household (7.8% of the total), in a two-room household (9.8% of the total), or in a three-room household, which is 11.8% of all. However, it is possible to see an intensity in the single-room and double-room households with 0-9 different appliances that correspond to 62.8% (Table 4.29).

*Table 4.29. Distribution of Number of Different Appliances by Rooms*

| | | | Rooms | | | | Total |
|---|---|---|---|---|---|---|---|
| | | | 1 | 2 | 3 | 4 | |
| Number of Different Appliances per Household | 0-3 | Count | 5 | 5 | 1 | 0 | 11 |
| | | % within Rooms | 35,7% | 27,8% | 8,3% | 0,0% | 21,6% |
| | | % of Total | 9,8% | 9,8% | 2,0% | 0,0% | 21,6% |
| | 4-6 | Count | 4 | 5 | 2 | 2 | 13 |
| | | % within Rooms | 28,6% | 27,8% | 16,7% | 28,6% | 25,5% |
| | | % of Total | 7,8% | 9,8% | 3,9% | 3,9% | 25,5% |
| | 7-9 | Count | 4 | 5 | 6 | 2 | 17 |
| | | % within Rooms | 28,6% | 27,8% | 50,0% | 28,6% | 33,3% |
| | | % of Total | 7,8% | 9,8% | 11,8% | 3,9% | 33,3% |
| | 10-12 | Count | 1 | 3 | 3 | 3 | 10 |
| | | % within Rooms | 7,1% | 16,7% | 25,0% | 42,9% | 19,6% |
| | | % of Total | 2,0% | 5,9% | 5,9% | 5,9% | 19,6% |
| Total | | Count | 14 | 18 | 12 | 7 | 51 |
| | | % within Rooms | 100,0% | 100,0% | 100,0% | 100,0% | 100,0% |
| | | % of Total | 27,5% | 35,3% | 23,5% | 13,7% | 100,0% |

*Source: Author*

According to Table 4.30, the primary school graduates have 17.6% (of the group with 7 to 9 appliances), followed by the ungraduated with 13.7% (of the group with 0-3 appliances) out of overall appliances. On the other hand, 11.8% (of the group with 4-6 appliances) were used by primary school graduates, and 9.8% were used by ungraduated. 7.8% of the respondents have 0-3 appliances used by primary school graduates, and the same ratio for 10-12 appliances, while 7.8% also apply 10-12 appliances used by tertiary school graduates. As a result, the higher the education, the more there is a tendency to have more appliances in the household and vice versa. However, the average is centered around the appliance groups of 0-3, 4-6, and 7-9. Furthermore, the distribution of different appliances by occupation, by average income, and by electricity expenses have been provided in Appendix C.

*Table 4.30. Distribution of Different Types of Appliances by Education*

| | | | Education | | | | Total |
|---|---|---|---|---|---|---|---|
| | | | No Education | Primary School | Secondary School | Tertiary School | |
| Number of Different Types of Appliances per Household | 0-3 | Count | 7 | 4 | 0 | 0 | 11 |
| | | % within Education | 43,8% | 17,4% | 0,0% | 0,0% | 21,6% |
| | | % of Total | 13,7% | 7,8% | 0,0% | 0,0% | 21,6% |
| | 4-6 | Count | 5 | 6 | 1 | 1 | 13 |
| | | % within Ed | 31,3% | 26,1% | 20,0% | 14,3% | 25,5% |
| | | % of Total | 9,8% | 11,8% | 2,0% | 2,0% | 25,5% |
| | 7-9 | Count | 3 | 9 | 3 | 2 | 17 |
| | | % within Education | 18,8% | 39,1% | 60,0% | 28,6% | 33,3% |
| | | % of Total | 5,9% | 17,6% | 5,9% | 3,9% | 33,3% |
| | 10-12 | Count | 1 | 4 | 1 | 4 | 10 |
| | | % within Education | 6,3% | 17,4% | 20,0% | 57,1% | 19,6% |
| | | % of Total | 2,0% | 7,8% | 2,0% | 7,8% | 19,6% |
| Total | | Count | 16 | 23 | 5 | 7 | 51 |
| | | % within Education | 100,0% | 100,0% | 100,0% | 100,0% | 100,0% |
| | | % of Total | 31,4% | 45,1% | 9,8% | 13,7% | 100,0% |

*Source: Author*

## Energy Appliances for Energy Practices of Everyday Life

The aforementioned different appliances illustrated in Table 4.30. Accordingly, the appliances used for lighting, cooking, cooling, communication, heating, entertainment, and washing practices have been shown on the descriptive statistics of the appliances; This suggests that the mean light per household is 3.33 and the mobile phone is 2.39, showing more than one of these appliances per household. Of the rest of the appliances not found in every household, the fridge has a mean value of 0.71, and the television has 0.69. Based on these means listed (Table 4.31), higher than one item per household suggest that the lighting and communication are significant when viewed from the local energy practices at the household level.

*Table 4.31. Descriptive Statistics for the Household Electric Appliances*

| | N | Minimum | Maximum | Mean | Std. Deviation |
|---|---|---|---|---|---|
| Lights | 51 | 0 | 15 | 3.33 | 2.951 |
| Fridge | 51 | 0 | 3 | .71 | .642 |
| Deep Freezer | 51 | 0 | 1 | .33 | .476 |
| Kettle | 51 | 0 | 1 | .41 | .497 |
| Stove, Hot Plate | 51 | 0 | 1 | .14 | .348 |
| Iron | 51 | 0 | 1 | .45 | .503 |
| Radio | 51 | 0 | 2 | .45 | .541 |
| Mobile Phone | 51 | 0 | 6 | 2.39 | 1.576 |
| Television | 51 | 0 | 2 | .69 | .547 |
| Computer Laptop | 51 | 0 | 2 | .22 | .461 |
| Fan | 51 | 0 | 2 | .39 | .568 |
| Washing Machine | 51 | 0 | 1 | .18 | .385 |
| Electrical Heater | 51 | 0 | 1 | .04 | .196 |
| Blow drier | 51 | 0 | 3 | .08 | .440 |
| Sewing Machine | 51 | 0 | 1 | .04 | .196 |
| Welding Machine | 51 | 0 | 1 | .04 | .196 |
| Music Player | 51 | 0 | 1 | .16 | .367 |
| Microwave | 51 | 0 | 1 | .16 | .367 |
| Shaving Machine | 51 | 0 | 1 | .02 | .140 |
| Total quantity of electric appliances counted per HH | 51 | 1 | 12 | 6.33 | 3.004 |
| Valid N (listwise) | 51 | | | | |

*Source: Author*

**Lighting in the Households**

In connection with the above-mentioned descriptive statistics, lights are used to enable the practice to be executed, such as providing light for the dark hours, enabling working or studying at night. Accordingly, "light" frequencies illustrate that 31.4% of the households have two lights, 25.5% have one light, and 11.8% have four lights (Appendix C, Table C.22). The bulbs that are used are mostly 60W, and some use 11W energy-efficient bulbs. The households that do not have any electricity, use candles, torches, solar bulbs, or mobile phones as a source of light.

## Cooking

Traditionally, the firewood that is collected from the bush is the main source of energy for cooking. According to the survey findings, 37.3% of the households still use firewood as the only source. It is followed by the gas (31.4%), gas and wood (13.7%), electricity (7.8%), wood and electricity and gas (3.9%), and gas and electricity (3.9%) (Figure 4.31).

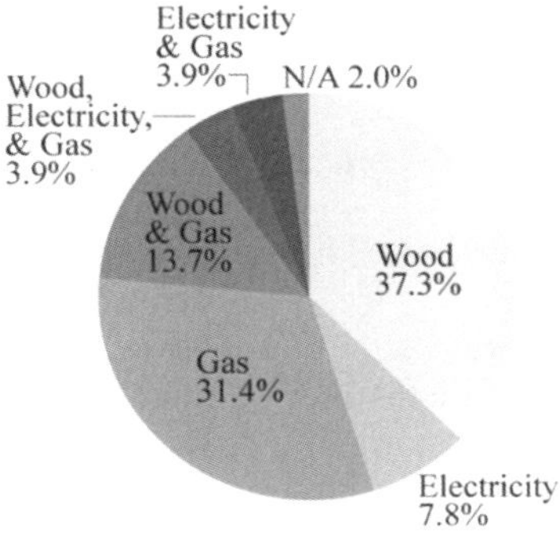


*Figure 4.31. Types of Energy Used for Cooking*

*Source: Author*

The author only heard of the use of some solar cookers but did not come across any of them during the field research. The survey also accounted for 13.7% of the households having a hot plate used for cooking. Some respondents discussed that buying the gas bottle for gas cookers has been challenging for them, since they were not always available, whereas electricity was always available. Therefore, they would prefer to use an electric stove. The firewood collectors from the bush stated that they were not struggling anymore to look for the wood to cook. In this way, they were saving time.

Furthermore, besides the shift from the traditional cooking method with firewood to gas cookers or hot plates, the respondents also used other cooking or kitchen items. The kettle was used by 41.2% of households for heating water, followed by 15.7% who use microwaves, and 11.8% who use toasters for cooking. All of this multiplication of kitchen items not only eases their everyday life, like "preparing tea with a kettle," but also highlights the specialized items that are used in the kitchen within the process of modernization.

## Communication and Media

Within communication practices, mobile phones, television, radio, and music have been identified as communication appliances. In this regard, 64.7% of the households have at least one television, 43.2% have at least one radio, 15.7% have a music player, and 19.6% have at least one computer or laptop in the household. Mobile phones and computers are invaluable appliances for communication with other people and acquaintances, as well as other purposes such as money transfers, social media and/or further purposes through the use of smartphones, which will be discussed further in the next subsection covering digitalization. According to the survey, only one household did not have a phone. The rest of the households had one to six mobile phones per household, making this appliance the second most preferred item, with 33.3% with one mobile phone, 31.4% with two, and 11.8% with three mobile phones per household (Table 4.32). Nonetheless, it is also important to note that households do not have any private internet connection. Access to the internet takes place with smartphones and GSM networks from the offices or the Community Development Center.

*Table 4.32. Quantity of Mobile Phones per Household*

| | | Frequency | Percent | Valid Percent | Cumulative Percent |
|---|---|---|---|---|---|
| Valid | 0 | 1 | 2.0 | 2.0 | 2.0 |
| | 1 | 17 | 33.3 | 33.3 | 35.3 |
| | 2 | 16 | 31.4 | 31.4 | 66.7 |
| | 3 | 6 | 11.8 | 11.8 | 78.4 |
| | 4 | 3 | 5.9 | 5.9 | 84.3 |
| | 5 | 5 | 9.8 | 9.8 | 94.1 |
| | 6 | 3 | 5.9 | 5.9 | 100.0 |
| | Total | 51 | 100.0 | 100.0 | |

*Source: Author*

The process of individualization defined by van der Loo and van Reijen (1992) suggests that the individual who breaks away from one´s collectivity of immediate environment has growing importance of the individual, whereby the development of the individual identity, as well as individual behaviors towards groups, society, and state, occurs. The individuals start dealing with social units different from their collectivity, thus reaching a certain level of independence. In the case of Tsumkwe, both the interviews and the surveys suggest that a certain level of individualization occurs with mobile phone and television use. The respondents often mentioned, "I don't visit the neighbors´ houses to charge my phone anymore." Referring to its reliability and availability, they stated that "It´s always available. No more asking other people for charging phones... please… please."

This empirical data is insufficient to claim anything about the people's individual identities. However, it brings to light changes in their behavior by charging the phone or watching television at their place. This self-efficacy reached at their premises also introduces a notion of comfort and happiness "I am happy to have it. Now I can charge my phone at my house" and "It brings light, and it is useful. It makes it easy for me to charge my phone and not going to the houses of the neighbors." At this stage, it is essential to note that possessing a mobile phone is one level of individualization and domestication while bringing the second level of *charging it at home and not going to the neighbors*. Following the mean of 2.39 for mobile phones, Table 4.33 shows the number of mobile phones, televisions, and fridges with the highest mean and median, excluding the lights. As a result, the individualization process shall be illustrated through the practices of communication and media and the cooling of households.

*Table 4.33. Descriptive Statistics of Individual Appliances*

| | | Adults | Rooms | Television | Mobile Phone | Fridge |
|---|---|---|---|---|---|---|
| N | Valid | 51 | 51 | 51 | 51 | 51 |
| | Missing | 0 | 0 | 0 | 0 | 0 |
| Mean | | 2.92 | 2.24 | .69 | 2.39 | .71 |
| Median | | 2.00 | 2.00 | 1.00 | 2.00 | 1.00 |
| Mode | | 2 | 2 | 1 | 1 | 1 |
| Sum | | 149 | 114 | 35 | 122 | 36 |

*Source: Author*

According to Table 4.33, there are more than two mobile phones per household, and the sum of all the mobile phones is nearly the sum of all adults, with 122 and 149, respectively. The distribution of mobile phones per size of the households, based on the number of rooms, one to three mobile phones per household occupy the majority with 33.3% (1 mobile phone), 31.4% (2 mobile phones), and 11.8% (three mobile phones), which is spread throughout different household sizes. However, households with one room and with mobile phones have the majority of the ratio, with 17.6%, followed by households with two rooms with two mobile phones (13.7%) and households with one room with two mobile phones (7.8%) (Appendix C., Table C.23). Furthermore, when the relation between the occupational and educational categories was analyzed, a significant difference was not observed. However, a distribution to overall categories was noted.

**Television**

Television is also among the most frequently mentioned communication and media items, accounting for 64.7% having at least one television, 19.6% in two-room households, 17.6% in three-room households, and 15.7% in one-room households (Table 4.34). It is mainly owned by primary school graduates (29.4%), followed by incomplete primary school degrees (11.8%), 9.8% secondary school graduates, and 9.8% tertiary school graduates. By occupation, state employees (23.5%), unemployed people (17.6%), and private sector employees demonstrate their television ownership in each household. This shall be interpreted as the lower education degrees using television as an item of access to information, but the occupation does not necessarily define a particular class of ownership.

*Table 4.34. Distribution of Television by Number of Rooms*

| | | | Number of Rooms | | | | |
|---|---|---|---|---|---|---|---|
| | | | 1 | 2 | 3 | 4 | Total |
| Television | 0 | Count | 6 | 7 | 3 | 2 | 18 |
| | | % within Rooms | 42,9% | 38,9% | 25,0% | 28,6% | 35,3% |
| | | % of Total | 11,8% | 13,7% | 5,9% | 3,9% | 35,3% |
| | 1 | Count | 8 | 10 | 9 | 4 | 31 |
| | | % within Rooms | 57,1% | 55,6% | 75,0% | 57,1% | 60,8% |
| | | % of Total | 15,7% | 19,6% | 17,6% | 7,8% | 60,8% |
| | 2 | Count | 0 | 1 | 0 | 1 | 2 |
| | | % within Rooms | 0,0% | 5,6% | 0,0% | 14,3% | 3,9% |
| | | % of Total | 0,0% | 2,0% | 0,0% | 2,0% | 3,9% |
| Total | | Count | 14 | 18 | 12 | 7 | 51 |
| | | % within Rooms | 100,0% | 100,0% | 100,0% | 100,0% | 100,0% |
| | | % of Total | 27,5% | 35,3% | 23,5% | 13,7% | 100,0% |

*Source: Author*

However, the future wishes of the interviewees show significant interest in the electrified futures, meaning future electric appliances they wish to have, as well as individual wishes or opportunities they see for themselves. The television, the fridge, and the music system (see section 4.6., electrified futures) are the high-ranking three items listed in the surveys, which also demonstrate the wish or the desire to supply individual electric appliances used in daily life. Thus, the residents have already embarked on the process of individualization and domestication. In connection to Maslow´s pyramid utilized for self-actualization in the hierarchical pyramid model, the residents are in the first level of their biological human needs for food, oxygen, and sexuality, and secondly, their physical needs, security, and safety in pursuing the respective steps.

**Cooling and Heating**

In continuation of the household appliances used, the fridge, deep-freezer, and fan have been identified. Based on the survey results, 56.9% of the households have one fridge, and 5.9% have at least two fridges in use (Table 4.35). Furthermore, 33.3% of the respondents use a deep freezer to store their food for extended periods. During the field survey, some respondents who were not connected to the mini-grid stated that they use their neighbors´ fridges to store food occasionally. Furthermore, 19.6% of the single-fridge-owners live in a two-room household, followed by 15.7% of three-room households, 11.8% in a one-room household, and 9.8% in a four-room household. 27.5% of the fridge-owning households have an average monthly income of 3,901 NAD or more. On the other hand, 27.5% are also owned by primary school graduates, followed by 13.8% with no education. This data set illustrates that a household's fridge ownership is significantly connected to higher income levels.

*Table 4.35. Distribution of Fridge by Number of Rooms*

| | | | Number of Rooms | | | | |
|---|---|---|---|---|---|---|---|
| | | | 1 | 2 | 3 | 4 | Total |
| Fridge | 0 | Count | 8 | 8 | 2 | 1 | 19 |
| | | % within Rooms | 57,1% | 44,4% | 16,7% | 14,3% | 37,3% |
| | | % of Total | 15,7% | 15,7% | 3,9% | 2,0% | 37,3% |
| | 1 | Count | 6 | 10 | 8 | 5 | 29 |
| | | % within Rooms | 42,9% | 55,6% | 66,7% | 71,4% | 56,9% |
| | | % of Total | 11,8% | 19,6% | 15,7% | 9,8% | 56,9% |
| | 2 | Count | 0 | 0 | 1 | 1 | 2 |
| | | % within Rooms | 0,0% | 0,0% | 8,3% | 14,3% | 3,9% |
| | | % of Total | 0,0% | 0,0% | 2,0% | 2,0% | 3,9% |
| | 3 | Count | 0 | 0 | 1 | 0 | 1 |
| | | % within Rooms | 0,0% | 0,0% | 8,3% | 0,0% | 2,0% |
| | | % of Total | 0,0% | 0,0% | 2,0% | 0,0% | 2,0% |
| Total | | Count | 14 | 18 | 12 | 7 | 51 |
| | | % within Rooms | 100,0% | 100,0% | 100,0% | 100,0% | 100,0% |
| | | % of Total | 27,5% | 35,3% | 23,5% | 13,7% | 100,0% |

*Source: Author*

On the other hand, another cooling appliance used besides food cooling is an electrical fan to cool the air in warm weather during the dry season. In this respect, 31.4% of the respondents mentioned having one fan, and 3.9% have two fans at their places.

In addition, according to the surveys, 96.1% of the respondents do not use any heating and the ones that use electric heaters (3.9%).

**Personal Care, Ironing, Laundry, Sewing, Welding as a Household Energy Practice**

For personal care, hair dryers (4%) and shaving machines (2%) have been stated by the respondents. Furthermore, ironing, laundry, and sewing shall also be counted. Accordingly, the survey results suggested that 45.1% of the respondents use iron, while 17.6% use a washing machine for laundry. 3.9% of the respondents have a sewing machine in use. Welding machines have also been mentioned by 3.9% of the households. In this respect, the iron is one of the electric appliances that is controversial for its use in the mini-grids but still practiced regularly.

### *4.3.3.2 Local Businesses*

The local businesses in Tsumkwe that were surveyed have diverse backgrounds. Therefore, the electric appliances used for energy practices have also been identified as more diverse than the households. In this regard, there are in total of 35 different appliances for all businesses, of which 19 appliances are the maximum different quantity per local business (Table 4.36), with a mean value of 6.53.

*Table 4.36. Descriptive Statistics of the Number of Different Type of Appliances per Local Business*

| | N | Min. | Max. | Mean | Std. Deviation |
|---|---|---|---|---|---|
| Category Total Appliances | 17 | 1 | 7 | 2.41 | 1.543 |
| Total Appliances | 17 | 1 | 19 | 6.53 | 4.460 |
| Valid N (listwise) | 17 | | | | |

*Source: Author*

Furthermore, the most significant quantity was found in three appliances (17.6%), six appliances (17.6%), five appliances (11.8%), and nine appliances (11.8%) (Appendix C, Table C.24).

Thus, they can be grouped under five categories as illustrated in Figure 4.26. Accordingly, 35.2% of the local businesses have four to six appliances in their businesses, followed by one to three appliances (29.21%), which constitutes the majority. The highest ratio has been reached by the private businesses having 1-3 appliances only; followed by private and public businesses having each 17.6% for 4-6 appliances. Moreover, the sum of the first two categories (1-3 and 4-6) in private businesses is 64.7% (Figure 4.32). Although thirty-five different appliances, there is a tendency to use fewer.

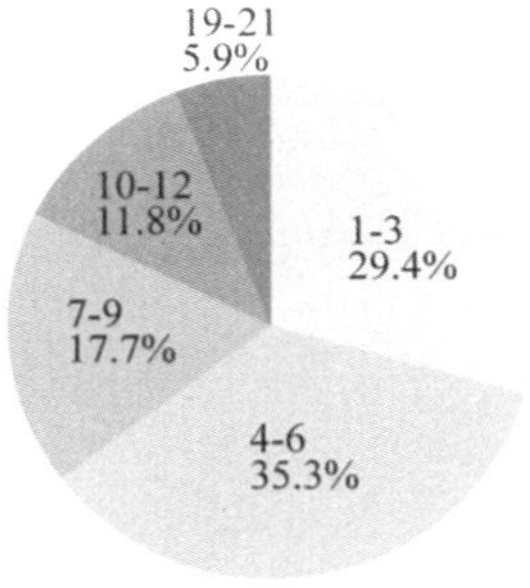


*Figure 4.32. Number of Different Types of Appliances per Local Businesses*

*Source: Author*

Thus, diversity comes with the private sector´s contribution to the matrix, with 58.8% private, while the quantification of the less diverse appliances results in the public sector, with schools, police stations, and clinics, that local small businesses that are private have been using fewer appliances than the public businesses.

From the list of all appliances used for different energy practices, fluorescent lamps, light bulbs (everyday and energy-saving), fridges, radios, televisions, mobile phones, computers, monitors, fans, printers, digital security cameras, and air conditioners are the most used items. These items provide the following energy practices: lighting, cooling, communication and media, and security within the surveyed respondents.

### Lighting in Local Businesses

Three different types of lights have been identified within the surveys: fluorescent lamps, light bulbs, and spotlights used in all local businesses, at least in one type. The fluorescent lamps are used in public offices and institutions, such as clinics, schools, and police stations, as well as private businesses, such as lodges and general dealers, which account for 58.8%. In comparison, 41.8% do not have any fluorescent lamps. The light bulbs are used in almost every local business, with a sum of 268 items. The spotlights are only used at the lodge. Thus, 55% of the light fittings are fluorescent, 44% bulbs, and 1% spots that are used for lighting in studying, working, and doing business.

### Cooling Appliances

The fridge is the first cooling appliance used in different quantities by 58.8% of the local businesses in Tsumkwe. The air conditioner and the fans are the other cooling appliances used to cool the air. According to the surveys, 41.2% of the respondent businesses utilize a cooling system (Table 4.37). Among them, 29.4% of the local businesses have used the air conditioner, while 17.6% have fans.

*Table 4.37. Quantity of Fridge Used in Local Businesses*

| | | Frequency | Percent | Valid Percent | Cumulative Percent |
|---|---|---|---|---|---|
| Valid | 0 | 7 | 41.2 | 41.2 | 41.2 |
| | 2 | 2 | 11.8 | 11.8 | 52.9 |
| | 3 | 2 | 11.8 | 11.8 | 64.7 |
| | 4 | 2 | 11.8 | 11.8 | 76.5 |
| | 5 | 3 | 17.6 | 17.6 | 94.1 |
| | 10 | 1 | 5.9 | 5.9 | 100.0 |
| | Total | 17 | 100.0 | 100.0 | |

*Source: Author*

Of these, 41.2% are local businesses, 11.8% are private businesses, and 29.4% are public businesses, which illustrates the differentiation between different businesses, and from households (Appendix C, Table C.25).

### Communication and Media Appliances

Computers, printers, and television are the devices for communication and media. According to the survey, 46.1% have been using televisions, 46.1% own computers, and 41.2% have printers. Furthermore, the central control (5.9%) used for broadcasting, projector (5.9%), copy machine (5.9%), and monitors (17.6%) have been the outcomes within this category. These items demonstrate a bundle of electric appliances used for different purposes in everyday life at work in schooling, working, broadcasting, or any other job related to communication and media.

#### *4.3.3.3 Digital Payment and Digital Security for Households and Local Businesses*

At this stage, it is essential to note that the residents of the settlement of Tsumkwe have been going through the processes of electrification and domestication at nearly the same time as digitalization. This means that people are introduced to several electric appliances and learn to use them while learning other tools, such as smartphones and computers, that connect them to digitalization. One of the best examples of this point is using digital payment methods provided by three national banks in Namibia. According to the survey, digital payment methods are available for 41.20 % of the respondents, and 58.80% do not have this access (Appendix C, Table C.26). In this sense, the residents are limited in using digital payment with the condition of having appropriate appliances. Nonetheless, 41.2% is a significant ratio to consider within this process.

Furthermore, the availability of digital payment methods through online banking has been used by most primary school graduates (19.6% of the total); this is followed by secondary and tertiary school graduates with 7.8% (Appendix C, Table C.14). Regarding occupational differentiation, state employees have the most significant portion, with 19.6% of the total, followed by private sector employees (11.8%) (Appendix C, Table C.15). In addition, the average income range higher than 3901 NAD/month has the most significant ratio with 25.5% of the total (Appendix C, Table C.26), followed by the range 1301-3900 NAD/month.

Digital payment was unavailable for local businesses, and the regional headquarters paid some of the expenses of public institutions. The local businesses have been paying in cash. However, digitalization has occurred beyond paying electricity tokens in security systems. Only two private local business general dealer/service stations and the lodge own these cameras to provide the security of their businesses, which account for 11.8% of the overall businesses, with a total sum of 21 digital cameras (Appendix C, Table C.16). In this sense, it is essential to note that they have not been employing any security personnel for the same purpose. However, they have been using technology to facilitate this service for themselves via digital security cameras.

#### *4.3.3.4 Energy Appliances wished in the Future – "Electrified Wishes"*

**Households**

The future energy appliances wished by the community, which were asked in the semi-structured surveys as an open-ended question, have been elaborated on in order of priority. Moreover, the same appliances were also analyzed thematically. According to the thematic categorization, domestic appliances are fridges, fans, lights, heaters, electric stoves, washing machines, microwaves, air conditioners, sewing machines, deep-freezers, kettles, and ovens. The items that can be found in the leisure category are television, music player, hi-fi, radio, and satellite. The office category appliances include laptops/iPads/PCs, wi-fi, phones, faxes, and printers. The appliances under the industrial/commercial are welding machine, water pressure, geezer, water, and grinder.

Table 4.38 demonstrates the electric appliances wished for within the community in Tsumkwe in the order of priority. According to this list, television, fridge, music player/system/hi-fi/radio, lights, heater, electric stove, washing machine, laptop, and electrical tools are the most wished items. It is important to note that the ratio of people who do not wish for any item for priority number 1 is 20 %, priority number 2 is 29%, and priority number 3 is 57%.

*Table 4.38. Future Appliances in the order of Priority at the Households**

| Future Appliances / Priority | No. 1 | No. 2 | No. 3 |
|---|---|---|---|
| Television | 16% | 5% | 11% |
| Fridge | 9% | 13% | 9% |
| Music Player/System/hi-fi/radio | 9% | 13% | 2% |
| Fan | 9% | 4% | 4% |
| Lights | 5% | 0% | 2% |
| Heater | 5% | 2% | 0 |
| Electric stove | 4% | 7% | 2% |
| Washing Machine | 5% | 5% | 2% |
| Laptop/IPAD/PC | 4% | 2% | 2% |

*Full table is presented in the Appendix C, Table C.18.

*Source: Author*

**Local Businesses**

The electrified visions of the local businesses have yet to result as diverse as the households. Table 4.39 demonstrates the cooling system appliances with fridge (18%) and the 12% for air conditioning and fans. In this sense, cooling the room temperature to store the food or to feel better in that space, has also been representing a process of domestication to utilize a process for cooling the air. In the second stage, the respondents prioritized communication and media appliances, with broadcasting equipment (6%), television (12%), and computers (12%). It shall be noted that the highest ratio with 47% did not provide any items for this category.

*Table 4.39. Future Appliances in the order of Priority at the Local Businesses**

| Future Appliances / Priority | No. 1 | No. 2 | No. 3 |
|---|---|---|---|
| Fridge | 18% | 6% | |
| Cooling Systems (A/C and fans) | 12% | 6% | |
| Internet servers | 6% | | |
| Broadcasting Equipment | 6% | | |
| Television | 6% | 6% | |
| Computers | 6% | 6% | |

*Full table is presented in the Appendix C, Table C.17.

*Source: Author*

## 4.4 Conclusion

This chapter investigated, reflected, and interpreted the local energy practices of everyday life, the visions of the community, and the changes that have taken place in the recent history of Tsumkwe and analyzed the demographic, socio-economic, and socio-technical data about RE governance. Thus, it provided a hermeneutical interpretation of the data about the local context, which will be further discussed in Chapter 7 with a comparison to Gam and the regional and national governance levels. Thus, it shall provide a temporal perspective by reflecting on the past twelve years of mini-grid and the historical standing point of Tsumkwe while considering current challenges and constraints that the community has been experiencing at the present in order to contribute to future possibilities and solutions for existing issues.

The visions of the community members provide a strong Ju/'hoansi or San town in the future, with all town-like characteristics such as tar-road, shops, banks, post offices, hospitals, malls, cinemas, vocational training centers, and computer centers, without being left-behind and all settlement having been electrified. These characteristics consisted of place-based infrastructural development expected from the government. In addition, community members expressed individual and collective wishes about living a "normal life" or a "better life" in harmony with equality among all people living there; this envisioned "better life" included future wishes, hopes, and desires about more opportunities in terms of development and growth, more electricity, and employment opportunities, especially for Ju/'hoansi. The education and training they wished to have, included computers, tablets, washing machines, and other household appliances. In this respect, they also wished to have co-management of such a center with the government or other potential partners in the pursuit of self-development, communication, and (life-long) learning. Some people mentioned their special wish to become business owners and be their boss with the autonomy they would like to have. Lastly, solidarity and respect among all the residents were among the overall wishes that were constructed by the community.

The process of modernization described by van der Loo and van Reijen (1992) looks at the processes crystallized over the centuries. The process of differentiation within the context of Tsumkwe analyses a relatively shorter period of thirty-two years, starting from the independence of Namibia in 1990 until 2022, with reflections on the pre-colonial and colonial times, which is approximately equivalent to seventy years that provides to observe the changes especially relevant to the electrification via the mini-grids. In summary, having a phenomenological look towards the phenomena of energy/electricity in everyday lives, investigating the past and the future visions with its hermeneutical interpretation, the study demonstrates the macro processes that have been occurring in Tsumkwe as differentiation, rationalization, domestication, and individualization. The "source of life" or "life" meaning associated with electricity and the overarching vision of becoming a town demonstrate the future tendency of these processes.

In this sense, there is not only differentiation from a hierarchical perspective between public, private, and community actors but also specialization in the differentiation and multiplication of these actors for specific, goal-oriented purposes in achieving their goals. The mini-grid ownership, operation, and maintenance, held by CENORED since 2017, is one of these differentiated works. A local branch office of the company is present in Tsumkwe to carry out electricity supply-related activities, such as connecting households or shops, operating and maintaining the mini-grid, providing customer relations, and providing technical support to clients. This office is also evidence of the changing structures within the settlement and its needs about the electrification needs and growing demand. Van der Loo and Van Reijen (1992) stated that technology is not only about the instruments and tools but also about the social networks that are formed around them and the mentality required to use these tools. Thus, introducing electricity to a wider population has also brought challenges and opportunities as well as its own networks.

The domestication of energy practices is growing, even though not all settlements have been electrified. Energy practices are defined as an "ingredient of dynamic social practices" (Shove & Walker, 2014, p. 51) and occur in constellations, bundles, and intersections of competences, meanings, and materials. The overarching meaning of electricity expressed as "magical" and interpreted as *a magical tool* that enlightens everyday life of life-worlds is a source of life for some others, with an analogy of sunlight enlightening lives in the form of electric light as an extension of a solar PV system illuminating at night. In addition, having electricity in households 24/7 without interruption creates autonomy for the individuals, which has the most crucial meaning.

The "light" enlightening lives, knowing that it is there in the darkness of the night, mobile phones that are charged at home instead of neighbors´ houses connecting them to loved ones and/or the world, payment of the prepaid tokens via online banking system, studying at night, cooking with electricity without the hassle of collecting firewood or waiting for the gas bottle, boiling the water in the kettle, storing the food in the fridge without wasting it, drinking cold water, selling cool drink, freezing the meat, doing laundry, watching television at own´s premises, sewing, ironing, and welding belong to the everyday lives. The most desired household energy appliances were televisions, fridges, and music systems, including hi-fi and radio. On the other hand, the businesses mentioned fridges, air conditioning and fans, television, and computers.

Thus, this situation of progressive consumption causes a potential limitation in the future of the mini-grid electricity use in Tsumkwe due to the growing demand for existing connections, followed by the demand for new connections. The growing population has also supported this situation since the 2000s, when Tsumkwe became an essential settlement attracting not only San people but also other ethnic groups. The census data suggested 1404 inhabitants with 290 households in 2011; the NSA estimation for 2021 was 3215 inhabitants with 664 households; for 2031, it is expected to be 5026 inhabitants with 1038 households. Furthermore, the students attending primary and secondary schools have been approximately one-third of the population in Tsumkwe as of 2021. Thus, Tsumkwe also attracts students from the village schools around Tsumkwe and frequent visits from the San people from the villages around to receive public services.

As a result, these ethnic and non-ethnic aspects contribute to the sense of the locality, the sense of the culture, and the "global sense of the local" with the electrification and all other changes that come with it. The affordability of the mini-grid connection with low-income, electric appliances with higher energy consumption but are more affordable than energy-efficient appliances brings the question of (un)employment conditions. The employment desires of San people are primarily due to their self-expressed "shyness" or low interaction with other ethnic groups step forward in this sense.

On the other hand, paradoxically, the process of modernization is not new to San people living in Tsumkwe. It has its roots in the colonial period when they defended and protected their communal lands and territory. San people belonged to a hunter-gatherer independent lifestyle living in their *n!ore*; they jumped to the dependent lifestyle of the colonial period. This drastic transition caused them sicknesses and death, called Tsumkwe, "a place of death." They concluded that their survival depended on food, water, and animals. In this way, they have revitalized their n!ore structure, water holes, and animals. They kept close ties with their acquaintances and supported each other. Tsumkwe´s subsequent development after the independence and recognition of communal land brought a new beginning, followed by the mini-grid connection in 2011. Then, the electrification again introduced a sort of "independence" for their everyday life by becoming dependent on electricity and technology.

In the meantime, different specialized services appeared in the settlement to carry out different services from public or private companies or institutions. Thus, through these new formations, a scale reduction has been occurring for specific services (water, electricity, Conservancy, agriculture, environment, gender equality, education, health, and security), and a scale enlargement has also been taking place at the same time. From Elias's figurational perspective, explained in Chapter 2, this multiplicity of actors or different purposes occurs within the same community; even though individuals change their positions or places, they remain there. They become more specialized, such as the CBNRM management by the NNC or NNDF supporting the communities' livelihoods, and become more goal-oriented at a community level. In the meantime, whether with/or without employment, they increasingly *and irrevocably* use electricity and electric appliances and become dependent on technology day by day while becoming autonomous in their daily activities. This autonomy also brings a type of individualization on an individual and/or household level.

# 5 Everyday Life Matters in Gam: Energy Practices and Future Visions

> "When we came here, I am a repatriator from Botswana. So, when we came here, it´s like animals in the bush. So now, we are human being! We can, even though it´s not all of them, we can have things that are being, people in town. Oh, there is a big difference (…). Not all things have developed. I am a teacher. Because I can even, I mean, I mean, because of the electricity, I am able to use my phone the way I want. Meaning that I can even google the way I can teach, using my phone. During that time, it was even difficult to charge your phone. When you come for cooking, I am able to cook. You can see that spot, electric pot. Now, I am able to use many things that... I am like a person in town. You can see that I am watching television, everything is normal" (Interview 6-G, Pos. 9-10).

This chapter outlines the precolonial, colonial, and post-colonial periods of Hereros and the everyday life of Gam residents by introducing mini-grid-generated electricity.

## 5.1 Pre-Colonial, Colonial, and Post-Colonial Gam and the Region

Gam Settlement is 100 km southeast of Tsumkwe with a relatively recent history dating back to April 28, 1993, the official date of the Herero people's repatriation; this is the day that Gam's Herero Community currently celebrates and commemorates their new beginning in Namibia. The literature on the early days of settlement is limited and primarily belongs to the official mission of the Ministry of Lands, Resettlement and Rehabilitation (MLRR) reports (Uytvanck, 1993) and studies conducted before and after repatriation (Uytvanck, 1993; Botelle & Rohde, 1995). The main aim of these reports was to collect the background data, develop operational strategies to receive the repatriated Herero, analyze the situation after repatriation in order to investigate the needs, and provide recommendations for the planning of a socially appropriate and economically/ecologically sustainable development program (Uytvanck, 1993, p. 4).

### 5.1.1 Pre-colonial Period for Herero

In the pre-colonial period, Herero people inhabiting northeastern and central Namibia were among the ethnic groups such as San, Nama, Damara, Ovambo, and Kavango people (Green, 2023). The early migration of Herero is believed to have occurred from eastern and central Africa to Southwest Africa from the north around 1550 and moved to central and northeastern Namibia gradually over the following years (until ca. 1750) (Vivelo, 1977). They belonged to the Bantu-speaking pastoralist cattle-herding groups.

These pastoralist Herero groups have placed the cattle as a focal point, playing an essential role in their economy and culture as livestock and traditional rituals. The "homestead" was their primary residential unit, which consisted of a group of huts or houses arranged in a circular form around the livestock (Vivelo, 1977). A distinct characteristic of Herero society is that they had a double descent structure comprising both matrilineality and patrilineality, linking their ancestral lineages upon their birth (Vivelo, 1977). They were autonomous in their clans, led by headmen (Green, 2023). The cultivation was adopted after the European contact (Green, 2023; Vivelo, 1977). The primary Herero crop is maize, grown for their consumption, and the excess is sold to neighbors (Vivelo, 1977). Ancestral worshipping has been part of the traditional religion. However, the arrival of colonials substantially impacted the Herero way of living at the end of the 19th century.

Gewald (1999, p. 8) suggested that Herero society, from the 1860s, has transformed from a decentralized transhumant pastoralist society into a centralized society in specific urban centers. Consequently, they have transformed their needs and adapted to the circumstances, political concepts, and structures that outsiders influenced (Gewald, 1999). In those years, the Herero society was led by five chieftaincies, who were closely related in kinship but tended to compete politically and economically (Gewald, 1999). Samuel Maharero, who was the son of the late Maharero Tjamuaha, was one of these chieftains who managed to receive the support of Germans in the struggle for political power in Namibia in exchange for selling land, cattle, or labor (Gewald, 1999). Thus, he became the most powerful Herero leader by having the support of the German Imperial. Following this situation, the natural disasters and the rinderpest outbreaks weakened the Herero communities, creating even more dependency on the German colonial state with medicines, vaccines, and food (Gewald, 1999).

### 5.1.2 Colonial Period in Namibia for Herero

The region was declared German South West Africa by Otto von Bismarck at the Berlin Congo Conference in Germany. Between 1904-1907, the German troops were involved in a war with Herero, which also triggered an uprising of Nama, which resulted in being genocidal (Bachmann, 2018). According to Gewald (1999, p. 8), the German-Herero war resulted from a set of misunderstandings that occurred in the events of the years at the end of the 19th century, which led to the destruction of Herero society. The Herero rising under the leadership of Samuel Maharero against the German military caused the death of 80% of the Herero during the four years (Gewald, 2000). The remaining survivors, primarily women and children, had to work in the prison camps (Gewald, 2000), the indigenous traditions and community structures were affected, cattle raising and access to education were stopped, or they fled to Botswana with their families during the war against the Germans (Katjavivi, 1989). The Conference of Peace (1919) in Paris introduced a mandate system (Katjavivi, 1990), and German South West Africa was put under the South African administration (Dugard, 2018). However, unlike the expectation, the Herero society re-organized and re-emerged by gathering their forces; the decease and funeral of Samuel Maharero strengthened this re-organization and unification of two main Herero groups (Christians and soldiers and the Herero groups who were in exile after 1904) under a single Herero identity in 1923 (Gewald, 1999). Between 1915 and 1923, in the transition period from German colonial rule towards South African rule, substantial changes took place in the way Herero groups lived; they obtained a degree of autonomy until 1940 by moving and grazing their cattle (Gewald, 2000). They could leave German employers, return to their land of origin, obtain stock, and re-settle into the farms left by the German settlers in central Namibia (Gewald, 2000). However, by 1940, this free movement and grazing have been limited by introducing a new set of legislation (Gewald, 2000). South Africa's administration continued colonialism until 1988 and agreed to new dialogues with the UN, which concluded with a ceasefire and UN resolution.

### 5.1.3 Post-Colonial Period in Gam

With Namibia's independence in 1990, the government started the repatriation program. In 1991, the Botswana Government recorded the first registrations from Namibian origin to move back to their home country (Uytvanck, 1993, p. 8). Although they were of Namibian origin, most people were born in Botswana. Before the repatriation, Gam had 200 inhabitants, of which 120 were Herero-speaking and 80 San-speaking (Uytvanck, 1993, p. 8). On April 28,1993, 1171 women and children arrived in Gam, whereas the men and the livestock were planned to arrive in the next two years. In total, 805 females and 366 males were registered, most of which were children (37,49% aged 1-10 years) and youth/young adults (30,83% aged 11-20 years) (Uytvanck, 1993, p. 8). The Ministry of Lands, Resettlement, and Rehabilitation (MLRR) provided tents, mattresses, blankets, paraffin lamps, soap, and food rations. As a result, the population in Gam with the arrival of the repatriation reached 1371 people in 1993.

According to the repatriation report, by that time, there was not any real shop in Gam. About 1 km from the campsite, a temporary shop has been set up and run by the people from Bophuthswana in the old Gam area. Moreover, a bottle store was established at one of the tents of government officials in the camp (Uytvanck, 1993, p. 10). The repatriates were accommodated in the tents temporarily until a more permanent structure was established (Uytvanck, 1993, p. 11). Ten staff houses, one single quarter, and one office have been built for government officials (Uytvanck, 1993, p. 11).

Regarding primary school education, 75 children from old Gam and 94 repatriated children attended the primary school with a principal and two teachers (Uytvanck, 1993, p.11). Of these 169 children (from Grade 1 to Grade 5), 49 have been Otjiherero, 26 are Khoisan, and 94 are repatriates (Uytvanck, 1993, p.11). In addition, 168 children in the camp had yet to attend the classes. The repatriated children have faced a change in the education language, from Twana to English, as well as in the curriculum. A secondary school was unavailable (Uytvanck, 1993, p. 13). up to Grade 9, the nearest secondary school was situated in Tsumkwe, about 100 km from Gam. From Grade 10 onwards, the nearest school was in Grootfontein, with the language of education in Afrikaans (Uytvanck, 1993, p. 13). The pupils who would like to continue their secondary school education in English had to go to Oshakati or Windhoek. On the other hand, in 1993, there was neither the possibility of a pre-primary school education for the 217 children aged 3-5 years in the Gam, nor in Namibia as a whole (Uytvanck, 1993, p. 14).

During the repatriation time, there was only one mobile (bus) clinic in the old Gam, with an appointed nurse. Due to the heavy workload, another nurse was sent a couple of months later (Uytvanck, 1993). Tsumkwe Clinic was another alternative. The only doctor available was in Mangetti Dune, which is approximately 200 km away from Gam (Uytvanck, 1993). However, the lack of transportation did not make any transfers available. The sanitation was composed of 4 prefabricated ablution blocks, consisting of 14 showers and basins, 7 toilets, and eighteen (18) come (Uytvanck, 1993, p. 16). Due to the lack of water and ignorance, the toilets, which were very close to the tents, were not used. The 1171 people continued to use the bushes. Four taps from boreholes provided the drinking water supply (Uytvanck, 1993, p. 16). The report states that only two taps were working during the mission. The surrounding environment has been observed to be very dirty, with rubbish scattered around (Uytvanck, 1993, p. 16).

According to the report, a significant change in the lifestyle of the repatriates took place with the resettlement program. Back in Botswana, people used to care for their own needs and live independently in the huts (Uytvanck, 1993). The children used to go to school. Upon arrival, however, they started relying on food aids and living in tents, becoming dependent on the MLRR (Uytvanck, 1993). The community committee, consisting of 8 people elected by the community, represented the local community in the meetings with the government officials and the local chief, Mr. K.Tjiho (Uytvanck, 1993). The report also estimates the cattle repatriation that was not realized in 1993. Finally, the short to long-term recommendations for basic needs, education, health, community development, and cattle repatriation have been expressed, including some activities to keep the repatriates busy and to fight the dependency syndrome. Based on a survey conducted during the mission, people showed interest in gardening, poultry, sewing/tailoring, and shoe making/repair (Uytvanck, 1993, pp. 18-19).

The integration activities to unite the San and Herero people have also been recommended (Uytvanck, 1993, p. 19).

From the period of repatriation to the census in 2001, there are nearly 20 years. The census data provided by the NSA shows the population in Gam as 1161 people, which indicates a slight decline in the population. The census followed this in 2011, whereby the population in Gam has been recorded as 1229 people residing in 293 households. Of these 293 households, 69 correspond to detached houses, 15 semi-detached houses, one guest flat, three mobile homes, 199 traditional dwellings, five improvised housing units (shack), and one other.

## Installation of a Mini-Grid

To accommodate the growing population and related needs of the Gam community, the Gam Development Organization asked for the electrification of Gam in order to strengthen the development activities (Hoeck et al., 2022). Gam's mini-grid was the largest in Africa at the time, and it was awarded to a Namibian company, HopSol (MME, 2014). The mini-grid was intended to supply electricity to the entire Gam Settlement with its 1680 inhabitants (MME, 2014). The hybrid solar-diesel power plant installation had been planned to consist of approximately 2000 CIS (thin film) solar modules with a battery bank of 2.5MWh capacity to meet the demand of the settlement for 24 hours (MME, 2014). As a result of these efforts, the hybrid solar-diesel mini-grid (Figure 5.1) with a 292 $kW_p$ solar PV capacity, along with 180kW, 3-phase inverter, two diesel generators, and lead-acid battery bank, was funded by the MME and commissioned in 2014 (Ileka et al., 2017; Hoeck et al., 2022). Upon the commissioning, the MME handed over the mini-grid to RC, and the operation and maintenance were done by the Ministry of Works, which is based in Tsumkwe (Hoeck et al., 2022). Like Tsumkwe, the ownership and the O&M were transferred to CENORED in 2017.

*Figure 5.1. Gam Solar Hybrid Mini-Grid*

*Source: Author*

The map below (Figure 5.2) illustrates the essential and non-essential lines providing electricity to Gam, indicating, at the same time, the locations of the households as per the 2011 census and some of the local shops. The extension of the mini-grid lines to an additional 400 connections underlines how the growing population has been settling in geospatially from 2011 until 2022, radiating from the center, with the local shops remaining in the central road.

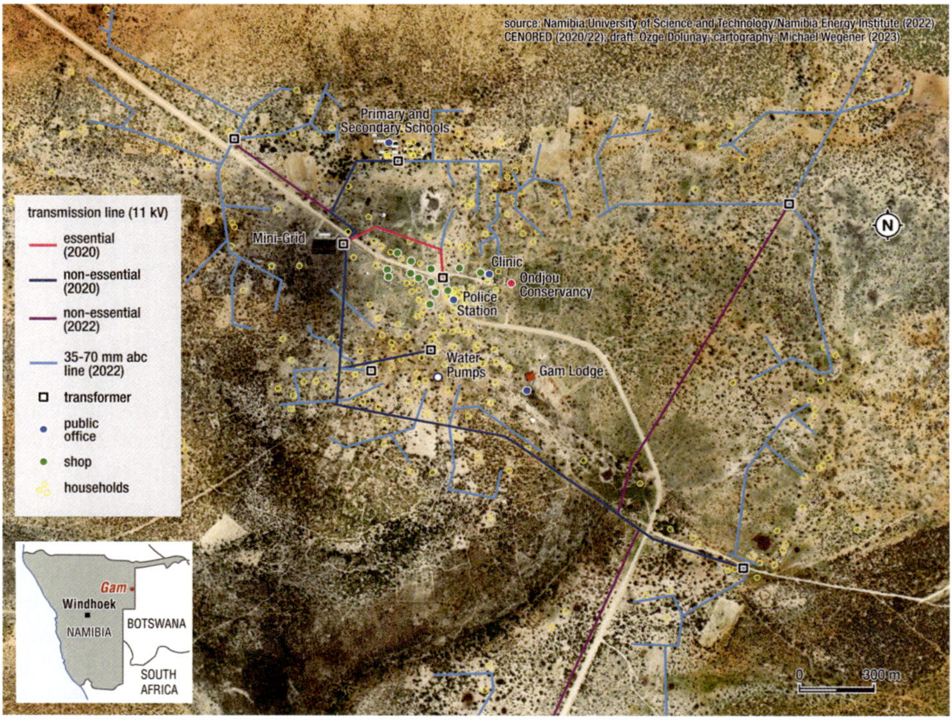


*Figure 5.2. Dwelling Units (2011) in Gam with the Electricity Distribution Lines*

In this regard, the mini-grid (Figure 5.1) was in poor condition due to the lack of maintenance while generating a surplus of electricity for Gam in 2020. Therefore, CENORED moved the generators to Tsumkwe at the beginning of their duty (Hoeck et al., 2022). However, this surplus was not due to the low electricity consumption but rather low connection points (Author field notes, 2020). In 2014, there were only 200 households that were connected to the mini-grid, excluding schools, the police station, and the clinic (Hoeck et al., 2022).

The community was told that there would be additional connections after the initial period but did not hear anything until 2020, although they made their applications to CENORED. Amidst the rising pandemic and the fear of the unknown during the first field research (March 2020), the community members showed a profound interest in the research by putting all their fear aside with the hope of having more information regarding the potential connection time. Phase 2 of the Gam mini-grid was completed in November 2020 with an additional connection of 400 households and local shops (Hoeck et al., 2021).

Notwithstanding this increase in the mini-grid connections, as of 2023, CENORED still does not have any office in Gam to establish O&M or customer relationship duties. In the final conference of the PROCEED project in September 2023, CENORED officials did not confirm or foresee any future local human resources allocation for this purpose.

### 5.1.4 Local Setting and Main Actors in Gam

The majority of the population in Gam is Herero, repatriated from Botswana, with a small San community. The settlement consists of households, local shops, Ondjou Conservancy, churches, Traditional Authorities, NGO, office of the ministries, Gam Primary School, Gam Secondary School, police station and the clinic (Figure 5.3). Gam has been appointed as a settlement and waiting for the administrative office to be opened (Field Notes, 2023).

*Figure 5.3. Map of the Local Actors in Gam*

*Source: Author*

#### *5.1.4.1 Community Based Organization - Ondjou Conservancy*

Ondjou Conservancy (OC) was established in October 2006 according to the Nature Conservation Ordinance (1975, no.4) under section 24A(2)(ii) (Official Gazette, 2006) and named after the Otjiherero word "elephant" (NACSO, 2023b). It covers an area of 8,730 km$^2$ with an approximate population of 3,068 inhabitants with traditional Herero culture (NACSO, 2023b). The area is characterized by diverse vegetation with grassland, woodlands, drainage lines, and an undulating landscape (NACSO, 2023b). The headquarters is in Grootfontein, and the field office is in Gam. OC oversees the management of natural resources, including forestry and wildlife. As of 2020, there were 20 staff in the OC. The primary wildlife resources are the elephant, lion, giraffe, kudu, leopard, wild dog, cheetah, gemsbok, black-backed jackal, and spotted hyena (NACSO, 2023b). The devil´s claw harvesting and trophy hunting are among income-generating activities.

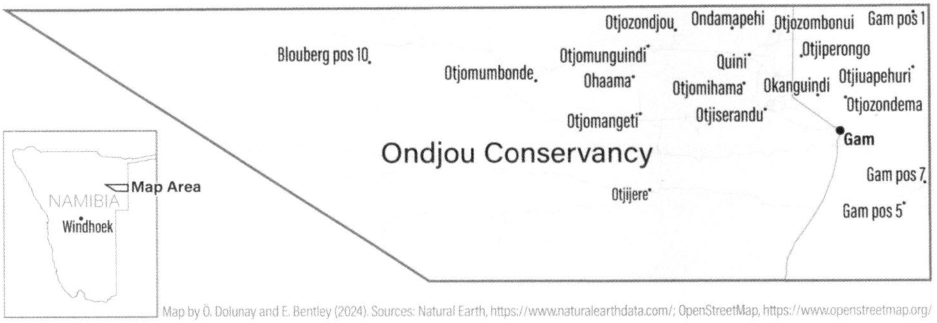


*Figure 5.4. Ondjou Conservancy Area*

In addition, Figure 5.4 illustrates the distribution of some of the 32 villages surrounding Gam, located within the Ondjou Conservancy area, including Omumbonda, Ohaama, Ondampeki, Otjipahewa, Otjomihama, Otjiserandu, Otjiminguindi, Okorute, Otjijene, Omangeti, Opekepeke, Otjiperongo, Onderajahorua (San), Otjozondema, Otjimingandu, Tjio Bonekole, Otjozombanjui, Post 5, Post 1, Post 3, Post 2, Otjotoveni, Otjivapekuvi, Otjozondjou, Oumanonake, Post 7, Oteku, Otura, Okehoro, Otjihamunika, Post 4, Otjongaravige.

### *5.1.4.2 Non-Governmental Organizations*

Tumba Tumba Association, which is not active anymore, was the predecessor of the Gam Farmers´ Association (GFA) in 2004. Currently, GFA represents the only NGO in the settlement. There are approximately 600-700 members in the GFA. The association is registered within the Otjozondjupa Regional Farmers´ Association, through which it is represented on a national level in the Namibia National Agriculture Union (Field Notes, 2023). GFA has an official board of members and an elected president. The election takes place every three years. The farmers´ representatives oversee the marketing. Two times a year, they organize Farmers´ Day, whereby they invite experts in the fields of farming and agriculture who listen the issues of the farmers.

The churches have been categorized under NGOs. Accordingly, there are in total eleven churches active as of 2023 in Gam: the Zion Council of Churches (ZCC), the Spiritual Healing Church, the Oruno Church, the Jesus Christ Church, the St. Amos Church, the Cross Healing Church, the Heaven Church, the Emmanuel Church, the Pentecostal Church, the St. John Church, and the United Congregational Church (UCC).

#### *5.1.4.3 Public Institutions and Parastatals*

The public institutions shall be divided into public services such as primary and secondary schools and their dormitories, police stations, and clinics. In the clinic in Gam, there are twelve employees in total. They have a generator and an old solar PV system, which is no longer used. The generator was planned for maintenance. However, running the generator is costly; therefore, they do not prefer to use it. The population is said to be around 2500 people. The total number of students in Gam Primary School (845) and Gam Secondary School (345) is 1169 (2020), equivalent to 46% of the overall population, constituting an essential part of society. Gam Primary School has 26 teachers, whereas Gam Secondary School has 12 teachers. Gam Secondary School only has a hostel with a capacity of 240; however, it is not operational. They also do not provide any meals. On the other hand, Gam Primary School does not have a kitchen. They prepare breakfast in the open fire at 10:00 for 824 students.

The office of the Ministry of Agriculture, Water, and Land Reform is located outside the main Gam center, and it is also connected to the mini-grid. Another public office is the Veterinary Office at the checkpoint with Nyae Nyae Conservancy area border.

Furthermore, Gam is home to four Traditional Authorities (TA): (1) Ovaherero Traditional Authority (OTA) (Figure 5.5), (2) Kambazembi Royal House, (3) Ovambanderu Traditional Authority, (4) Ju/'hoansi Traditional Authority (Field Notes, 2022). The OTA represents all the Herero, among others, as its name refers to. At the same time, Kambazembi TA is a branch of Kambazembi, another TA from the Otjozondjupa region based in Okakarara. Ovambanderu TA is another traditional group in Namibia that has its lifestyle; thus, they have their traditional authority.

In addition, although there are water pumps belonging to NamWater, no office of NamWater or other parastatal is present.

*Figure 5.5. Ovaherero Traditional Authority Representatives*
*Source: Author*

### *5.1.4.4 Private Businesses*

The main road in Gam has several local shops (Figure 5.6) that provide food or other items along both sides of the road. They usually have few employees and provide for communities' needs. Savannah Market, Maharero´s Market, Maria´s Market, Maharero William Market, shebeens, welders, Gam Lodge, car mechanics, Karumundu Shopping Center, tailors, restaurants or take-aways, and herbal medicine shops are some of the local private businesses. The settlement does not have a specific butcher, but the shops provide cooked food.

*Figure 5.6. Food Shop*
*Source: Author*

Besides these local businesses, there is also a Trophy Hunting Camp, which is located 6km outside of the Gam center and is used for trophy hunting activities. It operates with a generator.

### *5.1.4.5 Households*

The households comprise shacks, concrete (Figure 5.7), traditional wood mixed, or single quarters spread around the main road.

*Figure 5.7. Households in the Center of Gam*

*Source: Author*

In summary, Figure 5.8 below illustrates the timeline of significant events that have taken place in Gam. There was no event until the repatriation, which marked the initiation and shaping of the following events. The government upgraded the schools from grades one to seven, followed by grades seven to 11, between 1993 and 2000. In 2004, the Gam Farmers´ Association was founded as an NGO. The Ondjou Conservancy was founded in 2006 as CBO. The mini-grid installation in 2014, its commissioning by the MME, and CENORED´s takeover in 2017, are some of the biggest highlights in Gam´s short history. The beginning of 2020s introduces more events, such as the PROCEED research project between Germany and Namibia (2019-2023), the COVID-19 pandemic in 2020-2022, and Phase 2 of the mini-grid with the extension to the rest of the unelectrified places in Gam in 2020. The tarred- road constructed in 2022 surrounding Gam and the main road going through the settlement and Gam having been appointed as a "settlement" officially belong to the series of the most important events in a short time. According to the community members' latest news, they expect to receive 70 families from Botswana as a part of the repatriation program.

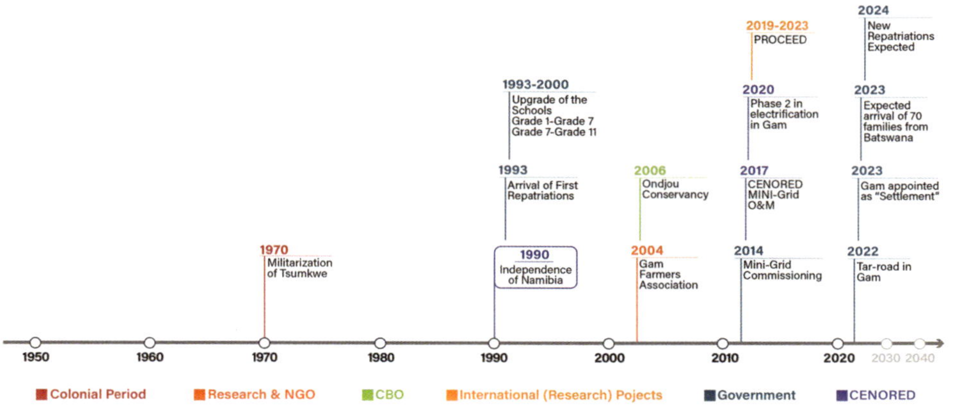


*Figure 5.8. Timeline of Major Events in Gam (1993 – 2023)*

*Source: Author*

## 5.2 Qualitative Findings

This section outlines the qualitative findings from the ten interviews conducted in 2022.

### 5.2.1 Process of Differentiation

#### *5.2.1.1 Place-Based Differentiation – Animal versus Human-being and Urban versus Rural*

Gam´s particular location, with its thirty-year relatively recent history, is still alive and active in the repatriates' memories. As Massey (1995, p. 186) quoted: "The past is present in places in a variety of ways At the arrival of the repatriates in Gam, the majority of the population was composed of 39.1% of the age group 1-10, 26.4% of the age group 11-20, 7.6% of the age group 21-30, and 6.9% of the age group 0-1. This population distribution suggests that 80% of the population had an age group of 0-30 at their arrival. Thus, they represent in 2023 a population range of 30-60 year-old residents residing in the settlement who have their childhood/adolescence memoirs. The quote below is one of them describing the conditions in their past as "like animals in the bush" and reflecting on their present life: "Now, we are human being[s]." MLRR´s efforts in welcoming "home" the 1911 registered Herero with Namibian origin before the Botswana authorities, the operational strategies that were developed seem to have quite a harsh impact on the arrival, albeit it was intended to realize a smooth transition home, by all means (MLRR, 1993, p. 1). Thus, only after thirty years do they feel like human beings, living under conditions designated for human beings and not like animals in the bush. On the other hand, it also underlines the profound changes of "big difference" or "development," as locals pronounce it, that they have been going through daily. The "things" that the people in the urban areas, in "towns," use are also available or usable by them, although not in their entirety in Gam.

> "When we came here, **I am repatriator** from Botswana. So, when we came here, it´s **like animals in the bush**. So **now, we are human being**! We can, even though it´s not all of them, we can **have things** that **are being, people in town**. Oh, there is **a big difference**" (Interview 6-G, Pos. 6).

The urban, peri-urban, and rural classifications depend on the methods and parameters used by the beneficiaries of these definitions, such as governmental institutions, academia, or statistical use (Cattivelli, 2021). The MME defines the rural areas in the Rural Electrification Distribution Master Plan (2010, p. 6) as "those communal areas that fall outside the proclaimed municipal areas and commercial farm"; thus, rural electrification has been planned. The national reports and statistics compare urban and rural areas in Namibia. Hence, interviewee six also refers to this dichotomy, mentioning/implying fewer differences. She also highlights that they have been informed that they are paying the same tariff for electricity as the people in town; however, they are restricted from using the appliances they would like to use, which creates a "real problem" for them.

> "We are **told but we are paying the same money**, been paid by the town people. But here, we´**ve been restricted not to use some of the things**. To us, it´s like a punishment. We **are paying the same amount and we are being restricted**. To us, it´s a really problem" (Interview 6-G, Pos. 19).

In the context of electricity, they feel an inequality compared to their fellow citizens in the towns, and they base this interpretation on the information provided to them verbally. Unfortunately, the information they are told is not accurate. Gam community has a subsidized tariff, paying less than the rest of the country; this shows how important it is to inform the customers of the mini-grid by CENORED, in this case, to have more conscious customers that will also eradicate the inequality consideration for the tariff, while proving and providing the specificities of the mini-grid and its (technical) differentiation as a user by the national grid.

#### *5.2.1.2 Differentiation in the Electricity Supply Structure*

**Mini-Grid versus National Grid and Customer Relations**

On the other hand, the (technical) difference between the mini-grid and the national grid in energy generation brings inequalities in terms of its use. The quote above also indicates the restrictions in the use of electricity: "We´ve been restricted not to use some of the things. To us, it´s like a punishment", which they interpret as a restriction as well as a punishment. But a punishment for what exactly? As a mini-grid user in rural Namibia? It would be difficult to interpret this. However, the real reason for this "restriction," so to speak, is due to the electricity generated by the solar modules during the sunlight hours of the day. The generation stops in the evening, and the batteries and diesel generators start operating at night hours. The first limiting or restricting factor comes from the sun itself and not by the operators. The second limiting factor is diesel consumption during the night hours when there is no solar energy production at night, and diesel consumption is costly and must be paid for. As a matter of fact, with the high electricity consumption of appliances in their households, with the addition that all community does similar things in their everyday life, electricity consumption increases more. Thus, it results in consuming more diesel, which the community does not pay extra; this is why the operators request a restriction and inform their customers via posters that are similar to Tsumkwe.

> "Like, first they told us that you can… Because this house of mine, I can **extend the electricity to that house**. But nowadays, we are been **prohibited not to do** that. They are the **same people**, but there **is different information**. Now, **which is which?** (…)
>
> They are not telling us. Only come and say, "don´t do this" or "do this". This one is telling you "**do this**", and when **this one goes, this one comes tell you** "do this" but **we don´t have their phone number**. And even, we are not, we don´t know them. Even to follow why they are doing that" (Interview 6-G, Pos. 43-47).

Furthermore, related to the technical concerns, interviewee six stated that before, they could extend the cables themselves to other houses, but now they are prohibited. She mentions the concerns of not having a contact person to address their issues, that the people come and go by saying different things, and that it is impossible to follow them. Since Gam has no CENORED Office in the settlement, the highlighted concerns inform the lack of place-based customer support, considering the large number of clients. Notwithstanding, it also informs place-based differentiation between mini-grid and national grid customers since CENORED´s website includes all the information necessary for the grid-connected customers and customer communication contacts. In Gam, no contact person delivers or disseminates accurate information and necessary updates for the mini-grid and its users. Currently, the operation and maintenance are managed by the office in Tsumkwe. The quote from interviewee one describes the situation and how CENORED responds to their everyday life issues regarding the mini-grid (in/mis)operation in Gam.

> "No, the problem it´s only that **CENORED, its place in Tsumkwe**. So, they used to **drive from Tsumkwe to here**. But at least, they tried, so near the distance between Gam and Tsumkwe for them **to drive everyday, to come and fix the electricity** or do something else. But they try at least. But when **we try to find out** what causes the electricity to go

> off. **They say, "no, the batteries are flat"**, so they need to buy maybe, to buy new batteries. They need the supporting machine, a generator.. to charge those batteries also" (Interview 1-G, Pos. 30).

This quote indicates that in the case of problems, CENORED is responsible for driving 100 km from Tsumkwe and back to fix the issue, creating more cost due to diesel and time with their limited personnel in Tsumkwe. On the other hand, "when we try to find out what causes the electricity to go off" shows the community's interest and willingness to understand the issues related to the mini-grid. Thus, it also illustrates a degree of ownership of the mini-grid. Interviewee five below articulates Gam's needs for water and electricity by criticizing the operation as very poor. He mentions that when the electricity goes off, they operate for long hours and cannot finish their jobs.

> "Water, electrcity. Because this electricity we are having fine, they can work. But the way they are **operating is very poor**. Because sometimes, even yesterday, they switch on at 3 o´clock since from morning time. **Which means we are not operating**. Even those people who are busy, those supposed to finish yesterday. **Because of the electricity, they didn´t manage to finish their job**. That´s the thing we are having" (Interview 5-G, Pos. 25-26).

This situation also translates to the reliability of electricity, which is closely connected to the process of rationalization, the predictability of electricity supply to run a business in a predictable and controllable way. On the one hand, there are many efforts from the government and funding agencies to increase capacity building and introduce productive use activities; on the other hand, in Gam, the unreliability of the electricity supply creates hurdles and prevents people from doing their daily business. Thus, they need help generating income.

#### *5.2.1.3 Differentiation in the Future Visions - Now and Future*

The differentiation in the future visions has been grouped under expectations, wishes, and hopes.

*Table 5.1. Expectations, Wishes, and Hopes for a "Better Future"*

| | | |
|---|---|---|
| Expectations, wishes, hopes for a *better future* | Expectations | "I **am expecting Gam to be**... a **developed town with hospitals**, **not clinics**, with a police station **with full staff appointed at the police station."** |
| | Wishes | Infrastructural : "it **needs a lot of development... We need training... Public toilet, water, service"** |
| | | **Educational - Training** **"For the future, we need training."** |
| | Hopes for a ***Better Future*** | "Opportunities...**we are left behind**. the government don´t care about us that much. So, **it hasn´t been that industrialized**." |

*Source: Author*

## From Settlement towards the Town of the Future

### Expectations

The previous chapter defined the visions as "a desirable future" or "actions towards a common goal." In the visions of the Gam community, as an overarching theme, there is an expectation and a wish to become a town and access the local towns' services in Gam to be developed. A town is associated with the development. Not only do they expect to have public services to the full extent, such as hospitals and a total capacity of police stations, but they also emphasize having buildings, banks, post offices, tar roads, and shops in towns like Shoprite and Agra. Only then will they know that the development has been achieved. In this sense, they expect to have a set of services both from the public and private sectors (Interviewee 1-G).

> "I **am expecting Gam to be**... a **developed town with hospitals**, **not clinics**, with a police station **with full staff appointed at the police station**. Hmm, what else?? Okay, we do have a Ministry of Environment and Tourism, Agriculture, Ministry of Education is here. What else? the tire road, around the town, that you can see it already, it´s here. It´s only the **construction of the buildings**, of the houses starting and also **expecting in the future**. But from these things are… Also, **shops like Shoprite, Agra**... It´s when development... it´s when the development, it´s when you are having these types of things**, bank, post office**... yeah. All those things. (…) These services... Either in Grootfontein or Gobabis" (Interview 1-G, Pos. 38- 40).

On the other hand, there is also a realistic skepticism about becoming a town because the electricity connection needs to be more stable and reliable for good shops to be established in the settlement. Interviewee six suggests that electricity might be both an advantage and a disadvantage in its current state for the shops or banks to invest in there. Thus, the development would fall into danger.

> "Gam in the future? In the future, Gam, I think, if the electricity switching off, like the way it is switching off, it **might develop**, **even good shops might come here**, **because of the electricity**. **Even the bank**. We might have **a bank because of the electricity**. But if it is switching off like this, I don´t think the people, they might come" (Interview 6-G, Pos. 14).

Furthermore, the visions in Gam have been realistically based on their existing struggles in their everyday lives, in which they articulate their needs to be met and their issues to be solved for their "development."

## Wishes - The Needs of a Town

### Infrastructural

This development they seek compares with the services they access in the towns they know or visit. Although the "little settlement" they call is not as small as they think it is, they are aware of their distance from those services and the difficulty of becoming a town without access to any of them. From this perspective of Interviewee 7, their needs are infrastructural, such as the public toilet, sewage system, and training. However, there is also an emphasis on the priority of the elders´ needs in asking for an especially easy-access toilet to prevent them from going to the bush and walking 50-70 meters. She keeps her needs or those of a younger generation after the elders.

> "…But it **needs a lot of development** to go around here. Because it´s a **little settlement** and it´s **very far from the town**. It **needs to be developed fast** to get the access in there. Things that we need here.
>
> O: What kind of things do you need?
>
> Interview 7 - G (2): We **need training**.
>
> Interview 7 - G: **Everything**! **Public toilet, water, service (filling station service**).
>
> Interview 7 - G (3): No filling station, no sanitation, **proper sanitation, like sewage**. There is **no sewage lines** here. People need to, **elderly need to walk for about 50 to 70 meters**. But then, the bush is a bit far. So, when there are a bit of toilets, for elderly it´s better.
>
> O: Okay. And for you? not`?
>
> Interview 7 - G (3): **For us, we also need, but it is first for the elderly**. Yes, yes" (Interview 7-G, Pos. 21-31).

Thus, their needs also highlight an expectation from the government to provide these services in an unspoken way. They also maintain their hierarchical community structure based on the priority provided for elders. On the other hand, for themselves, the younger generation considers "training" as a need. As a result, they envision a future with trained people living in a developed town.

### Educational - Training

Starting from a general wish and a considered need for training, the third person in the group focuses on the technical issues they are having with the unreliable and interrupted electricity they are having in Gam. As indicated in the previous section, regarding the place-based operational issues that CENORED has been going through, the interviewee clearly states the need for a staff dedicated 24/7 to the mini-grid in Gam. He also suggests that this person might not be part of CENORED but still take care of the mini-grid. Furthermore, he mentions he is an electrician and a specialized refrigeration technician who would like to receive training in electrical engineering to address the technical issues they face with the mini-grid. However, he needs more documents to prove that he is qualified to handle electrical work, and he observes that those who have documents fail to solve the problems. He proposes surge protectors for the households, enabling them to prevent damage to appliances caused when the electricity fluctuates or switches on and off.

> "O: And what kind of wishes do you have for the future?
>
> Interview 7 - G (3): **For the future, we need training**, like the plantation (solar power plant) here. People that used to service everything here. For CENORED, they come here from Tsumkwe. Or the whole day, the power is off until by the time they reach here. **So, we need somebody to be 24/7, yes**. Even if **it´s not somebody who is falling under CENORED, the one falling under the plantation itself**, yes. At least that person can be paid later via from the foundation if it doesn´t have to be paid by CENORED. **I don´t know where CENORED will fit in there.**
>
> O: Okay. So, it is not clear for you how?
>
> Interview 7 - G (3): It´s not clear! Not very clear.
>
> O: I see, I see. You mentioned trainings. What kind of trainings?
>
> Interview 7 - G (3): **Training in electrical engineering**. Because for me, **I am an electrician**. But then, **I need papers**. My experience is there. For me to present myself tell to the people, I can do this**, at least, they need certain paper**. But the job they are doing**, the ones who are said they are qualified, they are failing. Me, who doesn´t need papers, I can do it**. But then**, I only need certain paper**.
>
> O: Okay. I see. So, you are an electrician.
>
> Interview 7 - G (3): I am an electrician and **refrigeration technician**.
>
> O: How do you see? when the electricity switch off, then, I heard that a lot of appliances are damaged.

> **Interview 7 - G (2):** Yeah, yeah, yeah. they are getting damaged.
>
> **O:** What do you do in such a case? They are completely broken or?
>
> **Interview 7 - G (3): Then, they are getting burned** but then, **we have to introduce certain system** whereby they can **get surge protectors**. **Surge protector for every household until the power is stabilized"** (Interview 7- G, Pos. 31-41).

The following sections discuss training opportunities for youth or individuals since they are closely related to the rationalization and individualization processes. In addition, they are strongly connected, such as training for the elderly, nursing healthcare services, or training dedicated to youth. In the previous quote, two main points should be focused on. First, there is already a person who is familiar with electricity and has worked with it, and he would be available and willing for further training as a staff member for CENORED-Gam. In the workshop and conference of PROCEED, CENORED officials often argued that they do not have trained staff for the O&M, have strict regulations that must be followed, and cannot employ unqualified staff. However, from the discussions, one can interpret that they have never been given a chance or an opportunity to look for a local staff to train for this position. They have found it easier to employ a qualified person from elsewhere and to send him there. Thus, finding an interested and capable local person would be impossible if they are not looking in the right place.

Moreover, the technical considerations for a better mini-grid optimization exist in this sense, but it seems "left behind" by CENORED due to place-based concerns. Secondly, this interviewee highlights an essential point in the cross-cutting process of rationalization by articulating his wish to pursue training to get a certificate to prove that he is qualified, even though he witnessed that the qualified people do not do their work correctly. In this way, he reflects a goal-oriented manner by considering the bureaucratic requirements. Thus, he ensures that he steps toward his potential future.

## Disappointment – Failed Expectation: "Left Behind" and Hopes for a *Better* Future

The following quote indicates how people consider themselves left behind by the government when developing industrial facilities in Gam (relating to Chapter 4's discussion of "left behind"). The quoted interviewee sees himself as qualified to manufacture fridges, and further, with education, believes anything would become possible for him and those in the community, because they have a mini-grid and a power plant.

> "Opportunities...**we are left behind**. the government don´t care about us that much. So, **it hasn´t been that industrialized**. Even, we have the plantation whereby we can start manufacturing fridges, **as long as we**

> **get that knowledge**. How to do? **the education, everything is possible**. Yes.
>
> O: What else is needed apart from the fridge, what do you think, the essential things needed for?
>
> Interview 7 - G (2): I don´t know, maybe doctors. **Few doctors**... I don´t know about the clinic who can take care. Patients have been transported **to go to better hospitals** and so on. So, **at least we can get proper doctor here**. Yes" (Interview 7-G, Pos. 63-65).

In this context, disappointment appears as a negative form of a failed expectation from the government, which influences them to use the term "left behind." This disappointing situation might be related to their perceived expectations linked with the repatriation and coming home to Namibia. Even though, over the past years, they see "development" in their terminology, they still think they have been left behind compared to other places, especially in the infrastructural, health, and education services. In Gam, becoming a "Herero Town" has not been mentioned *per se*, since the Herero population is spread around Namibia, but town-like characteristics have been pronounced. In this sense, having more doctors so that the patients will be treated there, industrial facilities for them to create employment, and products they can use there become prominent.

As a result, these wishes and expectations depict the need for strong governmental support in order to reach and obtain infrastructural public services. The interviewees show an interest in the "development" of the Gam Settlement to make it a better place and town in the future, and they are ready to take responsibility when the opportunities are provided. Despite their struggles with electricity, water, and sanitation, they remain in Gam, attached to their places.

## 5.2.2 Process of Rationalization

### *5.2.2.1 Re-enchantment of the Worldview and the Meaning of Electricity*

*Table 5.2. Meaning of Electricity*

| | |
|---|---|
| Meaning of electricity | From the Optical Visibility to "***video ergo sum***" (I see; therefore, I am) or "***scio ergo sum***" (I know; therefore, I am) |
| | "**It´s life**!", "**It means life**!", "**Electricity is life**!" |
| | "Electricity is life, apart from water. Electricity comes second." |

*Source: Author*

### From the Optical Visibility to "*video ergo sum*" (I see; therefore, I am) or "*scio ergo sum*" (I know; therefore, I am)

The re-enchantment of the worldview appears as light "brightening the mind" and with the indescribable words used repeatedly: "it means a lot," emphasizing its importance. The enlightenment of the mind through electricity has been perceived and recognized by its use by human beings. There is neither darkness outside nor in the mind and/or in the brain. Reflecting on the relationality of the human and its environment (Relph, 1970) explained in the previous Chapter, the interviewee tries to articulate his experience(s) through the perception of his bodily senses and brings (re-)cognition into his overall experience. His repetitions draw attention to its prominence in everyday life when the words fall short in describing and explaining its meaning with "it means a lot." From these bodily experiences, it might be possible to explore and do further research on the connection with the "*enlightenment,*" defined and described by Kant (Humphrey, 1992, p.1), as "having the courage to use one´s understanding without the guidance from another" with a motto of *Sapere Aude* (as cited in Humphrey, 1992, p.1; Kant, 1784, p. X). In a simplistic understanding of this meaning of electricity, the interviewee gets in connection with its environment in a broader sense with electricity, explores the benefits of it by oneself without needing someone else, and creates space for a further (self-)exploration, knowing that their other electrical tools that he can use himself. Therefore, "it means a lot". Furthermore, it is used in other daily life practices, such as cooking.

> "It´s mean a lot, it´s kind of... **it´s mean a lot**! It´s mean a lot! It´s mean a lot! Now, **brighten our mind**, **you know**, we are **no more staying in darkness**, so it´s mean a lot. And we are using it for purpose **cooking and everything**. It means a lot" (Interview 4-G, Pos. 18).

From this space of convoluted experience and the knowledge gained through these practices, the interviewee recognizes the world or the reality he lives in; he recognizes it and possibly recognizes traits of himself in this enlightened and illuminated world with electricity. Further respondents describe it as "life" and "part of life".

### *"It´s life!", "It means life!", "Electricity is life!"*

The integration of electricity in the everyday, with its vitality, become very prominent and outstanding with the clear, genuine, and straightforward answers as in the following quote: "It´s life**."** It is as simple as it is profound. One could ponder the meaning of "what is life?" in many different contexts to understand the depth of its meaning for a word that we all use. The second interviewee takes the answer further, stating that "it´s life," as the second most important after water, "water is life," and combines it with light availability. In this way, he brings his reflections and explains that when there is no light, there is no vitality. In this way, he is also reflecting on the optical visibility as well as the light that brings life into their everyday life, everyday perceptions of everyday life that they live every day. Furthermore, he related it to the ongoing developments in the settlement; he admitted the development in a way that they would be left behind if they had not had any electricity from the mini-gird. The "plantation" means the "solar power plant" or mini-grid they have been benefiting since 2014.

> "O: For you, what does electricity mean?
>
> Interview 7-G (3): **It´s life!**
>
> O: It´s life!
>
> Interview 7-G (3): ha ha.
>
> O: and for you?
>
> Interview 7-G (2): the question was?
>
> O: What does electricity mean to you?
>
> Interview 7-G (2): **Electricity is life, apart from water**. **Electricity comes second**. There is no light, there is **no vivid. Development will be left behind that much**. Electricity **plantation came to help us very well**" (Interview 7-G, Pos. 54-61).

The vitality of electricity in their lives has been admitted with all its profoundness. They accept electricity not only as a tool that most of developers use when planning, but they live an improved life by using electricity and witnessing the developments they have in the settlement attached a meaning of life that comes, right after water. Hence, the optical meaning of life used by physicists or engineers is only a small part of what indeed they live, perceive, and experience in an overall holistic meaning of "light" and "life" for the residents in Gam.

**What is life?**

Chapter 4 included Box 1 and explored different definitions of "What is light?" as a second reflection on the hermeneutical cycle of interpretation. Since the convoluted, intertwined, intersectional, and cross-sectional meanings of light with life, the author included Box 2 and explored the definitions of life under different contexts. Within the context of electricity/energy or light or life, as the words very frequently used and taken for granted in our everyday lives in Global North, the introduction of "electricity" (or electricity access) in the off-grid areas brings a synergetic impact to the inhabitants. Therefore, the basic trajectory of electricity, light, and life are worth exploring to understand the issues in the localities better hermeneutically.

*Box 2. "Electricity is Life," What is Life?*

With respect to the hermeneutical interpretation in the light of phenomenology (*Noema* and *Noesis by Edmund Husserl*) moving towards the concept of *lifeworlds* (Schütz) and *everyday life*, the author finds it e*nlightening* to consider "*What is life*?" and its meanings under different contexts.

The first and foremost researched context is the biological perspective. Biology defines life as living and non-living organisms. It starts life from a cellular level up to an organism or a cluster of systems of organisms (Sagan et al., 2023). On a cellular level, cells also need energy to grow. Life from the lenses of biology includes the capacity for homeostasis, growth, reproduction, adaptation, response to stimuli, and metabolism (Molnar & Gair, 2015). NASA (2024) defines "Terran life" as "a self-sustaining chemical system capable of Darwinian evolution" and considers the specific features of the one life we know. As an institution conducting research in (outer) space, they consider the chemical transformational, biomolecular reactions to determine the characteristics of life. Starting from this definition, scholars have created further definitions, including thermodynamical aspects of an equilibrium system by considering the flow of information from the environment to the living system (Vitas & Dobovisek, 2019) or investigating the self-sustaining systems (Liu, 2020). The energetic life theory brings insights into the microcellular structure and its process (Wei et al., 2022).

Then, the list of diverse definitions continues in the philosophical debates starting from antiquity, relating to the consciousness, the purpose, and the nature of living beings. Life appears in Aristotle´s definition of a soul, "the actuality of a body that has life," where life means "the capacity for self-sustenance, growth, and reproduction" (Kenny & Amadio, 2023). Theology has considered human life to be rational, involving a spiritual, eternal soul with a conscious life (Mariscal, 2021).

On the other hand, the AI has the following answer to the same question.

> "I don't have personal experiences, feelings, or consciousness, so I don't have a concept of life in the way humans do. I'm a computer program created to provide information and assist with tasks. However, I can provide information and engage in discussions about life and its various

> aspects from a factual and conceptual perspective. If you have any questions or topics related to life that you'd like to discuss, please feel free to ask, and I'll do my best to provide information and insights" (ChatGPT by OpenAI, personal communication, 2023).

The above-mentioned definitions from different perspectives depict a very complex definition of "what is life?". But the overarching meaning that is accepted for human "beings" is that physically, chemically, and biologically, we have a beating heart to start with. However, for doctors, this should mean the basics, and then, comes our "healthy" cognitive abilities and the overall balance of these characteristics to pursue our life. The extent to which this life has been reflected, investigated, and explored through philosophy or spirituality, contributes to the multiplication of the definitions. The search of life in macro-cosmos in other planets or the search in the micro-cosmos at the cellular and atomic levels enrich the discussions regarding the vastness of its definition. However, in a further step, the definition provided by AI and linking personal experiences, feelings, and consciousness reinforces the philosophical debates of our era and carries it over to the debates of the robotization of humans and the humanization of robots (Habere Güven, 2020).

As a result, when these definitions have been reflected holistically, one can see that the living beings´ definition of life is powered by "light" as a source of energy to initiate life. Whereas AI is powered by electricity. In the robotization process of humans, electricity, AI, and even technization have been utilized. While in the process of the humanization of robots, a combination of digitized and digitalized data received directly or indirectly from humans has been used. The robotization of humans has been taking place by using the ever-increasing number of AI-based tools or devices that enable or promote "an easier or a more comfortable *life.*" Thus, in the case of Tsumkwe and Gam, electricity has been perceived and interpreted as *life* and as *light*, attempting to bring together different domestication processes in the same era and by marking a different model of the modernization process.

#### *5.2.2.2 Rationality of Purpose*

**"Electricity is everything", "Nothing to do without electricity!"**

Moving forward from the meaning as profound and as indescribable as "life," interviewees five and eight, with the quotes below, combine the meaning of "life" and "everything." Interviewee five makes an analogy with water while showing their competing equivalence with one another with his hands. It is a necessity that they have, and they cannot do anything without it. Those are the basics known for their vital characteristics for human life, which also pave the way for the rationality of purpose and how they are used in everyday life. Therefore, electricity within the "everything" meaning also includes comfort in their lives.

> "**Electricity everything**! We can do **nothing** without electricity, yeah! that´s why **we need it**! yeah, **that´s why we need it**. (…) **They are together**. They are more or less, they are going like this, they are going like this (showing together with his hands the parallel situation)" (Interview 5-G, Pos. 38-43).

Interviewee 8, while repeating that "it means a lot" and "it´s life," grounds her argument by explaining "because" the reason for her answers is "it is everything." The frequent repetition of similarities between water and electricity come together as an impactful "survival" tool coupled with each other. Recalling the times without electricity and the changes over time makes her conclude that they cannot live without it. Furthermore, the problem-solving aspect of electricity, such as charging phones, makes life easier for them.

> "The life **has changed much**. First, there is no electricity. We had **problem** with **charging cell phones**. **Right now**, I **can use it**. It **means life** become **more easier**" (Interview 8-G, Pos. 4).
>
> "Electricity means a lot! Electricity... Maybe it´s a life! **Because** it´s everything! She says, according to her, she takes the electricity as like the **helpful of water**, electricity is the **same**. **We can´t survive without it**" (Interview 8-G, Pos. 60-62).

**"It´s unfair"**

Within the rationality of purpose, for paying an electricity tariff in return for using the appliances they would like, interviewee six expresses her concerns about its unfairness. According to her, only paying a tariff to use lights is neither acceptable nor fair. She understands this situation as a punishment for them. For paying the same amount of tariff as others, they should also be able to use all the appliances they would like.

> "Now, this switching off, we should not use the electric things for cooking. When are we going to be allowed to do that? When? (…) But, what is like ... **it´s unfair**! We **are paying the same amount**. But we are still told **not to do some of the things using the electricity**. Well, people they are gaining. If they have to do things for us to use the electricity, instead of telling us pay the money, and just using for light. No, that one is not fair. To us, it´s like a punishment.
>
> O: So, it´s, it should be for everything? It should be allowed to use?
>
> Interview 6-G: **For everything**! **It should be allowed** like the way we are paying our money. **Or if it´s like that**. **We should not pay the same amount of money like other"** (Interview 6-G, Pos. 33-39).

As mentioned earlier, interviewee six lacks updated information regarding the subsidized electricity tariff they are paying. They do not pay the same amount as the other places in the region. However, her argumentation and reasoning are representative of the process of rationalization. The limitation they have been going through due to the utilization of high-consumption appliances provokes her to ask for a lower tariff, which they have. As a result, there is a demand for the reliability of electricity and unlimited use of the appliances by providing independence in choosing the appliances they would like to use and paying a fair amount for this service. In addition, they expect to be informed by the service providers.

#### *5.2.2.3 Reliability and Autonomy*

The reliability and availability of electricity contribute to the process of rationalization by bringing controllable and calculable aspects into daily life. Therefore, it also does not realize any productivity. Later, it also contributes to their independence as a part of the process of individualization or domestication. The quotes below from interviews seven and nine provide these aspects.

> "If you don't have electricity, **you don't know**, you **don't do your business**. It cannot work if you don't have electricity. **Nothing to do without electricity**" (Interview 7-G, Pos. 66-67).

> "The electricity means a lot of things. Because without the electricity, we cannot work, **we cannot do anything, without electricity**. But **if you have the electricity, you can do anything that you want**" (Interview 9-G, Pos. 24).

In these quotes, electricity is expressed as a precondition for doing anything they would like to do and conducting their business. When it is interrupted, and one does not know when it will be on again, it is impossible to continue working and generating income.

#### Opposing Thoughts – Skepticism with its Unavailability - Unreliability

Therefore, this situation creates opposing thoughts and some skepticism related to its (un-)availability and (un-)reliability, which still belong to the rationality of the community members. Besides being unable to do business, interviewee two complained about its unreliability and having a good operation only a few days a week. The rest of the time, the unreliabiltiy of the electricity causes problems with the rotten meat in the fridge and the warm beers that they are obliged to serve to their customers due to the lack of reliable electricity connection. Therefore, this is not a big change for her and her business since her issues continue to persist. As a result, they feel limited and stuck in this problematic situation.

> "Electricity here to her does **not mean anything**. Because it went off, even once **you put your meat in the fridge**. It ...out, **it becomes rotten**. Your beers that you have to **serve to your customers are hot**. So, there is no any changes between the older days and now. Okay, **maybe two days in a week, it operates well**, but after **that it started to get stuck, stuck, stuck**" (Interview 2-G, Pos. 24).

Gam, being part of the Kalahari Desert, might be very hot during the dry season. Meat and beer are important parts of the Gam community's daily diet. Therefore, neither rotten meat nor warm beer are attractive food and beverages to sell or buy.

### *5.2.2.4 Future Expectations*

The rationalization in visions contributes to the overarching wish to become a "town" in the future. Their rationalist visions have already demonstrated a pathway for this wish to come true. Under these visions, youth and training, infrastructural development, and business development or improvement have been foreseen (Table 5.3).

*Table 5.3. Future Expectations*

| | |
|---|---|
| Future expectations | Youth & Need for Youth Center (Training): "the youth can also keep them busy, on their own" |
| | Town-like infrastructure: "We don´t have the banks" |
| | Business Opportunities with the Electricity as a Pre-condition |

*Source: Author*

#### Youth & Need for Youth Center (Training)

Further discussed in the quantitative section, the growing number of children and youth is another component of the idea of the future. The youth has been involved in the community and OC activities. However, interviewee one suggests that they also need a center to keep them busier and lack a place and capacity to train them. In this way, he visualizes a trained youth for the future in Gam instead of wasting their time here and there for nothing. Therefore, he thinks of a rational use of time and youth to develop Gam.

> "They are all involved in Conservancy and community activities. But we **need also a youth center** where the youth can also **keep them busy**, on **their own**. **Now**, like now, **we don´t have any experience** where we can, where the **youth can gather together and do something**. They are just gathering around here, just ... us, because of nothing they can do during the day" (Interview 1-G, Pos. 66).

#### Town-like Infrastructure

The lack of banks to deposit the income generated from the business creates a life-threatening insecurity for the interviewee three. In his interview, his complaints about how to save his funds until the truck comes, with every rising fear of theft and sleepless nights, make him anxious and worried about his situation.

Eventually, this also prevents his business from growing. Furthermore, his complaints continue with the rocky soil structure in Gam and its unsuitable nature for digging, making the construction a challenging process. The lack of toilets also causes big trouble. Nevertheless, with the availability of electricity and water, he is confident that they can make their toilets without expecting it from the government. In this way, they would also save time by not going to the toilet, which is 2 km away, in the bush, and use this time to do their business. The lines of this quote demonstrate rational thinking in overcoming the challenges of a businessperson moving towards the process of individualization by taking responsibility in dealing with his issues as well as intersecting community issues. However, he cannot make the banks come to Gam, nor the supply of water, electricity, and some necessary tools to work with. In this, he expects the government´s support for their supply to enable them to work further to meet their needs.

> "People do safari here because of the ... **We don´t have the banks**, **e-wallet**, what.. **because if there will be the bank, the people they weren´t withdraw** the money even to buy. No e-wallet. What what ... **I am still having money here**. **Where can I go and save those one**? Means that each and every time I´m not sleeping well because the people **they might come and rob me or beat me**, because of taking their money. If there was a bank somewhere, where I can save my money, I supposed to have them somewhere already. Now, I used to keep the money until the.. when the truck will come here next week or after two weeks, to send the money to go and buy the goods. If there was a proper way of saving my money, **the business will also grow.** Because I can know that I will save this and then, I will keep this and I´d put this in stock and I´ll do this. Now, I´ll keep the money for the whole bank for two weeks, with the money here. Now, also, check how the house itself. Even to go to the bush, now you see. There is no toilet here, in Gam. Because we try, even if when you are going to decide that you are going to find corrugated iron. I wanted to make it but because of this stone. This stone, which you can see here, to dig is problem. **If the government is going to provide tools, water...** Now, **I think electricity is better, is finished**. Water and other machines, **whether we can hire for trailing or we can make our toilet**. Now, how many kilos (km) should I go to the bush? **Maybe 2 km, to go to the bush**. **Later on, have to come back also to attend to work**. Means that also that one contributing to some time because when the customer is here, I am not around. Because I went somewhere" (Interview 3-G, Pos. 32).

## Business Opportunities with the Electricity as a Pre-condition

The following quotes offer different business opportunities. At the same time, nearly all of them state the reliability and availability of electricity 24/7 as a precondition for having such businesses in the town they are designing. Interviewee two, in the quote below, describes that if CENORED does not improve its operation and provide constant and reliable electricity, it cannot start new businesses, such as restaurants or supermarkets, and selling cooked food.

> "There are supposed to be other business opportunities; there are. But **because the lack of this electricity, unless** when **CENORED upgrade the operation** of the electricity, that **will be very good for other businesses**. Even we want to start the business of **restaurant or mini-market** or what is the need of it? Because later on, the electricity will be going off. So, **first, we have to upgrade the electricity**, so that **it can work 24 hours without any problem**. **Then, we can start other businesses** also. (…) She was talking of starting a restaurant where she starts selling cooked food" (Interview 2-G, Pos. 27-30).

Interviewee 10 expresses similar concerns about the unreliable, fluctuating electricity supply. He thinks that in Gam, there should be banks, a post office, and shops like Shoprite, which would be very useful for the community.

> **"If, the electricity is constant**, you see, because the people here, you see, they are farmers, most of them, and if they are talking to farmers, **they are having money**. You see Gam, **we didn´t have a bank**, **we didn´t have a post office**, and sometimes it is… There must **be shops, Shoprite**. I think with the electricity**, if electricity is constant**, it´s not running up and down, **I think that things will come in and it will be helpful for the community**" (Interview 10-G, Pos. 22).

Interviewee one brings up butchers, dry cleaning, and wholesales as promising opportunities for Gam Settlement. He explains that many unmarried men need laundry and dry-cleaning services in their daily lives. He also sees the supply of shops as problematic, and a wholesaler in the settlement supplying all shops would also be very beneficial.

> "Hmm, even when you open a business **of a butchery**. There is an opportunity here in Gam for you to open a butchery here. Even **a dry clean.** There are many mans, **not married, they are not washing by themselves**. You can open a dry clean. hmm, you also need, coming to **the supplier of the shops**, what do you call it? hmmmm ... The supplying of our, the delivery of a stopping a shop is also very poor. We are hoping if it would have **a wholesale here**, that can supply to the shops also" (Interview 1-G, Pos. 60).

The business opportunities provided by the interviewees belong to the characteristics they envision in their future town, including specialized services like banks, big shops like Shoprite, dry cleaning, wholesales, restaurants, and butchers. They are aware of their needs and how they could meet them. However, they also admit that the lack of stable, constant electricity is a hurdle in achieving their goals as a community. Thus, all these visions contribute to the process of rationalization.

### 5.2.3 Domestication of "Energy Practices" of Everyday Life

> "**Very much** because before without **the mini-grid life** was **just impossible**, we could **do very less**. Now that we have electricity, **we can connect lots of appliance** and even expand **or upgrade our small business** to the standard level of a **modern town life.**"

The short response to the open-ended questions above describes precisely how the adaptation has occurred in Gam with the start of the "mini-grid life." It combines the process of rationalization in becoming more productive by doing "more things," upgrading small businesses, and obtaining "modern life standards." Through electricity and appliances, community members are now doing the things that were impossible in the earlier days of Gam. On the other hand, electric appliances also cause technical issues due to fluctuations and interruptions in the electricity supply. Thus, their appliances are damaged and not usable anymore.

> "Lots of challenges. This electric ... that we bought, we don´t know when are they going to last. Because it switch off any time without even noticing that at this time **you have to remove your things**, because the electricity will go. **We might even lose our things that we thought we might use for some years**, because of the switching off of the electricity. Really, that one, it´s a challenge" (Interview 6-G , Pos. 12).

With rationality in mind, they buy appliances that will last several years. However, they are usually challenged with technical complications they did not know earlier. This results in broken, unusable appliances and lost money for them.

#### *5.2.3.1 "Energy Practices"*

As indicated and explained in Chapter 4, local energy practices have also been classified according to the three-element social practice network concept by Shove et al. (2012, p. 14), which includes materials, meanings, and competences. Based on this practice concept, the following energy practices have been identified, explained, and visualized in Figure 5.9 for Gam:

*Figure 5.9. Energy Practices in Gam*

*Source: Author*

## Lighting

The lighting practices have changed many lives in Gam. They are not buying candles, diesel, or paraffin anymore. "My life has become now. I never had lighting and used candles, paraffin, and lanterns, but now I have electric light". Another respondent confirmed with "No more using candles and lanterns".

> "There **is no more need to buying diesel/paraffin** to use for lighting,"
>
> "I got electric lights and **don't need to buy candles on a regular basis**"(Quotes from open-ended questions [QOQ]).

The mini-grid provides electricity for the community members who have connections, and the unconnected households have indirect access to some services. The respondents expressed their satisfaction with the lighting for its availability 24/7: "I have 24 hour-lighting now!"; for having lights indoors, "we have light inside," and "It has changed the lives, the system of drinking cold stuffs and lighting in our houses." As a result, they are very content with having lights for the community: "We are happy because it gives us light in the community."

The competencies mentioned above signal the availability of electricity and having light for their places, linking it to the process of individualization while continuing the community practices. They also find electricity very helpful for their lives. The quote below highlights a prominent aspect of having "street lights," introducing a meaning of night safety in the streets of Gam. People can walk without having anyone else following them, and they can see this with the streetlights.

> "Electricity mean to me.... is....the electricity, for us here, **is light.** It´s **very helpful**, when **we are having some street lights here**, which is sometimes**, help to combat, people who are doing walk things, you can walk at night, without having someone behind you,** something like that" (Interview 10-G, Pos. 9-10).

The aspect of "light" and lighting of the households and businesses have appeared in this research. However, street lighting, being helpful for them, combating, and ensuring that they are safe in the streets is an aspect that does not directly appear. During the PROCEED Final Conference (2023), the community members brought up this issue as a complaint of not having streetlights on at night. After some consultation, the team was informed that the expenses of the streetlighting were on account of CENORED. Among other operational issues, this was critical for the safety of people in Gam. As a result, it has been reported during the conference discussions.

## Communication Practices

Mobile phones and televisions are the most used communication items in Gam. Mobile phones include smartphones and related digital applications as well. MTC telecommunication company has strengthened its telecommunication and network capacity over the past few years. During the first field research in Gam, making a phone call over an internet connection to send and receive internet-based messages was nearly impossible. However, the coverage was extended to the second field research in 2022, which enabled us to use it continuously. In 2023, the connection improved; even phone calls could be made with the internet. However, the community members stated that the network coverage has also fluctuated depending on their use.

One of the first competences is the individual charging mobile phones in own-household. "I can charge cellphones **at my own place,"** which is followed by using it to communicate with people. The quote below indicates how a person liberated oneself, needing other people and neighbors to use the mobile phone.

> "I have **my own electricity** now and can charge my phone at home instead of sending my sister to charge elsewhere" (QOQ).

The life-changing impact continues with the television and how they can follow the news, hear the happenings elsewhere, and connect to the country and the world at the same time: "I can watch news."

> "My life changed in a sense; I **can connect TV** and **watch daily happenings**; I can **charge my phone and communicate**" (QOQ).

Furthermore, television has also been used as an educational material beyond the meaning of individualization. "It **is like I'm now in Windhoek**. Also, my kids do **learning from TV**". It is an educational tool for children to learn, making them consider themselves as they live in an urban place like Windhoek. In this way, the similarities between the urban and rural/peripheries rise while the differences diminish.

Therefore, communication practices enable the community to attain town-like conditions in Gam. The quote from interview six strengthens this view with a more detailed description of a teacher's daily life. She explains events from searching for information on the Google database to charging her phone, cooking in an electric pot, using the electric appliances she would like, and watching television. She depicts this life as town-like, considers herself a person in town, and interpretes everything is normal.

> "….I am a teacher. Because I can even, I mean, I mean, **because of the electricity**, **I am able to use my phone the way I want**. Meaning that, I can even **google** the way **I can teach, using my phone**. During that time, **it was even difficult to charge your phone**. When you come for cooking, **I am able to cook**. You can see that spot, **electric pot**. Now, I am able to use many things that.. **I am like a person in town**. You can see **that I am watching television**, **everything is normal**" (Interview 6-G, Pos. 10).

The "normality" has been defined by the availability of tools (materials), by the way they have been integrated into her everyday life (competences) in work, cooking, communication, charging her mobile phone, and with the meaning she perceives by living in an urban-like condition. This quote presents communication practices and emphasizes the overall energy practices in Gam, in the periphery of Namibia.

### Cooking Practices

The traditional way of cooking is with firewood collected from the bush. Following up the electricity connection, firewood becomes an alternative cooking fuel: "I can cook on stove **less need of** firewood," "I don't have to fetch firewood **daily,**" in addition to the electric stove or microwave. People no longer need to collect firewood or spend time and effort cooking. Instead, they prefer using an electric stove, cooker, or microwave. "We are happy because we can watch TV and drink cold water. Cooking became easier with **no more firewood**". Although these electrical cooking appliances do bring comfort to people, the high energy consumption threatens the mini-grid's operation, as highlighted earlier. On the one hand, these high consumptions occur at the peak load of the day, like in dinner hours, increasing the necessity for battery use and/or generators, which might also cause supply interruptions due to the (over-) loading capacity. On the other hand, this is reflected in damaging overall appliances with added struggles that one needs to cope with, such as unfinished cooking, an entire family waiting for their meal, or a fridge that cannot store the meat for extended periods and eventually got rotten.

> "Challenges... A LOT of challenges. Not a lot of challenges but it´s only when challenges... Because when electricity it goes... and went down, so, whereby we are expecting, no… while I´m checking at my TV, electricity is off. You don´t know what damages it will be caused to my fridge, my Tv, my microwave, everything which was on, **you don´t know what will be the damages**, yeah… But that´s the main problem always. **While you are busy cooking, the electricity is off, without no... while you do not expect the electricity to go off.** Yeah… That´s a very big challenge, here" (Interview 1-G, Pos. 26).

An overarching meaning for multiplying cooking practices is "no more suffering" when everything functions properly: "We are happy because we **don't suffer anymore** for cooking." It is an important energy practice, which is in the course of transformation from traditional ways towards modernized cooking with the use of electric appliances.

### Cooling Practices

In Gam, household fridges, freezers, commercial refrigerators, and commercial freezers have been used for food storage and cooling. On the other hand, air conditioning has been used for space cooling.

Household fridges and freezers enable people to drink cold water: "I am drinking cold water and ironing in my own house. It changed my whole life," which has a life-changing impact for the respondents. Another respondent confirmed this: "I can connect my fridge to have cold water. I used to lose my animal carcasses due to the lack of a freezer. Now I can freeze my meat". He does not need to worry about the meat anymore because he has his freezer and can store not only meat but animal carcasses, keeping the meat fresh longer. "My life changed in a sense that I can **keep my meat** fresh **longer**". The quote from interview two explains the changes after the mini-grid connection. They used to use fridges with gas. However, electricity remains expensive for several refrigerators for business purposes.

> "When she started there was no electricity. And almost, how many years ago, since she started operating here, she started operating here in 2009 or 2010 around. But there was no electricity at that time around. So, after that, three years from there, the electricity came. Then, she started using electricity. She start **buying fridges to put the cool drink and other stuffs**, so whereby she sees there is a little bit change, comparing from the time when she started operating here. yeah.( …) That

> time when we were not having electricity, **she was using these fridges that are using gas**, and secondly, when she started using the electricity a bit, what she realizes **that the electricity is too expensive**, the units that we are buying, they are not enough but it´s expensive" (Interview 2-G, Pos. 3-6).

Moreover, another business that uses freezers, aside from keeping the meat, is making ice blocks. Interviewee nine has found a way of generating income by selling these ice blocks, which has also contributed to her life; this was not possible for her during the unelectrified times.

> "Like for me, I have changed a lot also. Because now, I am selling the ice blocks, in this. Because of the electricity. If there was no electricity, I couldn´t use it. Even this smug (?) relief, we use it because of electricity. Even to cook, we used to cook here" (Interview 9-G, Pos. 10).

During the PROCEED Final Conference (2023), the community members from Gam articulated another dimension of the lack of stable electricity for their mortuary. The challenges related to this have neither been "visible" nor expressed beforehand. They have explained their concerns about the interrupted and unreliable electricity and when they do not know when electricity will be "on" again. This situation results mainly in the early and sudden funeral services that need to be handled in place.

The cooling practices using cooling devices do not only develop the process of domestication and the use of electric appliances to store food and water or cool the space. Furthermore, they also have their own degree of individualization when people practice them in their private places.

### (Space) Heating

There have not been any heating practices recorded either in the interviews or in the open-ended questions recorded.

### Leisure/Entertainment/Enjoyment Practices

As indicated earlier, in communication practices, television is a part of everyday life as a means of communication to stay connected with the world and to receive news. However, one of the teachers explains below how it is used as an assisting hand for herself and her students. The teacher also uses it to improve her skills in teaching "singing." The students can learn from television. Moreover, television is used in the hostel during study time, after school, for studying, or for leisure. When there is no electricity, they are simply left in darkness.

> "They are growing. Because some of the things we are taking them, we are using the TV to show how to sing. Because I am not good at singing, so **I can put the TV, then they look, they observe how**. Other teachers can teach. It´s a way they are helping me. Other teachers are

> helping me because I don´t know what they are doing. Even **reading**... The **study time**, even the hostel day. People, they are suffering. Because the electricity is switching off any time. Then, I am left in darkness. So, at school, it´s affecting me a lot. During the test time, we are testing the learners. **The electricity is off**. It´s **an obstacle** also" (Interview 6-G, Pos. 23).

Television has invaluable meaning in leisure, communication, and teaching, as indicated. It enables an uninteractive connection between the world and the people in Gam but helps them contribute to the missing areas of their lives instead of sitting and staying in darkness.

### Sewing, Ironing, Welding

**"Just press the button and the things going on!"**

Welding machines, irons, and sewing machines were mentioned during the interviews as electric appliances used for additional energy practices.

The Herero community has been known for its traditional costumes, which require much fabric, and consequently, sewing or tailoring has been one of the critical occupations in the settlement. Interviewee eight describes how his wife can use her sewing machine to sew more clothes. She has developed it as a business, and her job got easier with the machines: "Just press the button and the things going on." Notwithstanding, tailoring, washing, and ironing are part of this bundle of sewing practices at home and in business.

> "We were struggling a lot before doing things. **My wife can use her sewing machine sewing more clothes**. We can connect **more appliances** for more **household needs**. Just like that. **even she has a business for sewing**. It means that right now she can use it **for machines, which is very helpful**. But the problem is that, it´s not like for the whole day. Sometimes you can say ...but had problem with it. (….) Sewing, yeah. It´s fine. It´s going well. **The job is very easier**, first, the old one with the Singer one, we can do things whole day. With the machines, **just press the button and the things going on"** (Interview 8-G, Pos. 6-14).

Interview 10 below highlights different bundles of practices of everyday life, including welding as another energy practice.

> "Electricity, **we are using for many things**. **Sometimes we are welding**, we are having freezer. We are watching, even the war between Russia and Ukraine, we are watching it on TV (laughters..) other kind of things, many things" (Interview 10-G, Pos. 12).

As a result, these energy practices outlined above have yet to be used individually. They have been used in bundles as a cluster of practices together. During the field research, many (business) women complained that they could not iron their clothes at their place; they were doing it in their shops. Therefore, although technical and economic constraints with ironing should be considered for the best operational practices of mini-grid, there appears to be a conflicting situation with the way of life of the Herero, especially women´s traditional clothing and its care; this would need to be explicitly addressed.

Furthermore, the local energy practices addressed tend to contribute to individualization, to the individual lifestyles at their own houses, that need to be considered within the planning.

#### *5.2.3.2 Visions in Domestication – Town-like Services*

The quantitative section will elaborate on the quantification of the wished electric appliances and their categories. Both of the quotes below depict town-like services that need to be developed and established in Gam. Some of these services were also mentioned in the process of rationalization. Apart from these services, Interviewee Nine´s vision is to have a fuel station, a bank, and an upgraded hostel to accommodate more people.

> "Yeah, like, in Gam, **we are looking for this service**, of **petrol or diesel** and what. And the banks, and what else? And the hostel, **a proper hostel here**. We have a hostel here but that one cannot accommodate the whole village. That´s all" (Interview 9-G, Pos. 16).

Interviewee four mentions the businesses needed in Gam, such as textile shops, bakeries, banks, and post offices, which will develop the conditions of their settlement. Thus, their life.

> "hmm a **lot of businesses are needed** in Gam, here. Such as **clothes, bakeries**, **everything,** everything. Most of them, they are needed here. For, in case of investing money, we need also **FNBs**, a lot of **banks** around here. Even **Nampost,** for those who are… we need a lot of businesses" (Interview 4-G, Pos. 15-16).

These visions of a town with its infrastructural and service upgrades are known to the community and expressed freely and intentionally. This is beyond a wish, a need-based expectation that might possibly be expected to be met by the residents or government.

### 5.2.4 Process of Individualization

In the previous sections of rationalization and domestication, the intersecting points of the process of individualization and the life-worlds of the residents of Gam have been explained. Within these processes, one can see how the individuals have been individualizing and shifting from the collectivity in certain everyday energy practices that they are living by also keeping community unity; this would suggest that although they have been experiencing a narrowed-down household life surrounded by the electric appliances they find beneficial for their everyday lives, they also envision themselves within the community and belonging to this future community as a whole. The quotes selected for this subsection outline independence, autonomy, and ownership.

#### *5.2.4.1 Autonomy*

The quote below from interview six has already been discussed in the previous subsections, which outlines not only the process of domestication but also the process of individualization with independence, autonomy, and a lifestyle like that of an urban area. Besides the electricity services, interviewee six can use her phone how she wants: "I am able to use my phone the way I want." The "I" emphasizes the individuality and freedom of doing this action with "the way I want" while making oneself autonomous for her work. Furthermore, since she works as a teacher, the time she has for cooking traditionally with firewood is limited, and she thus prefers the electric pot.

> "…I am a teacher. Because I can even, I mean, I mean, because of the electricity, **I am able to use my phone the way I want**. Meaning that, I can even google the way I can teach, using my phone. During that time, it was even difficult to charge your phone. **When you come for cooking, I am able to cook**. You can see that spot, electric pot. **Now, I am able to use many things** that… I am like a person in town. You can see that I am watching television, everything is normal" (Interview 6-G, Pos. 10).

When she reflects on her everyday life and the appliances she uses, she compares urban life with the availability of many appliances: "Now, I am able to use many things that… I am like a person in town". She watches television, and "everything is normal". The definition of "normal" indicates "conforming to a standard, usual, typical, or expected." In this sense, the "normal" conditions refer to normality with the possession and utilization of the electric appliances that are used in towns. Furthermore, with the freedom of using the appliances they prefer using, they break away from their dependency on others or other tools or discover new ways of finding/reaching information for their use or receiving information from the events that are going on in their country or in the world they live in; this also indicates a transition from analog searching methods and tools such as books and libraries, which they cannot reach in Gam, at their premises, towards digital searching/learning methods to fill this gap. This situation also demonstrates the intricate conditions whereby the lack of infrastructure shall be filled in with the digital tools and devices available.

#### *5.2.4.2 Ownership of Electricity "My Own Electricity"*

In line with independence and autonomy achieved with electricity in several parts of everyday life, the notion of ownership becomes visible in words. The quote below illustrates that, prior to the electricity in their premises, the interviewee already had a phone. One could argue whether this was one of the first ownerships or not that prompted individualization by linking the person to other people directly, by sending/receiving money, and being reachable. Although the Herero community has strong communal ties, with the electricity connection, they access light at their place, the electricity to charge their phone, becoming more independent from others and developing a sense of autonomy.

> "My life changed in a sense that I was using diesel to illuminate my house; now I have electricity light. I used to charge **my** phone **elsewhere**. But now I can connect in **my own house** using **my own electricity**" (QOQ).

With electricity, they also have access control of their possessions and may use them as self-expression, which was not part of this study. Furthermore, its contribution to self-esteem, with its link to Maslow´s pyramid, provides further autonomy on an individual level.

### *5.2.4.3 Wishes for Private Businesses*

Individualization continues as individualized visions for having their "own" businesses; this has been closely related to the process of rationalization in a goal-oriented approach. This was witnessed already in the first field research to Gam in 2020 when people heard that there was a researcher in the field of electricity. Regardless of the fear of Covid 19, people, one after the other, each of them was uttering their will for an electricity connection; a lady was running after, "I want to have **a bakery**, I want to have **my own business**," or another lady, "I **want to iron my cloths** at home," or others were complaining about the connection application that they had made one or two years ago and that they had not heard anything back. A goal-oriented commitment to having one´s own business, a specialized business such as a "bakery" lacked the electricity to run it. Another woman expressed her willingness to iron her clothes at her place, which was again specialized in their interest for their purpose and lacked the electricity to do so.

On the other hand, in interview nine below, the respondent would like to have a catering business to serve the visitors coming to Gam. This range of business also specializes in the gastronomy business. Currently, the food shops in Gam also sell cooked food. In this case, there appears to be an interest in selling cooked food by bringing it to the visitors as a form of specialized activity within the branch of hospitality that contributes to the process of differentiation.

> "Interview 9-G: **For myself**, ah, for myself, I just want to open **my own business.**
>
> O: Okay, which is?
>
> Interview 9-G: Which is, like I **want to open** the business **for catering**. I like the catering.

O: Okay nice. So, you will do this catering for all the people in Gam or for very specific groups of people? For whom will you be doing this catering?

Interview 9-G: Yeah, **For the visitors**" (Interview 9-G, Pos. 17-22).

Furthermore, interview seven highlights a healthcare and elderly service in Gam as a potential work for herself that is in close connection with community service. In this way, even though she is willing to have an individual job, she remains within the community boundaries. On the other hand, the need she foresees for taking care of the elders might also signal a change in the community, signaling a transition towards a process of individualization, whereby the elders need "other" people than their immediate family, simply because they are not available anymore, or have other duties to attend. From an individual perspective, the respondent may be willing to have a socially and officially "approved" job identity, enabling her social acceptance within the Gam community. This situation would also translate to rationalization by contributing to the bureaucratic administrative affairs while making it a differentiated part of the community, along with individual employment.

"For myself, for me... **I want to be a nurse**. Especially here, around. So, that´s what I am working on. (…)

Interview 7-G (3): (…) [Ovaherero translation]: she wants to **care for the elders**… once week or so. Even **houseworking**" (Interview 7-G, Pos. 47-51).

### 5.2.5 Paradoxes of Modernization

As indicated in the earlier chapters, the paradoxical part of the differentiation process appears as a scale reduction and enlargement. It allows new formations in society to be divided into specialized services or groups of the original group, re-organizing according to their needs. In this sense, in the case of Gam, Gam Farmers´ Association (GFA) and Ondjou Conservancy (OC) have been formed to carry out specified tasks within the community. Gam repatriates, coming from a pastoralist and cattle herder background, are represented under the GFA on a local, regional, and national level. At the same time, Ondjou Conservancy is the lead CBO for the region in establishing the management of the community land.

Furthermore, from an ethnic perspective, although the Gam population is nearly 95% majority Herero from Botswana with Namibian origin, there are four Traditional Authorities, one for Ju/'hoansi and three for Herero. This scale enlargement by creating smaller, specialized, dedicated units of an organization also suggests a goal-oriented approach to their specialization. Thus, remaining within the same community contributes to the differentiation process.

On the other hand, the "narrowing of vision" of the independent entities in achieving specific purposes while "widening of vision" belongs to the paradox of rationalization. In Gam, the paradoxes appear with the OC, or GFA, in pursuing their own goals with nature conservation, improving and supporting farmers´ activities, and people owning different specialized shops to generate income. On the other hand, although they specialize, they keep a vision of a town in the future with the multiplication of services. This situation also contributes to the pluralization of services or activities.

Furthermore, the paradox of domestication creates independence from the natural forces or physical barriers of people through the use of electric appliances in Gam while making people dependent on these technologies and electricity. As a result, since everyday life becomes more accessible, and easier, they tend to acquire more appliances, which need more energy and more connection points for the mini-grid. The vision of becoming a town and having access to the services provided in cities also requires more power, in which the current mini-grid demonstrates constraints. Thus, CENORED takes demand-based management measures in order to control and balance these demands arising from the Gam community. In long-term planning, considering the tendency of growth and development in Gam, the operators and owners of the mini-grid would need to take future-oriented visions according to the visions of the communities since the process of domestication under these conditions is irreversible, as the process of modernization itself.

The interrelatedness of the processes is linked in multiple ways, including individualization, rationalization, and differentiation, but all have their unique influence on humans. On the one hand, independence arises while changing its form to different dependencies, such as technology, administration, or other people, as in the case of elderly care. On the other hand, a human who is a part of a traditional societal structure under leadership as a subordinate transitions to an equal, impersonal position under a bureaucratic, hierarchical structure while self-experiencing a first-equal citizen. From the detachment of traditional ties, individualization starts by confronting them to the public structures. Thus, this process also demonstrates how independence from a specific type of relation, obligation, or struggle presents and how it creates a dependency paradoxically.

## 5.3 Quantitative Findings

This section presents the quantitative findings of the surveys, the data collected from the community, the libraries in Germany and Namibia, and the Namibia National Archives.

### 5.3.1 Quantified Local Setting

#### *5.3.1.1 Population in Gam*

Looking back nearly 30 years ago to the beginning of life in Gam for the repatriated Herero, including the Ovambanderu clan, sheds light on the present issues and the ongoing growth and development that have been taking place. Before 2000, Botelle and Rhode (1995, p. 197-198) recorded 60 inhabitants in 1980 and 139 in 1991. The repatriation of Herero from Botswana increased the population to 2000 in 1993, with the arrival of the first group in April 1991, 1171 people, followed by a second group of 544 people in September 1993. Gam was the first arrival point of the repatriates, from where they could also move to other parts of Namibia after living in a "tent city" with 700 tents. Half of the new Gam settlers expressed interest in remaining in the Gam area and its surroundings, whereas the other half thought about moving to Okahandja, Otjiwarongo, or Omaruru (Botelle & Rhode, 1995). Otherwise, the government had no concrete plans for developing or evolving this repatriation program (MLRR, 1993).

The census in 2001 recorded the population of 1161 inhabitants. The author assumes that the decrease in the population is due to the people leaving Gam to move to other places. The next census, held in 2011, recorded 1229 people. Since the planned census could not take place in 2021, no recent census data is available. However, the National Statistics Agency (NSA) calculated an estimate for 2021 and 2031 based on the population growth data recorded in the previous years. According to the estimation in this projection, the population in Gam has been estimated as 3195 for the year 2021 and 5162 for the year 2031, which could support understanding current energy practices and consumption as well as developing future scenarios. Starting from these figures and based on the data provided in the repatriation report, it is also possible to assume that the children at the time of repatriation have now become adults aged 31-40 years (37,49%) and 41-50 years (30,83%). Moreover, the following table estimates the household level based on data from the 2011 census, where the average number of people per household was calculated at 4.19. This figure has been, therefore, used to estimate the total housing in Gam for 2021, 761.70, and for 2031, 1230.65 households (Table 5.4).

*Table 5.4. Gam Population and Household Estimation**

| Years | 1980* | 1991* | 1993* | 2001 | 2011 | 2021** | 2031** |
|---|---|---|---|---|---|---|---|
| Population | 60 | 139 | 2000 | 1161 | 1229 | 3195** | 5162** |
| Total Housing | | 25 | | N/A | 293 | 761.70*** | 1,230.65*** |

*: Botelle and Rhode, 1995 (p. 197-198). **: Estimated data provided by NSA in 2022.
***: Average number of people per household, calculated as 4.19 (2011).

*Source: Author*

Thus, starting with these present estimations and the historical data, Table 5.4 was prepared to illustrate the population growth over the past 30 years. After the fluctuation in the first years of repatriation, the existing census data and the population estimation suggest an increase of 159.97% from 2011 to 2021 and an estimation of 61.56% from 2021 to 2031. In the PROCEED Conference (2023), Gam community members announced that 70 families would go from Botswana to Gam upon completing their administrative work in 2023. This increase over two decades indicates population growth unless there is any Force Majeure.

#### *5.3.1.2 Social Stratification and Division of Labor within the Local Actors in Gam*

The growing population in the past decades has also formed a division of labor and social stratification within society. The description of different groups within the Gam community has been shown in Table 5.5. Public institutions, national and regional representations, churches, community representations, and local private shops constitute different strata within the Gam community. Currently, the following public offices are situated in Gam: the police station, the clinic, the OTA Traditional Authority, Kambazembi Royal House, Ju´/hoansi Traditional Authority, Ovambanderu Traditional Authority, the Gam Primary School, the Gam Secondary School, the checkpoint Veterinary Office of Nyae Nyae, and the office of the Ministry of Agriculture. The Ondjou Conservancy and the Farmer´s Association are active NGOs. The private shops are generally composed of food shops such as Savannah Market, Maharero´s Market, Come All Shopping Centre, Maria´s Market, Maharero William Market, Karumundu Shopping Centre, as well as the car wash, textile shop, car mechanics, welders, *shebeens* distributed around Gam (Table 5.5). As a result, there are five public institutions, eleven churches, two representations from the ministries, four Traditional Authority, two community representations, and local private shops whose quantity is unknown. There is no parastatal or national/regional company present in Gam.

The number of TAs suggests further ethnic Herero differentiation within the community and the public, private, and community-based services operated in Gam. The eleven churches that are active in the settlement should also not be underestimated. According to Google Maps' geographical search, there are ten churches in Grootfontein (2023), with a population of 23,793 (2011).

*Table 5.5. Local Actors in Gam*

| | | |
|---|---|---|
| **National and Regional State Representations** | **Public Institutions** | Gam Primary School,<br>Gam Secondary School,<br>Clinic,<br>Police Station, Veterinary |
| | **Ministerial Representations** | Ministry of Agriculture, Water, and Land Reform, Ministry of Environment, Forestry, and Tourism (to be appointed) |
| | **Regional and Local Representations (all belong to Ministry of Urban and Rural Development)** | Ovaherero Traditional Authority (OTA)<br>Ju/'Hoansi Traditional Authority<br>Kambazembi Royal House Ovambanderu Traditional Authority<br>Gam Settlement Office (to be appointed) |
| **Churches** | | Zion Council of Churches (ZCC)<br>Spiritual Healing Church, Oruno Church, Jesus Christ Church, St. Amos Church, Cross Healing Church, Heaven Church, Emmanuel Church, Pentecostal Church<br>St. John Church, United Congregational Church (UCC) |
| **Parastatals** | | N/A |
| **Community Representations** | **Community-Based Organization** | Ondjou Conservancy |
| | **NGO** | Gam Farmers´ Association |
| **Private Companies** | | MTC Tower - No official representation |
| **(Local) Private Shops** | | Fuel Station / General Dealer<br>Food Shops and supermarkets<br>Groceries, Butchers, Bar / Shebeens, Hairdressers, Gam Lodge<br>Car wash, Tailors |
| **Households** | | 762 (estimation based on the NSA data) |

*Source: Author*

### *5.3.1.3 Electrified Households and Businesses*

In 2017, after CENORED, took over the operation and maintenance of the mini-grids in Tsumkwe and Gam, they established an office in Tsumkwe with one person in charge and one security guard for the hybrid solar PV power plant. This CENORED employee used to be the only one responsible for the mini-grids in the region and has been attending to customer relations, mini-grid connection applications, technical failures, and other duties related to the mini-gird. In February 2022, a second person has been assigned to support the service. They have both been residing in Tsumkwe, none in Gam. As described in the qualitative section, the lack of CENORED presence in Gam causes an addressable set of issues when a responsible person has been appointed to the settlement.

Receiving the actual of number clients from CENORED has been complicated throughout the research process. However, courtesy of CENORED officials, the number of clients was obtained in June 2020 and June 2022. Boamah's (2020) field research in 2019 indicated that the number of collected clients was 214 for Gam and that the system was underused. No change was recorded during the author's field research in March 2020. The PROCEED project reports and discussions mentioned underutilization of the system. However, the entire settlement was connected to the mini-grid. The additional connection of 400 new clients took place toward the end of 2020 and the beginning of 2021, which completely changed the discussions around the "underutilization" context. These 400 mini-grid applications were already on the list, waiting to be connected for a very long time, a year or two years, and had been frequently mentioned during the field research in March 2020. Thus, the total connection reached 611 connections, of which 556 were residential, 51 were business-prepaid, and four were business-institutional, with 11 waiting applications. With this differentiation in the connection to the mini-grid, three different types of clients have been found with three different electricity tariffs in Gam.

*Table 5.6. Electrified Households and Businesses in Detail (2019 – 2023)*

<table>
<tr><th>Years</th><th>2019 (Boamah, 2020)</th><th>2020*</th><th>2022*</th><th>Increase in Electrification</th><th>2023**</th></tr>
<tr><td>Total Connections</td><td rowspan="5">214</td><td>214</td><td>611</td><td>185.51 %</td><td rowspan="5">N/A</td></tr>
<tr><td>Residential (Prepaid)</td><td>194</td><td>556</td><td>186.59 %</td></tr>
<tr><td>Business (Prepaid)</td><td>16</td><td>51</td><td>218.75 %</td></tr>
<tr><td>Business (Institutional)</td><td>4</td><td>4</td><td>0</td></tr>
<tr><td>Awaiting Applications</td><td>9 (7 prepaid, 2 institu-tional)</td><td>15</td><td></td></tr>
<tr><td colspan="5">Unelectrified Households and Shops</td><td>246 HH, 5 shops, 3 church-es + 70 families expected from Botswana</td></tr>
</table>

*Data received from CENORED in 2020 and 2022.
**Counting realized by the community members (PROCEED Conference, 2023).

*Source: Author*

Moreover, this increase meant a 185.51% increase in total connections, 186.89% in residential prepaid connections, and 218.75% in business-prepaid connections without increasing the mini-grids capacity. Thus, the interpretation of this electrification increases between 185% - 220% and would increase electricity consumption drastically while placing the mini-grid capacity and sustainability in jeopardy. As a result of this situation, after their connection, the load profile of the mini-grid changed, and the fuel consumption at night increased. Occasional blackouts started occurring in Gam, and manual operations have been realized by the CENORED employees, causing many complaints by the community in Gam.

In conclusion, counting the unelectrified places with a total of 255 connection points would provide us with 866 connection points, of which 802 belong to households. There must be more than fifteen awaiting applications, and eight businesses must be counted to provide exact figures. However, it is possible to calculate and state that the electrification rate in Gam is 69% only when considering households. Moreover, 802 households would also surpass the estimation provided by 764 households. On the other hand, the prepaid electrification for businesses would be 86% and for business institutions 100% (excluding the waiting for applications). Thus, the electrification rate in Gam is above Namibia's national and rural electrification rates, with 55.2% and 33.2%, respectively (WB, 2021).

**Survey Data – Sample Size**

Although there was a substantial increase in the mini-grid connection from the beginning of 2020 until the end of 2020, the surveys were conducted and completed under very restrictive conditions in 2020. Then, electrified clients were chosen to continue with the study. Accordingly, the quantitative data was analyzed based on the figures before the increase. At the time of data collection, there was no accurate information regarding the detailed structure of CENORED´s clients. Different pieces of information were brought together during the analysis, and the results can be seen in Table 5.6.

According to the CENORED data (2020), there are 20 business clients, of which 16 are prepaid and four are business institutional. The surveys represented 17% of the overall electrified clients in 2020, 14% of residential prepaid, 38% of business-prepaid, and 75% of business-institutional (Table 5.7).

*Table 5.7. Detailed Electrified Households and Businesses*

| Years | Surveys | Connection in 2020* | % of the Survey with total connection |
|---|---|---|---|
| Total Connection | 37 | 214 | 17% |
| Residential (Prepaid) | 28 | 194 | 14% |
| Business (Prepaid) | 6 | 16 | 38% |
| Business (Institutional) | 3 | 4 | 75% |
| Awaiting Applications | 9 (7 prepaid, 2 institutional) | | |

*Source: Author*

Due to the COVID-19 restrictions described previously, only a sample size of 28 prepaid households, which is more than 26 samples as per the MaCorr calculator (MaCorr, 2023), provided a confidence level of 90% with a 15% confidence interval (MaCorr, 2023). Under normal conditions, a 95% confidence level with a 5% interval would have been presented. Therefore, although the analyzed confidence level and interval have limitations, they still provide a comprehensive overview of the settlement. On the other hand, since the sample sizes of six prepaid businesses and three institutional businesses were small, the confidence levels and intervals were not calculated. However, some critical data have been shared in the quantitative section to provide an overview of the local businesses.

#### *5.3.1.4 Local Businesses*

The local prepaid and institutional businesses (n=9) surveyed were 44.4% public and 55.6% private. The total number of employees fluctuated between one to thirty employees. 100% of the surveyed private businesses responded to having one to five employees, whereas public institutions, such as schools and clinic, had more than ten employees, reaching up to thirty. The author also observed many women operating the shops in Gam.

#### *5.3.1.5 Households*

In this section, quantitative data from the households in demography, education, employment, and income distribution have been analyzed (n=28).

**Household Demography**

From the 28 electrified households that the surveys conducted, 71.43 % were used only as a household, and 28.57% were used for commercial activity. These business activities were sales of clothing, general dealers, or food and drink sales. According to the descriptive statistics utilized, the number of residents for the total households was 163 people, with 73 females (45%) versus 90 males (55%) division as well as 98 children (60%) with 65 adults (40%). The mean value for the rooms was calculated as 1.8214 (Table 5.8). However, 50% of the residents live in 1-room households. In comparison, 32.14% live in 2-room and 10.71% in 3-room (Appendix D, Table D.1). The average number of people per household has been calculated as 5.82, placing Gam with a larger household size than the average size in rural areas, with 4.6 people and 3.9 people in the Otjozondjupa region (NIDS, 2016, p. 89).

*Table 5.8. Descriptive Statistics for Household Demographics in Gam*

| | | Residents | Females | Males | Children | Adults | Rooms |
|---|---|---|---|---|---|---|---|
| N | Valid | 28 | 28 | 28 | 28 | 28 | 28 |
| | Missing | 0 | 0 | 0 | 0 | 0 | 0 |
| Mean | | 5,82 | 2,61 | 3,21 | 3,50 | 2,32 | 1,82 |
| Median | | 6,00 | 2,50 | 3,00 | 4,00 | 2,00 | 1,50 |
| Mode | | 5[a] | 2 | 3 | 4 | 2 | 1 |
| Sum | | 163 | 73 | 90 | 98 | 65 | 51 |

a. Multiple modes exist. The smallest value is shown

*Source: Author*

## Educational Background

The educational background of the respondents has also been categorized based on their responses to an open-ended question for their highest degree. According to the Namibian educational classification, 21.4% of the respondents do not hold any degree completion, 46.4% have primary school graduation, 17.9% have secondary school graduation, and 14.3% have tertiary graduation (Figure 5.10). With these rates, the educational background of the respondents represents slightly above the average with the national statistics (NIDS, 2016, p. 75) with no education and incomplete primary (18.9%), slightly below the average with completed primary (49.7%) and 22.6% completed secondary schools. As a result, 78.57% of the respondents have at least a basic level of education.

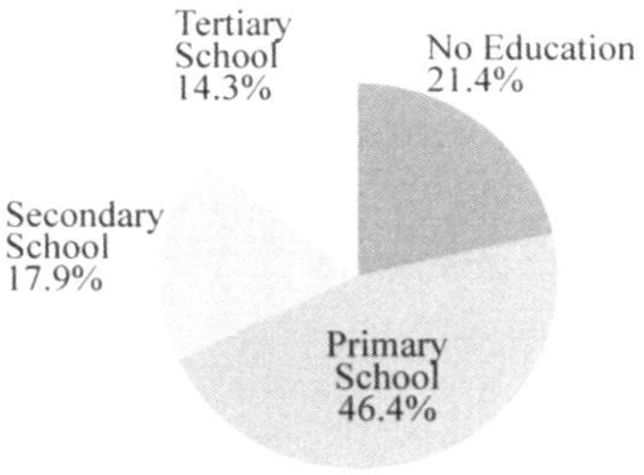


*Figure 5.10. Educational Background*
*Source: Author*

## Socio-Economic Background

The differentiation process with education has also been seen in the socio-economic backgrounds. Figure 5.11 displays that 46.4% of the respondents have been unemployed, 10.7% are state employees, 10.7% are private sector employees, 21.4% are self-employed, and 10.7% are retired.

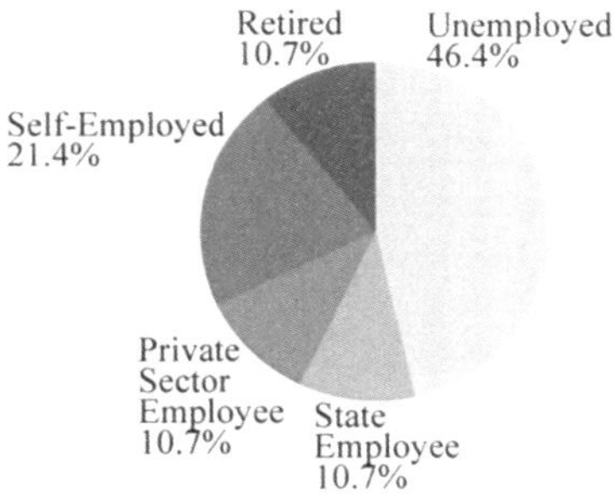


*Figure 5.11. Distribution of Occupation*
*Source: Author*

The unemployment rate (46.4%) has been recorded as more than double the national unemployment rate of 21.3% (WB, 2021). The state employees reflected here belong to the schools, the clinic, and the police station. The salaries and wages in Gam, which are 21.4% in total, are lower than the national statistics of the primary source of income, 53.6% (NHIES, 2016, p. 98). On the other hand, 21.4% of self-employed people describe the local ownership of shops or household businesses.

## Occupation and Education

Furthermore, in relation to occupation, Figure 5.12 indicates the distribution of different occupational categories with the educational background. All educational backgrounds appear among the unemployed cluster, with primary school graduates at 61.5%, 40% secondary school graduates, and 33.3% without educational background. 50% of the tertiary school graduates are private sector employees. Meanwhile, secondary school graduates are mainly self-employed (40%) or unemployed (40%). 50% of the people without any educational background are retired. As a result, it can be deduced that the educational background does not have a direct and absolute connection with the occupation in Gam. However, self-employment and private-sector employment are linked to educational backgrounds.

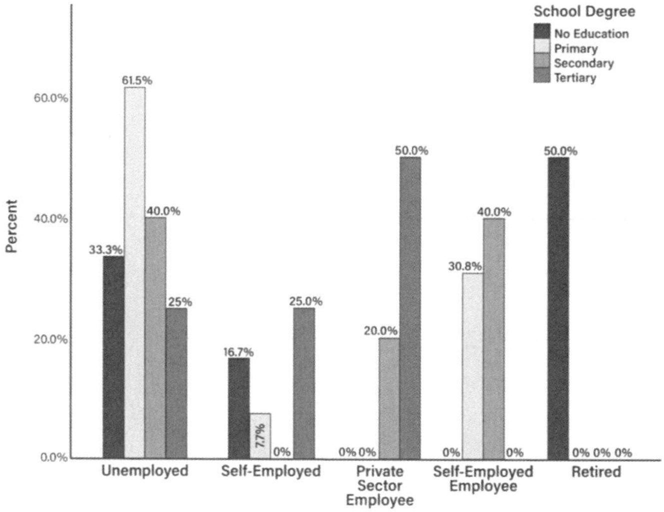


*Figure 5.12. Distribution of Occupation by Education*

*Source: Author*

## Income

As a part of the socio-economic conditions, the average income distribution for the households has been demonstrated in Figure 5.13. The average income range has been formed according to the minimum wage in Namibia at the end of 2019/beginning of 2020, being approximately 1,300 NAD. According to the findings (Figure 5.13), 50% of the respondents have an income range between 1-1,300 NAD, 7.1% did not want to respond to the question, 14.3% declared no income, and 17.9% responded more than 3,901 NAD.

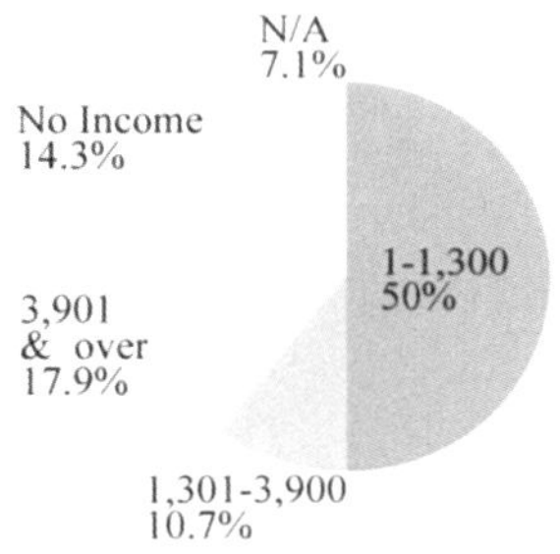


*Figure 5.13. Average Income per Household*
*Source: Author*

Furthermore, the mean income range was calculated as 1.25, corresponding to 1,301-3,900 NAD/month. These figures suggest a yearly income of 15,603 NAD – 46,800 NAD, which stands at the average income of Otjozondjupa of 34,897 NAD and the rural area average of Namibia of 27,792 NAD (NHIES, 2004, p. 36).

When the mean of adults per household for this study of 2.32 people is considered, this will result in a range of 565.65 – 1681.03 NAD/adult/month that is beyond the upper bound poverty line (UBPL) (NHIES, 2016, p. 104). However, 50% would still fall below UBPL.

## Income and Occupation

In continuation to the income ranges, Figure 5.14 illustrates that the lowest income range of 1-1300 NAD/month has been shared by 64.3% of the unemployed and 21.4% of the retired. The income range of 1301 – 3900 NAD/month has been evenly distributed among state, private, and self-employed employees. The income range of more than 3901 NAD is 40% for self-employed and 40% for state employees. This distribution suggests that nearly half of the unemployed have an income they do not consider a job.

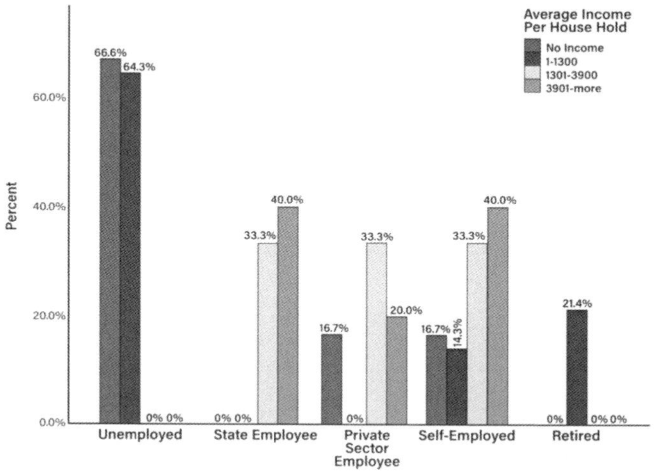


*Figure 5.14. Distribution of Income by Occupation*

*Source: Author*

## Ownership of Land, Livestock, Plot, and Car

In addition to the individual appliances, land ownership and livestock ownership have been investigated. Accordingly, 57.1% of the respondents have ownership of livestock or land, out of which 21.4% own both land and livestock and 35.7% own only livestock (Table 5.9). Based on the information provided, the livestock usually consisted of cattle, goats, and sheep. Of the livestock owners, 32.1% had small and big cattle, 21.4% had big cattle, and 3.6% had small cattle. Furthermore, the cultivated crops were maize, melon, cowpea, ground nut, calabasin, tomato, cabbage, and onions, with maize being the constant in any crop combination per household. This ownership structure suggests that although The Herero people are known to be pastoralists, they also started cultivating land after they settled in Gam.

*Table 5.9. Distribution of Ownership of Land and Livestock*

| | | Frequency | Percent | Valid Percent | Cumulative Percent |
|---|---|---|---|---|---|
| Valid | Land & Livestock | 6 | 21.4 | 21.4 | 21.4 |
| | Livestock | 10 | 35.7 | 35.7 | 57.1 |
| | No | 12 | 42.9 | 42.9 | 100.0 |
| | Total | 28 | 100.0 | 100.0 | |

*Source: Author*

When the crosstabulation was studied, there was a nearly even distribution (10.7%) of land and livestock owners among all educational backgrounds, with only 25% of primary school graduates (Appendix D.12). This distribution is also the same for average income per household (Appendix D.14). However, the occupation illustrates different distribution patterns, in which state and retired employees have small ratios of 7.1% and 3.6%, respectively (Appendix D.13). As a result, this suggests that the unemployed, private sector employees, and self-employed own land and/or livestock, regardless of their average income and educational background.

**Financial Aid**

In the quest to understand the financial situation of the households, their need for financial aid was asked. Accordingly, 64.3% of the households stated that they do not need any financial aid from family, friends, or financial institutions. 35.7% were affirmative to this question, which might be assumed to be part of the "unemployed" occupation cluster and might also belong to the no-ownership (42.9%) from Table 5.10 who need financial aid as a result of this analysis.

### *5.3.1.6 Continuity of Education and Wished Employment*

The respondents´ visions about their children´s education are classified as their future and their desired occupations. According to the surveyed households, there are 98 children, with a mean of 3.50 and a median of 4 children (Table 5.10). Of these, 60 children attend to school, which corresponds to 61.22%, with a mean of 2.14.

*Table 5. 10. Descriptive Statistics for School-Children*

| | | Children at School | Number of children per Household |
|---|---|---|---|
| N | Valid | 28 | 28 |
| | Missing | 0 | 0 |
| Mean | | 2.14 | 3.50 |
| Median | | 2.00 | 4.00 |
| Mode | | 1 | 4 |
| Sum | | 60 | 98 |

*Source: Author*

Moreover, 82.1% of the respondents added that they would like their children to pursue their studies depending on their finances, their success, and their circumstances (Table 5.11). Accordingly, the affirmative answers included continuing education in towns like Windhoek and Gobabis, and in institutions such as UNEMA and UNISA. This also shows the prominence of education in their life and the future they foresee. The fact that they already consider finance and success also provides another aspect of the ongoing rationalization process.

*Table 5.11. School Children's Potential Education in the Future*

| | | Frequency | Percent | Valid Percent | Cumulative Percent |
|---|---|---|---|---|---|
| Valid | No or N/A | 5 | 17.9 | 17.9 | 17.9 |
| | Yes | 17 | 60.7 | 60.7 | 78.6 |
| | Yes, depends on finance | 2 | 7.1 | 7.1 | 85.7 |
| | Yes, depends on success | 1 | 3.6 | 3.6 | 89.3 |
| | Yes, dep. on circumstances | 3 | 10.7 | 10.7 | 100.0 |
| | Total | 28 | 100.0 | 100.0 | |

*Source: Author*

**Wished Occupation**

From a future occupation perspective, 39.3% stated they would like a job, of which 21.4% stated that they would like to do any work. The remaining portion specified their future occupation wish as a cook, driver, teacher, trade, or own project (Table 5.12).

*Table 5.12. Occupation Wished in the Future*

| | | Frequency | Percent | Valid Percent | Cumulative Percent |
|---|---|---|---|---|---|
| Valid | No or N/A | 17 | 60.7 | 60.7 | 60.7 |
| | Any work | 6 | 21.4 | 21.4 | 82.1 |
| | Cook/Matron | 1 | 3.6 | 3.6 | 85.7 |
| | Driving | 1 | 3.6 | 3.6 | 89.3 |
| | My project | 1 | 3.6 | 3.6 | 92.9 |
| | Teaching | 1 | 3.6 | 3.6 | 96.4 |
| | Trade, fast food, clothing | 1 | 3.6 | 3.6 | 100.0 |
| | Total | 28 | 100.0 | 100.0 | |

*Source: Author*

### 5.3.2 Electricity Tariffs, Affordability, Expenses, and Supply Satisfaction

As mentioned earlier, electricity tariffs are differentiated within the interwoven processes of modernization. Table 5.13 below summarizes these tariffs, which were explained in the previous chapter. The government has subsidized these tariffs.

*Table 5.13. Mini-Grid Electricity Tariffs*

| Type of Client | Tariff |
|---|---|
| Prepaid Residential | 2.22 NAD/kWh (2.08 plus fixed charges) |
| Prepaid Business | 4.07 NAD/kWh (3.48 plus fixed charges) |
| Business Institutional | 2.60 NAD/kWh |
| Fixed Charges | ECB levy, NEF levy, LA surcharge |

*Source: Author (CENORED Correspondence, 2020-2022)*

#### *5.3.2.1 Households*

**Affordability of Tariffs and Perception of the Electricity Bill**

Considering the differentiated and subsidized tariffs (Table 5.13), 50% of the respondents stated that electricity is expensive. Whereas 32.1% mentioned it is affordable, and 17.9% found it cheap (Figure 5.15). This response is equivalent to 50% of the households with a monthly income of 0-1300 NAD/month. The data does not provide sufficient information about the salaries and wages income group, often mentioned in the national statistics.

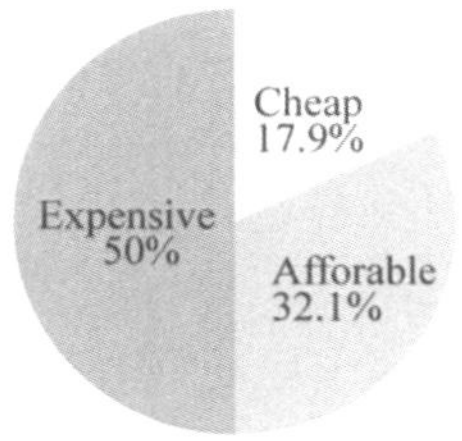


*Figure 5.15. Affordability of the Tariffs*
*Source: Author*

In addition to the affordability, 75% of the respondents stated that they do not know what they pay for their electricity bill. In comparison, 10.7% mentioned CENORED services, and 14.3% stated electricity and appliances. The recognized services of CENORED are essential to note that they do know that they are not only paying for electricity, despite the absence of CENORED office in the settlement. However, there is a need to increase community awareness and dissemination activities for mini-grid electricity, as was already realized before mini-grid installation.

**Electricity Expenses and the Perception of the Electricity Consumption in Units**

Moving forward, from the respondents who find electricity expensive, but they are not aware of what they pay, to the factual electricity consumption, which provides similar information as Figure 5.15 above. It's important to note that a significant portion of the respondents, 10 out of 28 respondents (35.71%) did not provide an answer to the question about the monthly electricity expenses (Table 5.14). For the respondents who answered the question, the mean of the electricity expenses is 253.06 NAD/month, with a mode of 50, 200, 300, and 600 NAD/month. Furthermore, when the known expenses are categorized, there is a range of 25% for the monthly expenses of 0-150 NAD, 25% for the range 151-300 NAD, and 14.3% for the range of 451-700 NAD with 35.7% without any response (Figure 5.16).

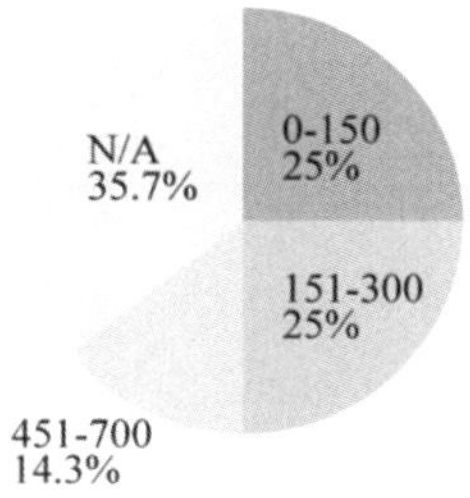


*Figure 5.16. Distribution of Electricity Expenses in Gam*
*Source: Author*

*Table 5.14. Descriptive Statistics of Electricity Expenses*

| | | |
|---|---|---|
| N | Valid | 18 |
| | Missing | 10 |
| Mean | | 253.06 |
| Median | | 200.00 |
| Mode | | 200[a] |
| Sum | | 4555 |

a. Multiple modes exist. The smallest value is shown

*Source: Author*

**Electricity Consumption**

The respondents also mentioned the units and the expenses. In this respect, the 200 NAD token would be equivalent to approximately 100 kWh/month, 300 NAD equivalent to 155 kWh, and 600 NAD equivalent to 300 kWh. The lowest consumption was 10.5 kWh, and the highest was 300 kWh monthly. In the latest report of the Africa Minigrid Developers Association ([AMDA], 2022, p. 52), the consumption in the mini-grids was recorded as 6.2 kWh monthly average, which they stated was lower than the 10 kWh average for the sites that were constructed before 2018. They made a remark that the new clients would consume less electricity than the longstanding customers, which brings an additional consumption tendency towards the design capacity of the mini-grids. Considering Gam mini-grid will complete its 10th anniversary in 2023, this also justifies the figures and findings that are reflected in the sub-section of domestication, with growing consumption patterns, placing Gam above the averages of mini-grid electricity consumption based on the report of AMDA.

On the other hand, the following tables indicate the distribution of the monthly expenses of the households with education (Table 5.15), income (Table 5.16), occupation Appendix, D.15), and rooms (Appendix, D17). Accordingly, the respondents who did not know about the expenses were the majority without education (10.7% of the total) or primary school graduates (17.9%). Moreover, primary school graduates had expenses of 0-150 NAD/month (14.3%) and 451-700 NAD/month (10.7%); this was followed by tertiary school graduates´ consumption of 151-300 NAD/month (10.7%). This scattered distribution of electricity expenses shows that primary school graduates do not have a single pattern in consumption. However, 75% of tertiary school graduates spend 151-300 NAD monthly.

*Table 5.15. Distribution of Electricity Expenses by Education*

| | | | Education | | | | |
|---|---|---|---|---|---|---|---|
| | | | No Education | Primary | Secondary | Tertiary | Total |
| Monthly Expenses | N/A | Count | 3 | 5 | 1 | 1 | 10 |
| | | % Education | 50,0% | 38,5% | 20,0% | 25,0% | 35,7% |
| | | % of Total | 10,7% | 17,9% | 3,6% | 3,6% | 35,7% |
| | 0-150 | Count | 1 | 4 | 2 | 0 | 7 |
| | | % Education | 16,7% | 30,8% | 40,0% | 0,0% | 25,0% |
| | | % of Total | 3,6% | 14,3% | 7,1% | 0,0% | 25,0% |
| | 151-300 | Count | 2 | 1 | 1 | 3 | 7 |
| | | % Education | 33,3% | 7,7% | 20,0% | 75,0% | 25,0% |
| | | % of Total | 7,1% | 3,6% | 3,6% | 10,7% | 25,0% |
| | 451-700 | Count | 0 | 3 | 1 | 0 | 4 |
| | | % Education | 0,0% | 23,1% | 20,0% | 0,0% | 14,3% |
| | | % of Total | 0,0% | 10,7% | 3,6% | 0,0% | 14,3% |
| Total | | Count | 6 | 13 | 5 | 4 | 28 |
| | | %Education | 100,0% | 100,0% | 100,0% | 100,0% | 100,0% |
| | | % of Total | 21,4% | 46,4% | 17,9% | 14,3% | 100,0% |

*Source: Author*

The respondents with an income range of 0-1300 NAD/month have distributed expenses in overall expense categories, with the unknown expense category having the highest ratio, with 21.4% of the total; this is followed by 14.3% of the income range of 0-1300 NAD/month in the 0-150 NAD category. The income range (3901-more) corresponds to 17.8% spending a total of 151-700 NAD monthly. The overall distribution demonstrates a tendency to state that the highest income levels spend more on electricity consumption. In contrast, lower income levels share lower expenses but are still distributed within all expense categories.

The unemployed category has 25% of the unknown expenses and shares the highest ratio of 14.3% of the total with 0-150 NAD/month; this is followed by 10.7% of self-employed people, with the highest expense category of 451-600 NAD. This situation and the overall distribution among different occupational backgrounds demonstrate that the unemployed have a certain income level to pay their electricity bills, and the self-employed have an income level that can afford to pay higher bills. (Appendix, D.15)Furthermore, 50% of the consumption in all expense categories has been realized in one-room households, followed by two-room households with 32.1 % (Appendix, D.17).

*Table 5.16. Distribution of Electricity Expenses by Income*

| | | | Average Income (NAD/month) | | | | |
|---|---|---|---|---|---|---|---|
| | | | No Income | 1-1300 | 1301-3900 | 3901-more | Total |
| Monthly Electricity Expenses | N/A | Count | 3 | 6 | 1 | 0 | 10 |
| | | % Income | 50,0% | 42,9% | 33,3% | 0,0% | 35,7% |
| | | % of Total | 10,7% | 21,4% | 3,6% | 0,0% | 35,7% |
| | 0-150 | Count | 2 | 4 | 1 | 0 | 7 |
| | | % Income | 33,3% | 28,6% | 33,3% | 0,0% | 25,0% |
| | | % of Total | 7,1% | 14,3% | 3,6% | 0,0% | 25,0% |
| | 151-300 | Count | 1 | 2 | 1 | 3 | 7 |
| | | % Income | 16,7% | 14,3% | 33,3% | 60,0% | 25,0% |
| | | % of Total | 3,6% | 7,1% | 3,6% | 10,7% | 25,0% |
| | 451-700 | Count | 0 | 2 | 0 | 2 | 4 |
| | | % Income | 0,0% | 14,3% | 0,0% | 40,0% | 14,3% |
| | | % of Total | 0,0% | 7,1% | 0,0% | 7,1% | 14,3% |
| Total | | Count | 6 | 14 | 3 | 5 | 28 |
| | | % Income | 100,0% | 100,0% | 100,0% | 100,0% | 100,0% |
| | | % of Total | 21,4% | 50,0% | 10,7% | 17,9% | 100,0% |

*Source: Author*

**Electricity Supply Satisfaction**

The survey results indicated that 82.1% of the respondents were happy with the mini-grid, and 17.9% responded: “Yes and no.” Among the affirmative category, 46.4% of the respondents were satisfied with its reliability and had no problems. 25% said they were happy with having light, drinking cold water, not buying candles, and using electric appliances (Table 5.17). On the other hand, the dissatisfaction was due to its high price, the damages to the electric appliances, and the inability to use the iron for clothes. As a result, their satisfaction is directly related to its reliability and the use of electric appliances, and their dissatisfaction is due to its price and technical difficulties they face in their everyday life.

*Table 5.17. Reasons for Supply Satisfaction*

| | | Frequency | Percent | Valid Percent | Cumulative Percent |
|---|---|---|---|---|---|
| Valid | No problems - Reliability | 13 | 46.4 | 46.4 | 46.4 |
| | Light | 3 | 10.7 | 10.7 | 57.1 |
| | Use of electric appliances | 5 | 17.9 | 17.9 | 75.0 |
| | Light & Appliances | 7 | 25.0 | 25.0 | 100.0 |
| | Total | 28 | 100.0 | 100.0 | |

*Source: Author*

### *5.3.2.2 Local Businesses*

**Affordability, Awareness of the Electricity Price, and Expenses**

In the differentiated tariff structure, local businesses have a different perception on the electricity price. Accordingly, 44.4% of the local businesses answered the question very expensive, 11.1% found it cheap, and another 11.1% expressed affordable. 33.3% did not respond to the question (Figure 5.17). This ratio is quite high considering the overall perception.

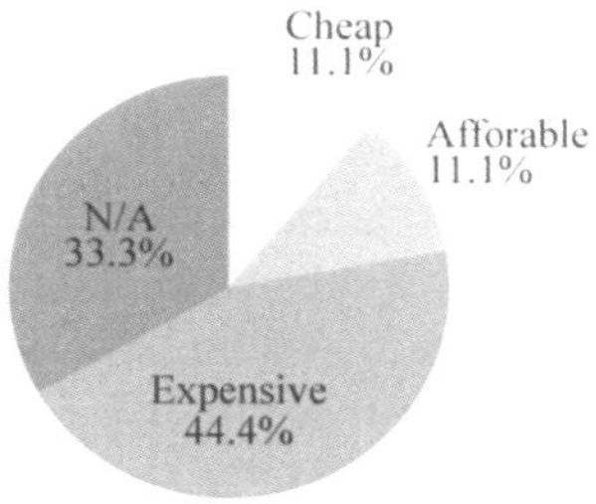


*Figure 5.17. Affordability of the Tariffs*
*Source: Author*

However, Figure 5.18 illustrates that 50% of the public businesses belong to this category, and the rest is distributed to affordable and cheap, which suggests that the respondents are not the payers of these bills. On the other hand, 100% of the local private businesses (prepaid) indicated that the electricity prices were "expensive" for them.

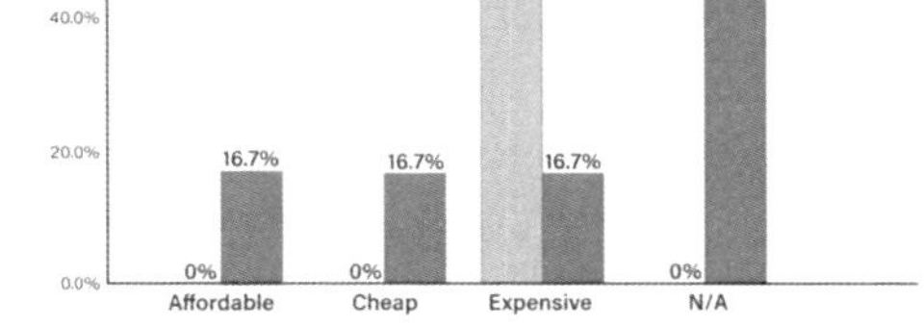


*Figure 5.18. Awareness of the Electricity Bill*
*Source: Author*

Unfortunately, the collected data is limited to fully describing the electricity units. The public clients did not know the bills since the regional centers directly paid it. On the other hand, only a few businesses mentioned the units. The lowest bill, with 200 NAD per year, is equivalent to 16.67 NAD/month (nearly equal to 1.6 kWh), and the highest for the prepaid is 234.5 kWh for 900

NAD/month. The lowest consumption would present the use of lights and phone charging only. The highest bill would also cover the fridge, freezer, and lights; this suggests that local private businesses know how much they pay. However, they are interested in something other than the technical details.

**Supply Satisfaction**

Furthermore, 66.7% of the respondents mentioned they were happy with the mini-grid electricity supply at the time of this survey, in 2020, before the connection of 400 additional points. Whereas 22.2 % stated they were unhappy, and 11.1% said: "Yes and no." The reasons for dissatisfaction are similar to those of households with technical issues related to damaged appliances, insufficient electricity to run several pieces of equipment simultaneously, and cost. Their satisfaction was due to the sales of cold drinks, having electricity 24/7 without interruption, and making things easy. This situation demonstrates that after ten years of using electricity, the business community has also accommodated to run the businesses with electricity. Despite its expensive nature, they still find it worth buying, investing, and buying more electric appliances, which will be elaborated on in the following sub-section.

### 5.3.3 Energy Appliances for Energy Practices of Everyday Life

The sub-section elaborates in detail what has already been described about the energy practices in the qualitative section, combining the quantitative findings at the household and local business levels.

#### *5.3.3.1 Households*

**Quantity of different appliances per household**

The total quantity of different appliances used in a household informs us about nine different electric appliances: light, fridge, deep freezer, kettle, hot plate, iron, radio, mobile phone, television, computer/laptop, fan, and washing machine. This list of appliances was gathered during the field research.

Accordingly, the mean number of different electric appliances found in a household is 5.32. The multiple modes that exist per household are three appliances, five appliances, seven appliances, and nine appliances (Appendix, D.19), corresponding to 14.3% of different appliances.

*Table 5.18. Descriptive Statistics of Different Appliances*

| | | |
|---|---|---|
| N | Valid | 28 |
| | Missing | 0 |
| Mean | | 5.32 |
| Median | | 5.00 |
| Mode | | 3[a] |
| Sum | | 149 |

a. Multiple modes exist. The smallest value is shown

*Source: Author*

Furthermore, every household has at least one light (92,9%); the mean for the light is 1.86 (Table 5.19). 50% of the households do not have any fridge, and the remaining 50% have at least one fridge. The deep-freezer holders have a total percentage of 25%. At least one hot plate is used in 53,6% of the households as a cooking appliance. The use of kettles is limited, with 32.1% and 67.9% of the households not owning any kettle. 89.3% do not have any (electric) fan. The mean for radio is 0.43, for television 0.39 (only 39.3% have a television), and for computer/laptop 0.11 (10.7% own a device). The mobile phone has the highest mean per household after the lights, with 2.21. Only two households do not own any mobile phones. The mean for iron is 0.68, equivalent to 74.3% of the households that count for at least one iron per household. There was no washing machine mentioned in the survey responses. Some of the appliances on this list suggest that diverse energy practices occur in the residents' daily lives, which is also linked with the differentiation of the devices, such as boiling the water in the kettle and using an electric stove or microwave.

*Table 5.19. Descriptive Statistics for the Household Electric Appliances*

| | N | Minimum | Maximum | Mean | Std. Deviation |
|---|---|---|---|---|---|
| Quantity of Different Appliances counted per old | 28 | 1 | 9 | 5.32 | 2.435 |
| Lights | 28 | 0 | 4 | 1.86 | 1.325 |
| **Fridge** | **28** | **0** | **2** | **.57** | **.634** |
| Deep Freezer | 28 | 0 | 3 | .32 | .670 |
| Kettle | 28 | 0 | 2 | .36 | .559 |
| **Stove/Hot Plate** | **28** | **0** | **5** | **.68** | **.983** |
| **Iron** | **28** | **0** | **2** | **.68** | **.548** |
| Radio | 28 | 0 | 1 | .43 | .504 |
| **Mobile Phone** | **28** | **0** | **7** | **2.21** | **1.686** |
| **Television** | **28** | **0** | **1** | **.39** | **.497** |
| Computer/Laptop | 28 | 0 | 1 | .11 | .315 |
| Fan | 28 | 0 | 1 | .11 | .315 |
| Blow drier | 28 | 0 | 1 | .04 | .189 |
| Sewing Machine | 28 | 0 | 1 | .07 | .262 |
| Microwave | 28 | 0 | 1 | .04 | .189 |
| Valid N (listwise) | 28 | | | | |

*Source: Author*

Within the individualization process, the mobile phone was the only individual item that was identified. Furthermore, fridges, electric stoves, irons, and televisions have been considered individual household appliances. In this sense, Table 5.19 illustrates the mean for mobile phones as 2.21 and the mean for adults per household as 2.32, suggesting that every household had an average of one mobile phone per adult. The mean for fridges with 0.57, electric stoves with 0.68, and irons with 0.68 falls behind one item per household. However, it also provides the perspective that it is not too far to become an individual appliance per household, which already suggests the beginning of individualized households as well. This could also be interpreted as self-sufficient households.

Furthermore (Figure 5.19), 28.6% of the households have one to three different appliances, 35.7% have four to six appliances, and 35.7% have seven to nine appliances, suggesting that the majority of the households (71.4%) have more than three appliances that are usually counted as lighting, mobile phone, and radio (Tier 1 of Multi-Tier Framework, Box 3). The increasing quantity of the different appliances demonstrates the domestication through the use of electric appliances in order to make "life easy" as the respondents often mentioned. Therefore, the connections according to household size, educational background, occupation, income, and monthly expenses show diverse insights.

When the different-sized households were analyzed, single-room households with one to three appliances accounted for 28.6% of the total, suggesting the potential group that is using electricity with the Tier 1 needs but spread in 24 hours. Moreover, 17.9% of the total belongs to two-room households and 10.7% to single-room households of the second largest group with four to six appliances. Lastly, in the group of seven to nine different appliances, 10.7% belong to single households and 14.3% to two-room households (Appendix, D.16)

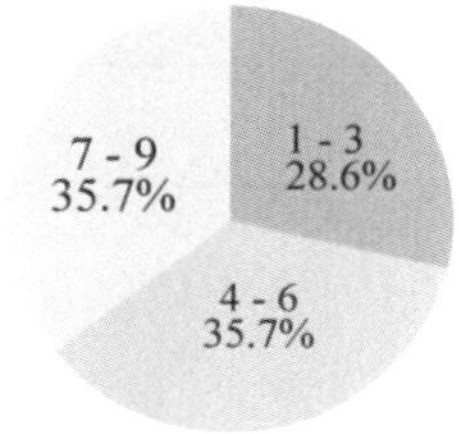


*Figure 5.19. Classification of Different Appliances per Household*
*Source: Author*

According to educational background, the distribution of the appliances reveals that 21.4% of the respondents have four to six different appliances, and 17.9% with seven to nine appliances are primary school graduates. They are followed by 10.7% with one to three without education and 10.7% with seven to nine appliances in the tertiary school graduates, who constitute 75% within this education group (Table 5.20). The rest is distributed to all educational levels; this suggests that the diversity of appliances is not directly related to the level of education. Although some data illustrate a strong link for a small ratio, it would not be significant to generalize it.

*Table 5.20. Distribution of Different Appliances by Education*

| | | | Education | | | | Total |
|---|---|---|---|---|---|---|---|
| | | | No education | Primary | Secondary | Tertiary | |
| Quantity of Different Appliances per Household | 1-3 | Count | 3 | 2 | 2 | 1 | 8 |
| | | % within Education | 50,0% | 15,4% | 40,0% | 25,0% | 28,6% |
| | | % of Total | 10,7% | 7,1% | 7,1% | 3,6% | 28,6% |
| | 4-6 | Count | 2 | 6 | 2 | 0 | 10 |
| | | % within Education | 33,3% | 46,2% | 40,0% | 0,0% | 35,7% |
| | | % of Total | 7,1% | 21,4% | 7,1% | 0,0% | 35,7% |
| | 7-9 | Count | 1 | 5 | 1 | 3 | 10 |
| | | % within Education | 16,7% | 38,5% | 20,0% | 75,0% | 35,7% |
| | | % of Total | 3,6% | 17,9% | 3,6% | 10,7% | 35,7% |
| Total | | Count | 6 | 13 | 5 | 4 | 28 |
| | | % within Education | 100,0% | 100,0% | 100,0% | 100,0% | 100,0% |
| | | % of Total | 21,4% | 46,4% | 17,9% | 14,3% | 100,0% |

*Source: Author*

Table 5.21 illustrates that 17.9% of the respondents who could not provide their electricity expenses have four to six appliances, and 3.6% have seven to nine appliances. On the other hand, 14.3% of the respondents have an electricity expense between 0 to 150 NAD and have seven to nine appliances. The same ratio with the same quantity applies to an expense range of 151 to 300 NAD/month. 14.3% of the total belongs to the smallest group with one to three appliances with unknown electricity expenses.

*Table 5.21. Distribution of Different Appliances by Electricity Expenses*

| | | | Expenses | | | | Total |
|---|---|---|---|---|---|---|---|
| | | | N/A | 0-150 | 151-300 | 451-700 | |
| Quantity of Different Appliances per Household | 1-3 | Count | 4 | 2 | 1 | 1 | 8 |
| | | % within Expenses | 40,0% | 28,6% | 14,3% | 25,0% | 28,6% |
| | | % of Total | 14,3% | 7,1% | 3,6% | 3,6% | 28,6% |
| | 4-6 | Count | 5 | 1 | 2 | 2 | 10 |
| | | % within Expenses | 50,0% | 14,3% | 28,6% | 50,0% | 35,7% |
| | | % of Total | 17,9% | 3,6% | 7,1% | 7,1% | 35,7% |
| | 7-9 | Count | 1 | 4 | 4 | 1 | 10 |
| | | % within Expenses | 10,0% | 57,1% | 57,1% | 25,0% | 35,7% |
| | | % of Total | 3,6% | 14,3% | 14,3% | 3,6% | 35,7% |
| Total | | Count | 10 | 7 | 7 | 4 | 28 |
| | | % within Expenses | 100,0% | 100,0% | 100,0% | 100,0% | 100,0% |
| | | % of Total | 35,7% | 25,0% | 25,0% | 14,3% | 100,0% |

*Source: Author*

## Energy Appliances for Energy Practices of Everyday Life

From the nine different appliances mentioned earlier with descriptive statistics, lighting, cooking, communication and media, cooling, personal care, and sewing, have been identified as the energy practices of everyday life.

### Lighting Appliances

As mentioned earlier, lighting in households is the first energy practice with the highest mean. Accordingly, 50% of the households have one light (bulb), 21.4% have four light bulbs, and 14.3% have two light bulbs (Figure 5.20). The households that do not have any bulbs still use the electricity for other purposes. As a result, 82.9% of the households use electric light.

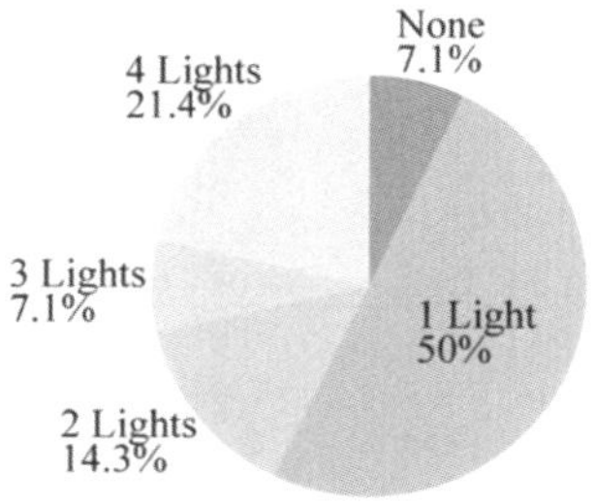


*Figure 5.20. Distribution of Light (Bulbs) per Household*
*Source: Author*

### Cooking Appliances

The traditional way of cooking is with firewood. However, the findings of the survey (Figure 5.21) illustrate that 28.6 % of the households use wood, and the same percentage uses electricity. 17.9% use wood and electricity in combination. Only 10.7% use only gas for cooking. If all of the electricity or electricity in combination accounted for cooking, 53.7% of the respondents used electricity solely or combined with other energy sources.

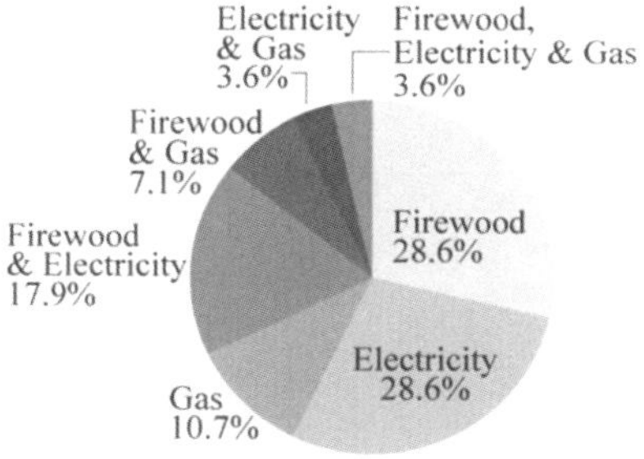


*Figure 5.21. Energy Sources Used for Cooking*
*Source: Author*

Thus, 53.6% of the households have at least an electric stove/hot plate, 3.6% have a microwave, and 32.2% have a kettle, recorded in the surveys. This situation suggests that the households' primary source of energy for cooking is still firewood or its combination with electricity or gas. Firewood is followed by electricity, and the appliances suggest a growing tendency to multiply devices in cooking practices.

The electric cooking distribution based on educational background was spread in all education groups, with 21.4% of primary school graduates and 14.3% of those without education having the highest ratios (Appendix, D.18). According to income range, the respondents with an income range of 0-1300 NAD have 25%, and those with "no income" have a 14.3% share in electric stoves (Table 5.22). Although these figures suggest a tendency for low-income and low-education people to have electric stoves, the rest of the categories also use them.

*Table 5.22. Distribution of Hot Plate/Electric Stove by Income*

| | | | Average Income | | | | Total |
|---|---|---|---|---|---|---|---|
| | | | No Income | 1-1300 | 1301-3900 | 3901-more | |
| Hot Plate/ Electric Stove | 0 | Count | 2 | 7 | 1 | 3 | 13 |
| | | % within Income | 33,3% | 50,0% | 33,3% | 60,0% | 46,4% |
| | | % of Total | 7,1% | 25,0% | 3,6% | 10,7% | 46,4% |
| | 1 | Count | 4 | 7 | 2 | 1 | 14 |
| | | % within Income | 66,7% | 50,0% | 66,7% | 20,0% | 50,0% |
| | | % of Total | 14,3% | 25,0% | 7,1% | 3,6% | 50,0% |
| | 5 | Count | 0 | 0 | 0 | 1 | 1 |
| | | % within Income | 0,0% | 0,0% | 0,0% | 20,0% | 3,6% |
| | | % of Total | 0,0% | 0,0% | 0,0% | 3,6% | 3,6% |
| Total | | Count | 6 | 14 | 3 | 5 | 28 |
| | | % within Income | 100,0% | 100,0% | 100,0% | 100,0% | 100,0% |
| | | % of Total | 21,4% | 50,0% | 10,7% | 17,9% | 100,0% |

*Source: Author*

## Communication and Media Appliances

The appliances used in communication and media are mobile phones, televisions, radios, and computers/laptops. According to the surveys, only 7.1% of the households did not have a mobile phone, whereas 92.9% had at least one mobile phone per household as a communication and payment tool when the phone characteristics allow it to be. Moreover, 39.3% had a television, and 42.9% had a radio. Only 10.7% had a computer/laptop. These figures suggest that the substantial use of mobile phones indicates an individualization process and domestication. On the other hand, the radio and television accounted for Tier 1-2 of the MTF and did not appear as substantial as the mobile phone.

As indicated earlier in the domestication process, television is not very prominent on a household level, with 39.3% ownership. However, it was included within the electrified wishes of the households. The radio (42.9%) and the computer (10.7%) also have a lower ratio. Therefore, they were not further analyzed.

**Mobile Phones**

However, in addition to the descriptive statistics, the tables (Table D.4, Table D.5, Table D.6, in Appendix D) also demonstrate that mobile phone ownership has been distributed to all households regardless of their educational background, occupation, and average income.

**Cooling Appliances**

The cooling appliances include a fridge, deep freezer for food storage, and fans for space cooling. In Gam, 50% of the respondents stated that they have at least one fridge, and 50% mentioned they do not have any. Furthermore, 25% stated they own at least one deep freezer. The fan owners accounted for 10.7% of the households. A fridge and a freezer have been considered for Tier 3 use, but only for 5 hours during the day and 3 hours in the evening (Box 3), allowing a yearly consumption of 365 kWh. In the case of Gam, considering the consumption trends of a mean of 123 kWh/month (253 NAD), it has already exceeded the Tier 3 category and would reach a yearly consumption of 1476 kWh; this would settle the energy consumption in between Tier 4 and Tier 5, in a consumption of approximately 4.04 kWh daily. The only difference is that in Gam, the residents would be using it 24/7, and it would be much appreciated.

**Personal Care Appliances, Sewing Machines, Irons, Washing Machines**

Within the personal care practices, 3.6% of the respondents stated that they have a blow dryer. Other house-related practices can also include ironing and sewing, which can be individually practiced or for commercial purposes. Although the surveys do not explicitly provide information on this commercial part, 64.3% of the households have an iron, and 7.1% have a sewing machine.

*Figure 5.22. Herero Traditional Dress*
*Source: Author*

These practices are closely linked to the traditional Herero costume, representing the cultural identity (Naughten & Marten, 2013). The traditional dresses (Figure 5.22) worn by the women are garment-intensive. Thus, it requires washing, ironing, and sewing as a bundle of practices for them to wear it properly. Therefore, it also presents one of the most essential local businesses in the settlement, very open to developing further equipment and laundry services. Moreover, of the 64.3% of the respondents with at least one iron in the household, 30.9% are primary school graduates, the highest ratio among the rest of the education groups that share 7.1% (Appendix, D.21). Furthermore, the distribution of iron owners according to the monthly income range reveals the lowest income (21.4%) and no income (17.9%) being the highest ratios. However, they are followed by the highest income category with 14.3% (Appendix, D.22. Both data tables show that the iron practices are prominent within the community.

#### *5.3.3.2 Local Businesses*

Electric appliances considered by local businesses have been classified under lighting, cooling, communication and media, and digital practices. Under these practices, a total of 17 different appliances have been identified, with a minimum of 3 and a maximum of 10 different appliances have been found available in the local businesses (Table 5.23). The mean was calculated as 6.67, and the median 7 (Table 5.25). The descriptive statistics of each appliance is presented in Table D.7 (Appendix D).

As a result, the overall different appliance types found in the local businesses have been classified into four groups. 44.4% of the businesses have seven to nine different appliances, 33.3% have four to six different appliances, and 11.1% share the minimum (1-3) and maximum (10-12) appliances each. The highest diversity, with 11.1%, indicates private businesses, while the lowest diversity indicates public businesses (Figure 5.23).

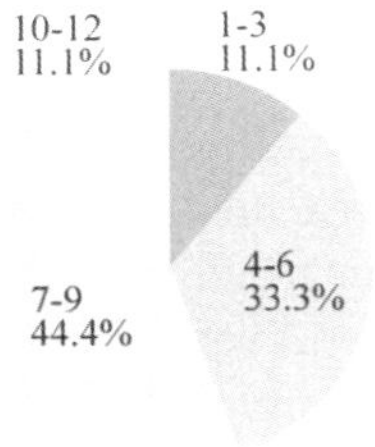


*Figure 5.23. Classification of Different Appliances at Local Businesses*
*Source: Author*

*Table 5.23. Descriptive Statistics of Different Appliances*

| | | |
|---|---|---|
| N | Valid | 9 |
| | Missing | 0 |
| Mean | | 6.67 |
| Median | | 7.00 |
| Mode | | 6[a] |
| Sum | | 60 |

a. Multiple modes exist. The smallest value is shown

*Source: Author*

#### Lighting Appliances

All businesses use electric lights, such as light bulbs or fluorescent lights. Accordingly, 66.7% of the businesses do not have any fluorescent light fittings. 33.3% of the businesses have only one light bulb, and 11.1% have two light bulbs. The remaining businesses have 7 to 23 light bulbs (Table D.8, Appendix D).

#### Cooling Appliances

55.6% of the businesses used the fridge as a cooling appliance. On the other hand, 66.7% of the local businesses confirmed having a space cooling system. 22.2% of the businesses stated that they had an air conditioner, and 55.6% had a fan. Of these space cooling systems, 44.4% belong to the public, and 22.2% belong to private businesses (Appendix, D.23); this suggests that cooling food and space has been practiced in both public and private businesses.

#### Communication and Media Appliances

Within communication and media practices, 22.2% of the respondents had a radio; in addition, television was not recorded. Furthermore, 22.2% confirmed that they do not have either computers or printers. 77.8% stated that they have both computers and printers, and 66.7% have monitors. These ratios mostly show the well-equipped schools, clinics, and some of the local shops.

### *5.3.3.3 Digital Payment*

#### Households

During the field research in 2020, the mobile network coverage from MTC that enables internet use was very limited. Under these conditions, 17.9% of the respondents confirmed using cash and digital payment methods to buy their electricity tokens, despite the constraints. These households belonged to the group of employed people (Table D.9, Appendix D) in all sectors, and they had an income of more than 1301 NAD (Table 5.10, Appendix D). Furthermore, they had all educational backgrounds (Table D.11, Appendix D). Although these figures do not represent any majority or significance, they still demonstrate the (beginning of) digitalization and the use of digital tools at a household level. Its significance is that despite the constraints of having a limited network, they still use it. Moreover, they have reached this utilization level only in *ten* years.

#### Local Businesses

In the local businesses, only one local shop uses digital payment methods. Moreover, two private shops have digital security cameras (22.2%). These brief findings suggest that business owners do not use digital payment methods widely. One of the reasons for this finding was the mobile network coverage of MTC, which did not allow the use of digital payment systems. The author expects that this will change from 2020 to 2023. Secondly, using digital security cameras shed light on the theft potential in Gam, where the owner decided to take such precautions to stop or identify the thieves. The fact that this only occurs in private businesses shows that public possessions are more secure than private ones.

#### *5.3.3.4 Energy Appliances Wished in the Future – "Electrified Wishes"*

**Households**

The future electric appliances the community wishes for have been asked as an open-ended question in the order of priority. The respondents' answers were analyzed in two different ways, thematically and in order of priority. Thematically, it is possible to differentiate the categories as domestic, leisure, office, and industrial/commercial.

*Table 5.24. Electric Appliances Wished in the Future in Households*

| Future Appliances / Priority | No. 1 | No. 2 | No. 3 |
|---|---|---|---|
| **Television** | **24%** | **19%** | 8% |
| **Fridge** | **14%** | **19%** | 0% |
| Music Player/ System/ Hi-fi/ Radio | 0% | 3% | 0% |
| Fan | 0% | 3% | 5% |
| Lights | 3% | 3% | 5% |
| Heater | 3% | 0% | 3% |
| **Electric stove** | **8%** | 11% | 16% |
| Washing Machine | 5% | 0% | 3% |
| Laptop/IPAD/PC | 0% | 3% | 3% |
| Microwave | 0% | 3% | 3% |
| A/C | 3% | 3% | 5% |
| Sewing Machine | 0% | 3% | 3% |
| **Deep-Freezer** | **16%** | 5% | 5% |
| Kettle | 0% | 3% | 8% |
| Wi-fi | 0% | 3% | 0% |
| Iron | 5% | 0% | 3% |
| Grinder | 0% | 0% | 3% |
| Electrical tools | 5% | 8% | 0% |
| Toaster | 3% | 0% | 3% |
| Vacuum Cleaner | 0% | 3% | 0% |
| Fridge/Deep Freezer | 5% | 0% | 0% |
| Nothing | 5% | 11% | 24% |
| Total | 1 | 1 | 1 |

*Source: Author*

The domestic category includes appliances such as fridges, fans, lights, heaters, electric stoves, washing machines, microwaves, air conditioners, sewing machines, deep freezers, kettles, irons, ovens, hair/barber machines, toasters, and vacuum cleaners. Televisions, music players, hi-fis, and radios are categorized under leisure. The office category appliances that are wished for are laptop/iPad/PC. The appliances that can be used for industrial/commercial purposes are welding machines, tire removers, grinders, generators, compressors, and drills.

Table 5.24 shows the electric appliances wished for the community in the order of priority. Accordingly, the items most wished for are a television, fridge, lights, heater, electric stove, washing machine, deep-freezer, and electrical tools. Televisions, deep-freezers, fridges, and electric stoves have been ranked as the most wished appliances with 24%, 16%, 14%, and 8%, respectively. On the other hand, it is essential to note that people who do not wish for any item for priority number 1 is 5%, priority number 2 is 11%, and priority number 3 is 24%. These electrified wishes demonstrate a will for further improvement in their media, cooking, and cooling practices that total up to 30%. Excluding television, all are high-consumption appliances for mini-grids, which need to run continuously. As mobile phones are not listed, this also illustrates a saturation in this appliance.

### Local Businesses

In local businesses, the priority of electrified wishes has a different listing. The fridge and computers have the same ratio of 22%; the rest of the appliances are air conditioners (11%), deep-freezer (11%), fax (11%), and infant baby warmer (11%) (Table 5.25). 11% did not express any electrical wish (5.48). In this way, 44% accounts for the cooling practices, and communication/digitalization with the computer follows this. The local businesses weigh high considering the ratio of cooling practices.

*Table 5.25. Electric Appliances Wished in the Future in Local Businesses*

| Future Appliances / Priority | No. 1 | No. 2 | No. 3 |
|---|---|---|---|
| Fridge | 22% | | |
| Computers | 22% | 11% | |
| Air Conditioner | 11% | | |
| Deep-freezer | 11% | | |
| Projectors | | 11% | |
| Fax | 11% | | |
| Infant Baby Warmer | 11% | | |
| Internet/Wifi | | | 11% |
| Generator | | 11% | |
| Nothing | 11% | 67% | 89% |
| Total | 100% | 100% | 100% |

*Source: Author*

In this way, households and local businesses are willing to get more appliances when they can; this is also linked to their affordability. However, when the domestication of lighting, cooking, or cooling practices is considered, there is a tendency for higher technization and, thus, higher energy consumption. On the other, the use of mobile phones and computers can bridge the gap between the peripheries and urban areas with the internet, enabling them to access some of the things they have never seen and wished for before. Certainly, one cannot discuss digitalization without its technology or technological appliances that are linked with domestication/technization. Therefore, their ownership equally suggests tendencies of growth and further use.

UN supports the use of technologies and digital advances in accelerating the achievement of seventeen Sustainable Development Goals to make our world a more just, peaceful, and fairer place. However, there are also studies on its adverse personal, interpersonal, organizational, inter-organizational, and socio-economic effects (Gimpel & Schmied, 2019). Considering that Gam has its recent thirty-year-old history, it has been going through the process of domestication and the process of digitalization at the same time. In Ghana, internet use in rural areas increased from 8% (2010) to 69% (2021) in recent years, and internet use among the age group of 15 to 40 has reached approximately 80% (Dabaren & Mensah, 2023). Furthermore, the high ratio of children and youth in the repatriation program suggests that children and youth from 0 to 20 years old corresponded to 72.4% of the repatriates. Therefore, in 2023, one can consider having an average age group of 30 to 50 (and their children), assuming the children remained in Gam after their arrival. Hence, the population in Gam is young and suitable to follow the growth patterns of digitalization and internet use, as in Ghana.

## 5.4 Conclusion

The Gam Community has an overarching vision to become a town. They know the importance of electricity as a pre-condition for creating or attracting more business opportunities, more employment, and better infrastructural conditions to have town-like services such as banks, hospitals, post offices, and extensive and specialized shops in their everyday life. They are disappointed with "being left-behind" but still carry the hopes for a better future. However, they support their children in their studies. They also look forward to working or creating their businesses to reach their goals.

The differentiation process has been identified from a place-based differentiation with an analogy to animal and human conditions upon their arrival in Gam, as well as the urban and rural differentiation that arose when it comes to the electricity supply first, followed by additional services provided in the towns such as hospitals, banks, big shops, post office. The prepaid tariff differentiation between households and businesses, as well as conventional institutional tariffs, is important. Furthermore, the existence of four Traditional Authorities representing different clans of Herero and San brought a different official public differentiation level in Gam. On the other hand, the differentiation between educational, occupational, and income levels has been found in the field.

“Electricity is life!” or “it´s life” have been expressed several times during the interviews, enabling a re-enchantment in their worldview by associating it with the rationality of purpose. In this sense, electricity did not only mean “light” or “life” but it was also considered “everything” and that “nothing to do without electricity.” However, they showed their concerns regarding its reliability 24/7 without interruption since, after 410 additional connections at the end of 2020, they went through frequent interruptions due to the peak loads during the day, when the generated electricity was insufficient to meet the consumption.

Although this increase in consumption has been partly related to the new connections, it also had to do with the energy practices they were doing in their everyday lives. These identified practices included lighting, cooking, communication, cooling (food and space), leisure and entertainment, ironing and sewing for households, food shops, butchers, tailors, car washing, welding, and supermarkets for businesses. These practices would enable them to live town life in Gam with all the electric appliances they can use at home. Thus, they would not need to go to other towns to acquire those services or experience that way of life. In addition, the digital processes also started when MTC placed a tower close to Gam, whereby the network coverage improved drastically between 2020 and 2022, enabling them to use internet services and social media as well as digital payment services that were not available in 2020.

Along with all these differentiation, rationalization, and domestication processes, the ownership of land, livestock, or energy appliances has also contributed to the process of individualization by bringing them a certain autonomy in their everyday lives. They can have their fridge, iron at their houses, charge their phones, or watch television in their places. They would like to continue learning, engaging in community services as employment, getting their own businesses, and living in Gam. Thus, after thirty years of their arrival there, they envision living in a town like Gam, providing all the necessary public services currently provided in towns, which will not make them travel to the towns close by to meet their needs.

# 6 Multilevel Off-Grid Renewable Energy Governance in Namibia

*"VISION 2030*

*A prosperous and industrialized Namibia, developed by her human resources, enjoying peace, harmony, and political stability*[15]."

Renewable energy is considered one of the critical tools in addressing the socio-economic development goals of the *Namibia Vision 2030*, established 2004. Following these case studies, this chapter will investigate off-grid RE governance from supra-national, national, and sub-national perspectives in addressing off-grid electrification strategies, policies, and visions surrounding electricity access with data sources (literature, reports, archives, policies, and interviews) explained in line with the modernization process.

As indicated in Chapter 2, the governance concept itself is a "new mode of governing" society and economy (Mayntz, 2001, p. 4) through the involvement of private, public, and semi-public actors and networks in policy-making (Lo, 2018, p. 1). It has also played a significant role in RE governance as countries strive to transition to RE sources and technologies to pursue energy access, supply, and security to reduce greenhouse gas emissions, mitigate climate change, and contribute to sustainable development. Furthermore, the emerging interdependency of the actors and networks combined with a significant authority of the state (Rhodes, 2007, p. 4) provide an MLG at supra-national and sub-national levels in a bidirectional way (Hooghe et al., 2020, p. 197). Thus, the design and implementation of effective policies play a crucial role in providing the necessary policy and regulatory framework for advancing and adopting RE technologies and shaping the trajectory of electricity access and energy transitions in a bidirectional way for its implementation.

In this regard, Namibia´s abundant solar and wind resources (GIZ, 2022) will help harness this RE potential within the energy mix and electrification plans. Therefore, RE presents a technological deployment and encompasses a complex web of actors, networks, and policies operating across multiple governance levels. At the supra-national level, Namibia has been engaged in international agreements such as the Paris Agreement and international frameworks such as the SDGs. Furthermore, Namibia is also part of the Southern African Development Community (SADC) initiative, which, together with a web of international agreements and partnerships, shapes RE governance and its RE policies within the country and paves the way for its subnational implementations.

In contrast, at a national level, the Namibian government plays a pivotal role in designing, formulating, and implementing RE policies aligned with the development visions and environmental commitments. The expertise acquired

---

[15] Namibia Vision 2030, 2004, p. 15.

through the transnational networks and knowledge exchanges has become operational in formulating strategic and development plans, RE policies, and master plans at the national and sub-national levels of RE governance. Therefore, the RE governance in Namibia presents two levels down the national state at sub-national, regional, and local levels and two levels up the supra-national level with continental and global actors. The Namibian government maintains the authority and power to conduct network and actor activities within this MLG framework. Furthermore, although Namibia´s population is small compared to its land area, the number of actors at different governance levels remains high, which is in contrast to Hooghe et al.'s (2020) argument. As a result, from a structural perspective, Namibia´s RE governance provides hierarchical governance, that is, type I governance, which is undergoing the process of modernization (Chapter 2, Figure 2.2).

## 6.1 Renewable Energy Key Market Actors

This section provides the background of the key RE electricity market actors in Namibia (Figure 6.1) within the MLG framework. It also covers the pluralization of the actors at all levels, from the public and private sectors. This multiplication contributes to the differentiation of energy resources, such as renewable energies, the specialization of the generation, transmission, and distribution of services as well as regional differentiations. All services within the MLG of energy in Namibia have been undertaken by the public and private actors that are specialized as a part of the process of differentiation in order to meet the rational goals within their area of jurisdiction. In this sense, Figure 6.1 illustrates the multitude of national, sub-national, and supra-national actors involved in the energy/electricity services, surrounding the MME´s leadership within the process of expansion with more specialized actors.

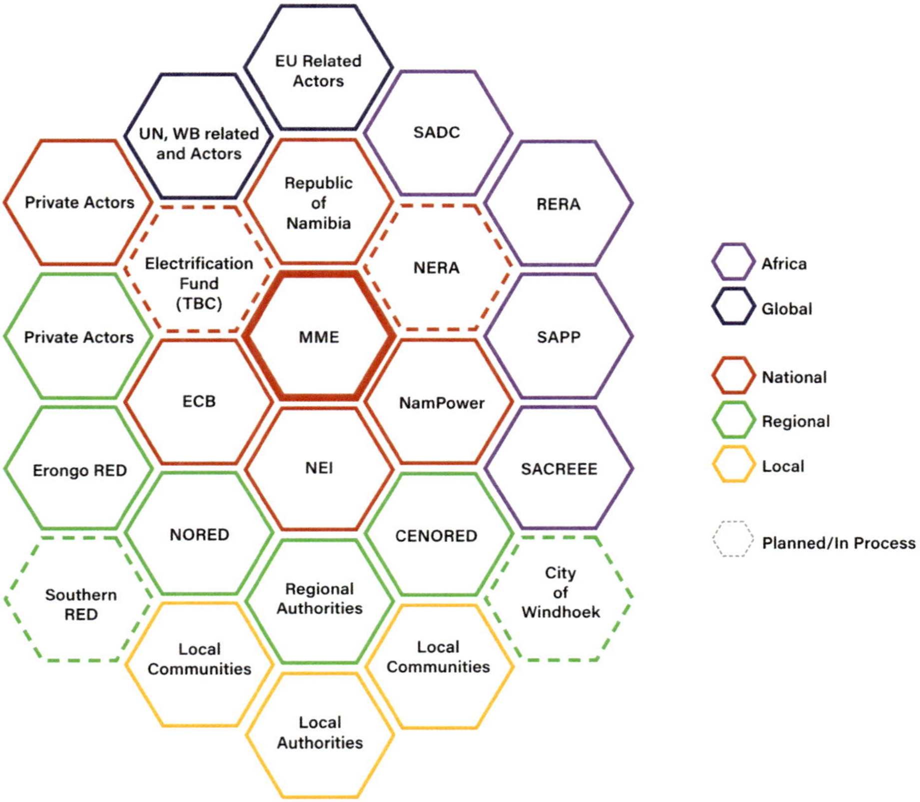


*Figure 6.1. Map of the Key RE Market Actors in Namibia*

*Source: Author*

### 6.1.1 Supra-National Level

At the supra-national level, Southern African organizations, international states, and international governmental organizations share the scene in the (renewable) energy sector with the Namibian energy market actors. This section outlines the main market actors in the energy sector at the supra-national level.

#### *6.1.1.1 Key Supra-Regional Actors*

The African countries declaring their independence from colonial rule in the 1960s marked a new era of forming newly independent states in Southern Africa. In the 70s, the earlier process of consultation for the Southern African Development Co-ordination Conference (SADCC) began; in 1980, it was established with nine member states and was a predecessor of SADC, which was formalized with the Treaty of the Southern African Development Community (SADC) in 1992. Namibia was the 10th member state (SADC, 2023). In the following years, SADC increased the number of members and became an intergovernmental organization as a Regional Economic Community player with 16 member states. From the differentiation process point of view, these freshly independent states supported each other in this Regional Economic Community to reduce their dependency, particularly on apartheid-era South Africa, and also to become effective and competitive players in the world economy with SADCC (SADC). They strengthened this cooperation with the regional cooperation under SADC with the support of the European Commission (Vanheukelom, 2017).

In this way, the initiative of politically regional cooperation transitioned into a broader agenda for political and economic development in the region in the areas of energy, transport, infrastructure development, environment, good governance, anti-corruption, and gender equality (Vanheukelom, 2017). As a paradox of differentiating (member) states and forming a scale reduction, a scale enlargement also occurred as a part of the regional community. At the same time, while forming a national identity, regional-community identity was also shaped to achieve the desired objectives. Therefore, the differentiation process was intertwined with the rationalization process by developing a network of relationships and cooperations in order to achieve the specific objectives of promoting peace and security, self-sustaining development based on collective self-reliance, achieving complementarity between national and regional strategies, increasing employment, the sustainable use of natural resources, protecting the environment, and consolidating the historical, social and cultural affinities in the Region (SADC, 1993). SADC formed preliminary intergovernmental regional cooperation in specialized fields, such as energy (SADC, 2023). The energy division led by the energy ministers consisted of the South African Power Pool (SAPP) in charge of power projects: The SADC Centre for Renewable Energy and Energy Efficiency (SACREEE) responsible for grid-connected RE and EE, off-grid, cooking, heating, cooling, and biofuels, the Regional Petroleum and Gas Association (REPGA), the Coal Association, and the Regional Energy Regulators Association (RERA) (SACREEE, 2023).

SAPP was created in 1995 as a regional organization of power utilities in SADC to undertake the energy field within SADC, optimize regional electricity, and enhance energy trading (SAPP, 2023a). As of 2023, SAPP comprises 12 national power utilities, five operating members, and three market participants, including NamPower (ECB, 2023a). SAPP has a vision of becoming a fully integrated, competitive energy market and a sustainable energy provider for the SADC region (ECB, 2023a). It is governed under four agreements and four working committees that operate under a Management Committee with the Environmental Sub-Committee, the Markets Sub-Committee, the Operating Sub-Committee, and the Planning Sub-Committee reporting to the Executive Committee (SAPP, 2023b). With these committees, SAPP coordinates the planning and operation of the interconnected electric power systems among member utilities and energy trading. Thus, SAPP undertakes a focal coordination point for the specialized energy services for its member states in a planned, organized, and coordinated way as a part of a rationalization process that acts on a regional scale. Figure 6.2 illustrates the extent to which SAPP operates in Southern Africa and the markets in which the national power utility companies operate daily.

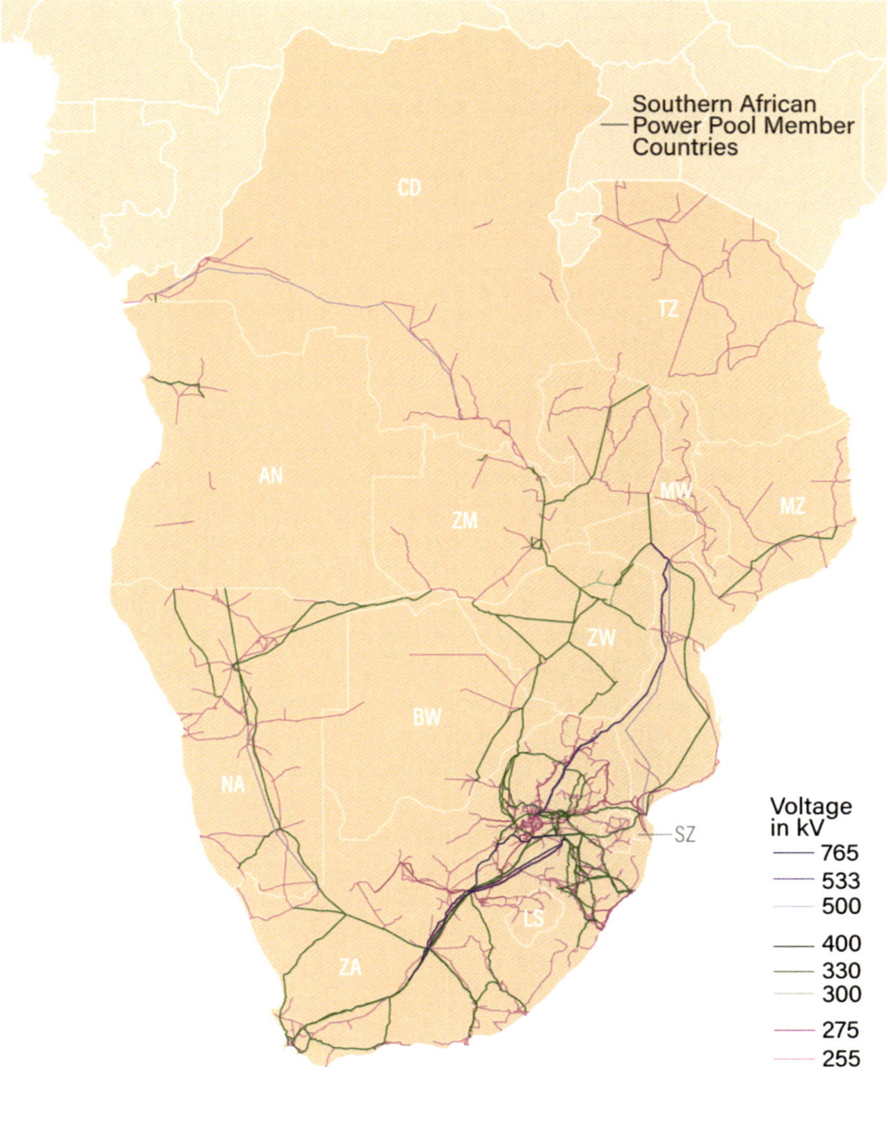


Map by Ö. Dolunay and E. Bentley (2024)
Sources: World Bank, https://datacatalog.worldbank.org/search/dataset/0040465;
International Renewable Energy Agency, https://mapre.es.ucsb.edu/wp-content/uploads/2015/07/botswana_interactive_MapRE.pdf
Natural Earth, https://www.naturalearthdata.com/; OpenStreetMap, https://www.openstreetmap.org/

*Figure 6.2. SAPP SADC Grid Map*

In the pursuit of regional support within SADC, responsible Ministers established RERA in 2002 as a formal Association of Energy Regulators, with an initial focus on electricity regulation only, which was expanded to energy regulation in 2019 (RERA, 2023). With its own constitution, the association aims to be a dependable and leading regulatory organization in order to advance the harmonization of electricity/energy regulation for developing a sustainable regional energy market (RERA, 2023). RERA´s governance and operational structure have been covered under four main categories and sub-committees, presenting a hierarchical level while working with a differentiated committee. These committees are run to build capacity and share information for the facilitation of energy policy, legislation, trade, consumer protection and communication services, and regional regulation and cooperation under the SADC Secretariat (RERA, 2023)

In continuation of the rational goals undertaken within SADC objectives, the SACREEE, another (specialized) subsidiary organization, was established in 2015 concerning the development of Renewable Energy and Energy Efficiency through the coordination of RE and EE-based projects within the SADC member states as a part of the regional coordination with its office in Windhoek. Its overarching objective is to increase access to modern energy services, improve energy security within the region, and support the sustainable development objectives of SADC (SACREEE, 2023). SACREEE focuses on the programs and projects that can be implemented on a regional level, such as harmonizing policy approaches, regulations, and standards, regional capacity building, knowledge management, and investment coordination. On the other hand, promoting market-based adoption of RE and EE technologies is also part of SACREEE's tasks. In addition, SACREE has been nominated as a key player in implementing the Southern Africa Renewable Energy and Energy Efficiency Strategy and Action Plan (REEESAP) (SACREEE, 2023). The secretariat in Windhoek works on long-term business plans and annual work plans, collaborating with 15 National Focal Institutions and a Technical Committee. SACREEE positions itself as a "regional RE & EE promotion agency rather than an implementer on micro-and grass-root levels" (SACREEE, 2023). Thus, this hierarchical structure is a follow-up of the hierarchy of the SADC and clearly states the position of a national and supra-national specialized actor.

As a result, the SADC-related supra-regional organizations serve as a think tank and supportive mechanism to national member states of SADC in the quest to develop and ensure sustainable, reliable, affordable, clean energy services for all. Although SAPP, RERA, and SACREEE are the specialized units for action on a regional level, a top-down approach has been implemented to achieve each unit's envisioned missions and goals. In addition to the SADC and related organizations, Namibia is also a member of the African Union and supported by the African Development Bank (AfDB) in the fields of economic governance for improved business enabling environment and infrastructure development and promoting value addition priorities in addressing the cross-cutting issues of capacity building, climate change, growth, youth, and gender including digital technologies and solutions (AfDB, 2023). Hence, the beginning of SADC and the pluralization of its specialized organizations in order to pursue and support narrowed-down goals demonstrate the paradoxical side of the processes. In the meantime, there is a multiplication of institutional actors for energy and its services at a supranational level, while the new national institutions also emerge to support supra-national agreements hierarchically.

#### *6.1.1.2 Key European Actors*

In 1993, the government of Namibia and the European Union signed an agreement to make some 40 million Euros available under the Sysmin scheme to support the Namibian mining industry. This support program aimed to reverse unfavorable developments in the mining industry by providing loans at reduced interest rates and establishing a revolving fund, the Minerals Development Fund (established in 1996). A major Mining Investment Forum was held in Windhoek in the same year and marked the beginning of cooperation with the mining industry (MME, 2023).

This beginning was followed by RE, European actors through various technical assistance programs, partnerships, and investments, contributing significantly to the development of renewable infrastructure and policies in Namibia. The European Union, the bilateral development agencies of the European States such as the Austrian Development Agency, German Development Agency, Danish Development Agency, to name a few, as well as the individual states, bilateral development banks such as KfW, European technology suppliers, and project developers actively engage in funding and implementing RE projects in Namibia. These collaborations aim to improve energy supply and security, mitigate climate change impacts, and foster a transition towards more sustainable development.

#### *6.1.1.3 Key Global Actors*

The global actors within the supra-national governance have been referred to as the rest of the international actors, excluding the European and the African actors, which include organizations such as the United Nations (UN) and the World Bank Group (WBG) that play a pivotal role in advancing RE agendas in Namibia. They support RE promotion and implementation projects through diverse programs in various agencies, such as the United Nations Development Programme (UNDP) and the United Nations Environment Programme (UNEP). Furthermore, Sustainable Energy for All (SEforALL) works specifically to act faster towards the achievement of SDG7 by working in partnership with the UN, governments, the private sector, civil society, financial institutions, and philanthropies. On the other hand, these agencies support capacity building, policy development, and project financing, directly or through funds the multilateral banks such as the International Finance Corporation (IFC), the WB, or climate funds such as the Global Environment Facility (GEF), and Green Climate Fund (GCF). In addition to multilateral agencies and funds, the Japanese International Cooperation Agency (JICA), the US International Development Finance Corporation, and the Korea Development Bank have been considered within this category (National Electrification Funding Portfolio [NELFP] published by MME, 2022a). As a result, all these multilateral and bilateral agencies and banks support Namibia´s efforts in RE-related activities to deploy RE solutions to address energy access, energy supply, and security challenges.

### 6.1.2 National Level

The supra-national actors are linked with the national actors directly or indirectly depending on the type of work they are involved in. In this section, the key energy market actors have been outlined in relation to the (off-grid) electrification within the RE governance framework (Figure 6.3).

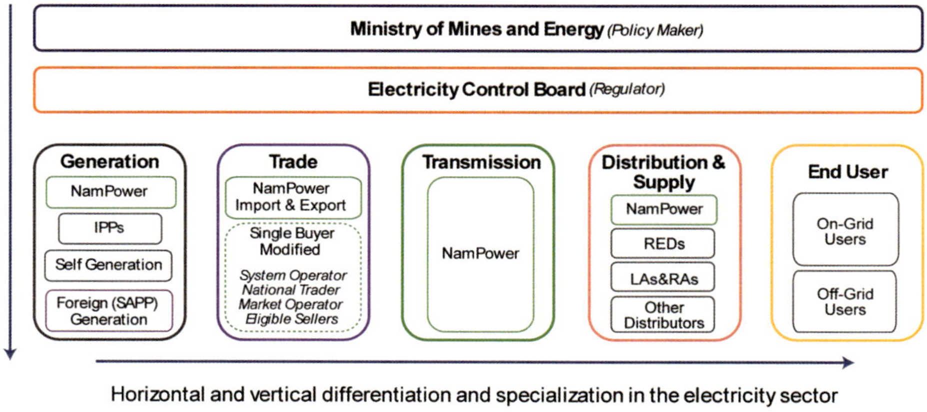


*Figure 6.3. Structure and Actors of the Namibian Electricity Market*
*Source: Author (adapted from MME, 2021; ECB, 2022)*

#### *6.1.2.1 Ministry of Mines and Energy*

The Ministry of Mines and Energy (MME) is the leading actor within the (renewable) energy governance in Namibia, which was established in 1990 after Namibia's independence (MME, 2023a). At its establishment, it had four directorates: the Directorate of Mining, the Directorate of Geological Survey of Namibia, the Directorate of Energy, and the Directorate of Administration and Finance (MME, 2023a). Besides these four directorates, the Diamond Board regulating and controlling the process of diamond production, evaluation, and security, the National Energy Council, NamPower, and the national oil company NamCor were under the auspices of the MME (MME, 2023a). MME had a set of objectives on land-based resources, mining, minerals, and energy. The very first legislation that was approved was the Petroleum Exploration and Production Act No.2 (1990) and the Petroleum Taxation Act No.3 (1991); mining and minerals, mine health and safety followed the petroleum legislation (MME, 2023b). With the restructuring in 1992 and 1993, The Directorate of Geological Survey was supported by three divisions: Regional Geoscience, Applied Geoscience, and Geotechnology, with 78 staff (MME, 2023b)

What was described by MME as a milestone in November 1994 was the three agreements that were signed between the Government of Namibia (GRN) and De Beers diamond company, which provided GRN 50% in the former Consolidated Diamond Mines and led with the establishment of Namibia´s leading diamond producer of Namdeb as a joint venture between GRN and the private sector (MME, 2023b). The Directorate of Diamond Affairs was established in 2001 within MME. On the other hand, the White Paper on Energy Policy (1998) restructured the electricity supply industry, established REDs, and introduced additional petroleum and mining bills in the 2000s (MME, 2023b). As a consequence of the new developments, the Energy Directorate was established. In 2006, they facilitated the establishment of the Renewable Energy and Energy Efficiency Institute (REEEI) within the Polytechnic of Namibia. Later, REEEI´s tasks were taken over by the Centre for Renewable Energy and Energy Efficiency (CREEE) within the Namibia Energy Institute (NEI) under the Namibia University of Technology in 2014 (NUST, former Polytechnic of Namibia). The Namibian Diamond Trading Company was formed in 2007, and the Off-Grid Energization Master Plan was prepared (MME, 2023b). With the uranium exploration reaching unexpected dimensions in 2008, MME undertook a geological survey on the uranium through the work of a particular purpose company, Epangelo Mining, registered by the ministry to explore and develop mineral resources fully (MME, 2023b). Furthermore, in 2010, a high-resolution airborne geophysical survey, which started in 1994, was completed (MME, 2023b).

As of 2012/2013, MME had 345 employees, of whom 48% are female and 52% male (MME, 2013). In 2023, MME had seven directorates: a Geological Survey, Energy, Energy Funds, Diamond Affairs, Mines, Petroleum Affairs, and Administration and Services. The Directorate of Energy, with the divisions of Electricity, Renewable Energy, and Energy Planning and Research (established in 2018), is entitled to implement energy policies and compliance with energy legislation (Electricity Act, 2007) as well as to support research on (new) renewable sources of energy (Electricity Act, 2007). The MME has a strategic mandate to contribute to Namibia´s socioeconomic development through its diverse geological, mineral, and energy resources.

The MME has been described as the "State´s leading agency in attracting private investment in resources exploration and development through the provision of geoscientific information on minerals and energy resources, and management of an equitable and secure titles systems for the mining, petroleum and geothermal industries" (MME, 2023a). Then, it is the institution in charge of regulating these extractive industries, overseeing them concerning the relevant State and Commonwealth legislation, and collecting royalties. The website and the information provided about the history and the development of MME highlight the intensity of the mining industry compared to other divisions, such as energy and electricity. There is a multiplication or pluralization of activities within the differentiation process in MME while specializing in different fields of activity.

Indeed, the description on the website does not present the main focus of MME in the 2020s, but it provides the roots of the development of MME. Considering the pluralization of services and activities undertaken, one may discern, that the MME gave priority to the mining and petroleum activities in the 1990s by partnering with the private sector initiatives. However, for the energy/electricity sector, a reluctance of the involvement of the private sector in renewable electricity/off-grid energy services appear. The inclusion of the private actors within the IPP scheme in 2018 is one of the indicators. On the other hand, the increase in the MME´s number of employees from 75 to 345 with a nearly equal ratio of men and women is remarkable in the observation of the multiplication of services/activities provided at the national level.

#### *6.1.2.2 Electricity Control Board*

The Electricity Control Board (ECB) is the statutory regulatory authority established in 2000 under the Electricity Act no.2 (2000), which was abolished with the updated Electricity Act no.4 (2007) and took effect the expansion of the ECB mandate and its core responsibilities (ECB, 2023a). ECB acts as the regulator of the electricity market and controls the supply industry in terms of electricity generation, transmission, distribution, supply, import, and export in Namibia by issuing generation licenses and setting electricity tariffs. It also plays an advisory role for all operators in the electricity distribution industry (EDI) and electricity supply industry (ESI) (von Oertzen, 2019). ECB Report (ECB, 2023a) summarizes the legal and regulatory instruments under eight categories: (1) legislation and draft bills, (2) government notices, (3) regulations, (4) policies and other national instruments, (5) rules, codes, standards, and guidelines, (6) ECB methodologies, (7) mechanisms, (8) ECB internal procedures. During the reporting period of 2022-23, ECB issued nineteen new licenses, of which 12 were for generation, six for export, and one for trading, with a total installed capacity of 699.354 MW. In addition, two-generation licenses were amended, one renewed, and twenty-one under review. As of 2023, ECB has 53 staff members with a 49% female and 51% male ratio.

#### *6.1.2.3 NamPower*

NamPower is Namibia´s national power utility, established initially as South West Africa Water and Electricity Corporation (SWAWEK) on 19 December 1964 under South African rule (NamPower, 2023). By then, it was a fully affiliated private company of the Industrial Development Corporation (IDC) of the Republic of South Africa, which was the critical entity for the development of the Ruacana Hydropower Station and the establishment of the transmission system to distribute the electricity throughout the country´s central districts to Windhoek. NamPower, taking over the duties and responsibilities of the existing facilities of SWAWEK in generation, transmission, and energy trading in Namibia, also contributed to the country's socio-economic development. NamPower is a state-owned enterprise that reports to MME and is regulated by the ECB. It is registered as a company operating under the Companies Act with a board of directors (NamPower, 2023). NamPower also reports to the Ministry of Public Enterprises (MPE) according to the Public Enterprises Governance Act 1 ([PEGA Act], Republic of Namibia, 2019). NamPower is the bulk electricity supplier for the Regional Electricity Distributors (REDs), mines, farms, and local authorities where REDs are unavailable.

## Grid Infrastructure and Investments

During SWAWEK´s term, the Van Eck Power Station was commissioned (between 1972 and 1976) with an installed capacity of 120 MW, and the Ruacana Power Plant (Figure 6.4) was put into operation in 1978 with a capacity of 240 MW (NamPower, 2016). Additionally, the Paratus Power Station was commissioned in 1976 at Walvis Bay with a capacity of 24 MW. Before SWAWEK became NamPower in July 1996, the intention to construct a 400 kV power line over a distance of 900 km between Aries (Kenhardt, South Africa) and Keetmanshoop to Auas (near Windhoek) was announced (NamPower, 2016).

*Figure 6.4. Ruacana Waterfall during the Rainy Season*

*Source: Author*

Following up on the intention of SWAWEK in extending transmission lines, the 900 km power line (400kV) was commissioned in 1997, and the first line was completed in 1999, being the longest line in the world at the time (NamPower, 2016). This line was followed by the Zambezi Interconnector transmission line with 350 kV and 300 MW High Voltage Direct Current (HDVC) line, which had a 951 km length in 2010 (NamPower, 2016). This strategic point was designed to ensure a reliable power supply to Caprivi Region and connected to the

Namibian grid. It also constituted a north-south link in the SAPP with the vision of interconnecting the region (NamPower, 2016). NamPower continued reinforcing the 220 kV transmission line in the economically growing Erongo Region to service the new mining activities under the West Coast Development project (NamPower, 2016). As of 2022, NamPower extended its transmission network of 66kV to 400kV overhead lines to 11,690 km, low voltage lines to 22,809 km with an 8,968 MVA transformer capacity, and is committed to continuously prioritizing the planning, development, strengthening, and maintenance of its generation and transmission infrastructure (NamPower, 2022).

In the years after the takeover, NamPower upgraded the Ruacana power plant´s capacity to 332 MW in 2012 by adding a fourth unit as it was initially designed. From 2014 until 2016, the first three units of the power plant went through the runner refurbishment process of the initial units, and the capacity was increased to 347 MW with this improved efficiency (NamPower, 2016). The only power station that was built by NamPower was the Anixas Power station in 2011 at Walvis Bay, with an installed capacity of 22.5 MW. The Omburu PV Power Station in the Erongo Region, with a capacity of 20 MW was put in operation in March 2022, marking the first wholly owned and operated RE project of NamPower. The ECB funded this power plant, which cost approximately N$340 million, expecting to serve 16,000 – 18,000 middle-income households (NamPower, 2022). With this recent investment, NamPower increased its generation capacity to 509.5 MW (NamPower, 2022). As a result, by 2022, NamPower´s energy generation investment accounted for 149.5 MW in 26 years, presenting 30% of the capacity realized during the apartheid period.

In 2018, the minister of MME determined that NamPower´s generation capacity would be increased by 150 MW, of which 100 MW were planned to be renewable resources, including the Omburu Power Plant. Within this framework, the Omburu 58 MW Battery Energy Storage System (BESS) Project, Soch Pinah 40 MW Wind Power Plant Project, 50 MW (Firm) Anixas II Power Plant, and the Otjikoto 40 MW Biomass Power Plant Project are in the pipeline (NamPower, 2022). In this way, NamPower aims to support COP21 initiatives by increasing RE share in the energy mix to at least 70%, as supported by the National Renewable Energy Policy (NamPower, 2022). Moreover, the Independent Power Producers and energy imports from SAPP constitute Namibia´s Energy Mix, which will be discussed further. Including these scheduled and foreseen investments, NamPower´s expected installed capacity is estimated at 639.5 MW with 58 MW BESS, increasing NamPower´s overall investment to 43.7% in approximately 30 years of establishment (NamPower, 2022).

Besides these large-scale national and SAPP-supporting energy generation projects, NamPower was also appointed by the MME to carry out RE projects on their behalf and provided with their funding. As of 2022, there are 32 schools to be electrified, and 25 projects from 2021/2022 are in the different stages of project implementation, which sums up to a total cost of N$ 96,522,105 (NamPower, 2022). This investment by the MME belongs to the rural electrification program for public buildings in rural areas within the framework of the National Electrification Plan. After the electrification, depending on the area of operation, the assets are handed over to the distribution and O&M licensee of the area/region (NamPower, 2022). According to the 2022 report, the construction and completion of the 60 km overhead line from Ivilivinzi to Impalila Island in the Zambezi Region intersecting point of Namibia, Botswana, Zimbabwe, and Zambia, was the highlight of the year. The electrification connecting the localities of Mbalasinte, Kasika, and Impalila Island cost N$23.5 million to benefit 300 households, schools, businesses, and public offices in three localities (NamPower, 2022). In addition, the electrification of the 125 km-Mangetti Dune line, electrifying 183 households, hostels, schools, health center, two police stations, and several businesses at Marulaboom, Mkata, and Mangetti Dune settlements cost N$40 million (NBC Digital News, 2022).

As a result, these electrification projects indicate the presence of the public sector and government, which are highly involved in the construction and electrification works, limiting the involvement of the private sector. However, they are involved in the IPP projects. Furthermore, they indicate strategic investment points aligned with SAPP or national political decisions. Furthermore, the growing energy market demand urges NamPower and the state to take measures under these conditions, such as the Modified Single Buyer model, which enables IPPs to sell electricity to the end users based on certain pre-conditions. This consists of another level of differentiation within NamPower's responsibilities, which involves the private sector. It is also part of the rationalization process in which a goal-oriented measure has been planned and taken into action considering the productivity in the energy sector as well as to have a more calculable and controllable energy (generation) future in Namibia. With these processes, NamPower is slowly moving away from traditional ties, but it also involves an intense process of domestication and rationalization. The inconsistency of the rainfalls in the rain season and the flow of the Kunene River has a direct impact on the energy generation of Ruacana HPP, in which the state and NamPower aim to secure the supply through the national energy investment and the energy trading measures from SAPP. Furthermore, using digital measures in power plants facilitates the authorities having more power for energy generation, transmission, and trading operations. However, the paradox is there: On one side, energy investments bring more autonomy and liberalization in one way, but on the other side, they create more dependencies on further investment, further technologies, and/or resources such as green hydrogen.

Furthermore, NamPower is the coordinating entity, responsible for the electricity supply operations and trades for Namibia within the SAPP. The above-mentioned 509.5 MW NamPower generation capacity is equivalent to 20% of Namibia´s energy mix, and it is dependent on the rainfall in the season flowing through the Kunene River to the Ruacana hydropower plant (HPP). In the 2021/2022 report period, Ruacana HPP´s generation was reduced from 24% of the previous to 20% (NamPower, 2022). In addition to NamPower´s generation capacity, the IPPs in Namibia supply 9%, long-term bilateral agreements from SAPP supply 65%, and 6% are supplied through short-term trades from SAPP (NamPower, 2022, p. 96).

In summary, as of 2022, the bilateral agreements (PPAs) held within the SAPP belong to 100 MW with ESKOM in South Africa, 180 MW with ZESCO in Zambia, and 80 MW with ZPC in Zimbabwe (NamPower, 2022, p. 96). Thus, this dependency on energy imports encourages Namibian authorities to invest more in RE-based energy infrastructures and energy storage solutions to provide the required power of supply for the citizens of Namibia. However, it is essential to underline that this current energy supply presented in the NamPower report is the necessary electricity to supply the grid-connected households of Namibia, only, consisting of approximately 55.2% of the entire population.

#### *6.1.2.4 Private Actors*

The private actors consist of national and foreign private companies or persons involved in the renewable energy business, such as business developers, investors, contractors, independent producers, or larger power users and producers. With the adoption of the Independent Power Producers (IPPs) Policy, they have become prominent within the energy market actors. They started their operations in 2015 (von Oertzen, 2019). As of 2022, ECB licensed 21 IPPs with an installed capacity of 699.35 MW (ECB, 2023, p. 72), of which 173.6 MW solar and wind power plants were operational (ECB, 2023a, p. 75). The IPPs operate under the REFIT Programme, which aims to accelerate the investment in RE technologies in order to reduce Namibia´s import share while ensuring supply security. According to NamPower's Annual Report (2022), IPPs accounted for 9% of the total energy generation in Namibia. In addition to the licensed IPPs, there are 15 generation licenses under review in ECB, with an installed capacity of 1,021.78 MW with solar PV technologies (ECB, 2023a).

### 6.1.3 Regional Level and Local Levels

According to the Namibian Constitution, there are three levels of government, central, regional, and local, which also lay out the structure of energy governance. Sub-regional actors also play an essential role in the energy sector, following the national actors who act on national and supra-national levels. The subregional level consists of the Regional Councils, Local Authorities, Traditional Leaders, REDs, private enterprises, communities, and people.

#### *6.1.3.1 Regional Councils – Local Authorities – Traditional Leaders*

The Boundaries Delimitation and Demarcation Commission (appointed by the National Assembly) initially divided Namibia into 13 regions, which then became 14 regions with the split of the Kavango Region into Kavango East and

Kavango West in 2013 (Mcgirr, 2021). According to the Constitution (Legal Assistance Centre [LAC], n.d.), each region consists of 6 to 12 constituencies, and Namibia has 121 constituencies in total. The regions have their Governors appointed by the President, who act as a link between regional and central governments and assist in matters related to their regions and a Regional Council per region. The governors report to the President.

On the other hand, the Regional Councils (RC) are formed by the elected people of each constituency of the respective region, having a management committee (LAC, n.d.). The size of this committee is determined by the number of its members, which will be three members for small (less than nine members) and four members for larger Councils (LAC, n.d.). The members have appointed a Chief Regional Officer to administer the Council´s affairs. In addition, three members of the RC represent the region in the National Council for five years (LAC, n.d.). They also collaborate with the National Planning Commission in developing plans and guides for the development and growth of each region and support local government within their region (LAC, n.d.). Additional duties can be assigned by the President, Parliament, or the Governor (LAC, n.d.).

Therefore, the classification of an LA is dynamic but might vary depending on the resources. An LA's duties and powers are also linked with its classification by municipalities having more services than the towns and the towns more than the villages (LAC, n.d.). All have a council consisting of five members for Village councils, seven to twelve members for town councils, and seven to fifteen members, depending on the size of municipal councils, elected every five years (LAC, n.d.). According to the LA Act 23 (Republic of Namibia, 1992), hey are responsible for the supply of water (Part VI), sewerage and drainage (Part VII), cemeteries (Part VIII), streets and public places (Part IX), electricity and gas (Part X), public transport services (Part XI), and housing schemes (Part XII). When they cannot perform their services, the central government may act accordingly (LAC, n.d.). Namibia has 14 RCs and 57 LAs, comprising 15 Municipal Councils, 26 Town Councils, and 18 Village Councils (Common Wealth Local Government Forum [CLGF], 2023).

In light of this hierarchical structure of governments, the case studies in Tsumkwe and Gam are settlements with Settlement Administration in place. Gam was recently appointed as a settlement. Therefore, as explained in the respective chapters, their access to the services outlined within the LA Act .23 (Republic of Namibia, 1992) is linked with the Local Authority in Tsumkwe Constituency. Moreover, at the local level, the Traditional Authorities (TA) are present to lead the traditional community as per their customary law and preserve the community´s traditions, culture, language, and values (LAC, n.d). A TA comprises senior and traditional councilors, who must live in the communal area they lead. They also support local authorities, RCs, and central government LAC, n.d. They have the authority to collect contributions for Community Trust Funds under the consent of the community in order to finance community projects and cover the administrative costs of the TA, who also receive allowances from the central government (LAC, n.d). The Council of Traditional Leaders has regulated the TAs according to the Council of Traditional Leaders Act 13 (Republic of Namibia, 1997), which defines its powers, duties, and functions and advises the President on using and controlling communal land. As of 2021, there are 51 recognized TAs in Namibia, along with 20 pending applications (Hitchcock et al., 2021, p. 270). Thus, this hierarchical local and regional governance including the TA´s contribution, demonstrates how structural and infrastructural activities of everyday life have been governed through the intersecting layers of political and community networks.

#### *6.1.3.2 Regional Electricity Distributors (REDs)*

After the White Paper on Energy Policy defined the goals and guided the energy policies in 1998, the Electricity Supply Industry (ESI) restructuring started, and studies were completed in 2000, in which REDs were proposed as the electricity distributors in five different regions (ECB, 2023b). With the approval of the ESI Restructuring Study in 2000, MME was appointed by the Cabinet for its implementation. Accordingly, Northern Electricity Distribution (NORED) was established in 2002, followed by Central-Northern Electricity Distribution (CENORED) in 2003, and by Erongo RED for coastal Namibia in 2005 (ECB, 2007b; Figure 6.5). The proposed SORED for southern Namibia and Central RED for central Namibia still need to be established (ECB, 2007b). Before REDs, RCs, and LAs were the leading distributors in Namibia, supplying electricity to households and businesses within their jurisdiction. NamPower supplied to farms and mines, and there were over 70 licensees in Namibia (ECB, 2007a).

The establishment of REDs was realized according to the Companies Act (1973) as a private company with mainly public shareholders such as LAs, RCs, and NamPower (ECB, 2007a; ECB, 2007b). This type of private company's structural and legal setting was considered to improve operational efficiencies, unify tariff structures, and ensure good customer service in a very specialized service area: the supply of electricity (ECB, 2007b). Thus, with the introduction of REDs, an increase in efficiency to decrease costs was expected to reach Vision 2030 (ECB, 2007).

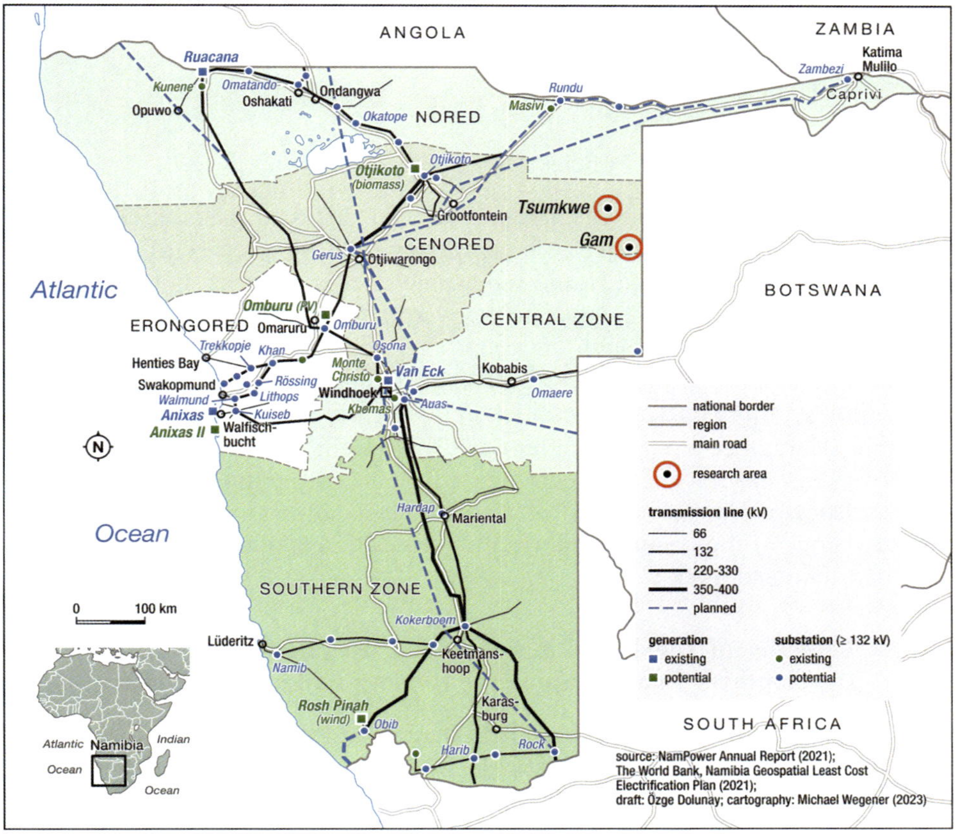


*Figure 6.5. REDs and the Power Plants with the Transmission Lines*

The formation of REDs on a specific regional level, while taking responsibility from the regional and local authorities, demonstrates the differentiation and rationalization in the operation of electricity networks. The restructuring of ESI also involves the differentiation and rationalization process, without any doubt, since the conformity of standards, tariffs, infrastructure, and operation are interrelated with the public actors involved in the SAPP network and supply of national customers and the involvement of private actors.

### *6.1.3.3 Private Actors at the Intersection of Regional and Local Levels*

Besides the private actors at the national level, there are also other private companies at the regional level that work in the intersection of regional and local, providing construction, installation, and O&M of the off-grid systems throughout the country, and supporting off-grid electrification financed through different donors.

In summary, this brief introduction of MLG electricity market actors developed and emerged starting from the independence, enabled a wide energy policy framework elaborated and adopted.

## 6.2 Renewable Energy Policies

The process of differentiation on supra-national, national, and regional levels with very clearly defined objectives complying with the process of rationalization has also been reflected in the renewable energy policies as a part of the modernization process of energy. The government's vision of reaching socio-economic development through the country's industrialization has led the decision-makers to shape the policies accordingly. Thus, "energy" and "electricity" were admitted to be critical necessities in taking key steps toward a prosperous Namibia (Harambee Prosperity Plan [HPP]/Republic of Namibia, 2021). The modernization process that the energy sector has been going through since the independence of Namibia has introduced new legislation and policies to pursue the set visions of the country and meet the needs and conditions of the national and supra-national market.

### 6.2.1 Timeline of the Key Energy Policy, Regulatory Framework and Development Plans Shaping the Renewable Energy Market

Consequently, several energy-related legislation and policies have been elaborated on in accordance with the National Development Plans. Figure 6.5 below presents the key legislation and policies, along with the main actors in the field of (renewable) energy, by showing the processes of differentiation and rationalization occurring subsequently in the restructuring of the electricity sector. The differentiating structures for specific necessary services interlinked and intertwined to each other have been illustrated from the levels of governance, the type of the legal entities, and their grid-connected/related infrastructures.

In the pursuit of becoming the 10$^{th}$ member of SADC, Namibia signed the SADC Protocol on Energy in 1996, which defined the objectives of energy cooperation in the SADC Region. Accordingly, it promoted the development of national energy policies for harmonious, balanced, and equitable energy development in the SADC region. The guiding principles outlined in the Protocol of using energy to support economic development and growth, alleviate poverty, promote self-reliance among Member States, encourage the development and transfer of science and technology related to energy through research and development, encourage direct participation of the citizens and communities in the development and use of energy, and to ensure sectoral and sub-sectoral policies and development (SADC, 1996). Thus, the SADC Energy Cooperation Policy and Strategy (1996), the SADC Energy Action Plan (1997), the SADC Energy Activity Plan (2000), and the Regional Infrastructure Development Master Plan and Energy Sector Plan (2012) were developed after this Protocol (von Oertzen,

2019). This supra-national cooperation also supported the development of the Namibian energy sector, its restructuring, and the specialization within the energy sector and its sub-sectors on a national level.

Figure 6.6 also illustrates how supra-national support has underpinned the policy development activities from the beginning, while the energy infrastructure remained public until the National Policy on Independent Power Producers ([IPPs], published by MME, 2018). Furthermore, the beginning of the 2020s brought new policy strategies with the (expected) addition of the NERA and the Electrification Fund to meet universal energy access by 2040. Thus, this temporal development of the energy market with its infrastructure, policies, actors, and networks highlights the differentiation, rationalization, and domestication in off-grid RE governance. However, it also demonstrates that the regional and local activities do not follow the same governance unfoldment, illustrating where the next steps should take place for universal access.

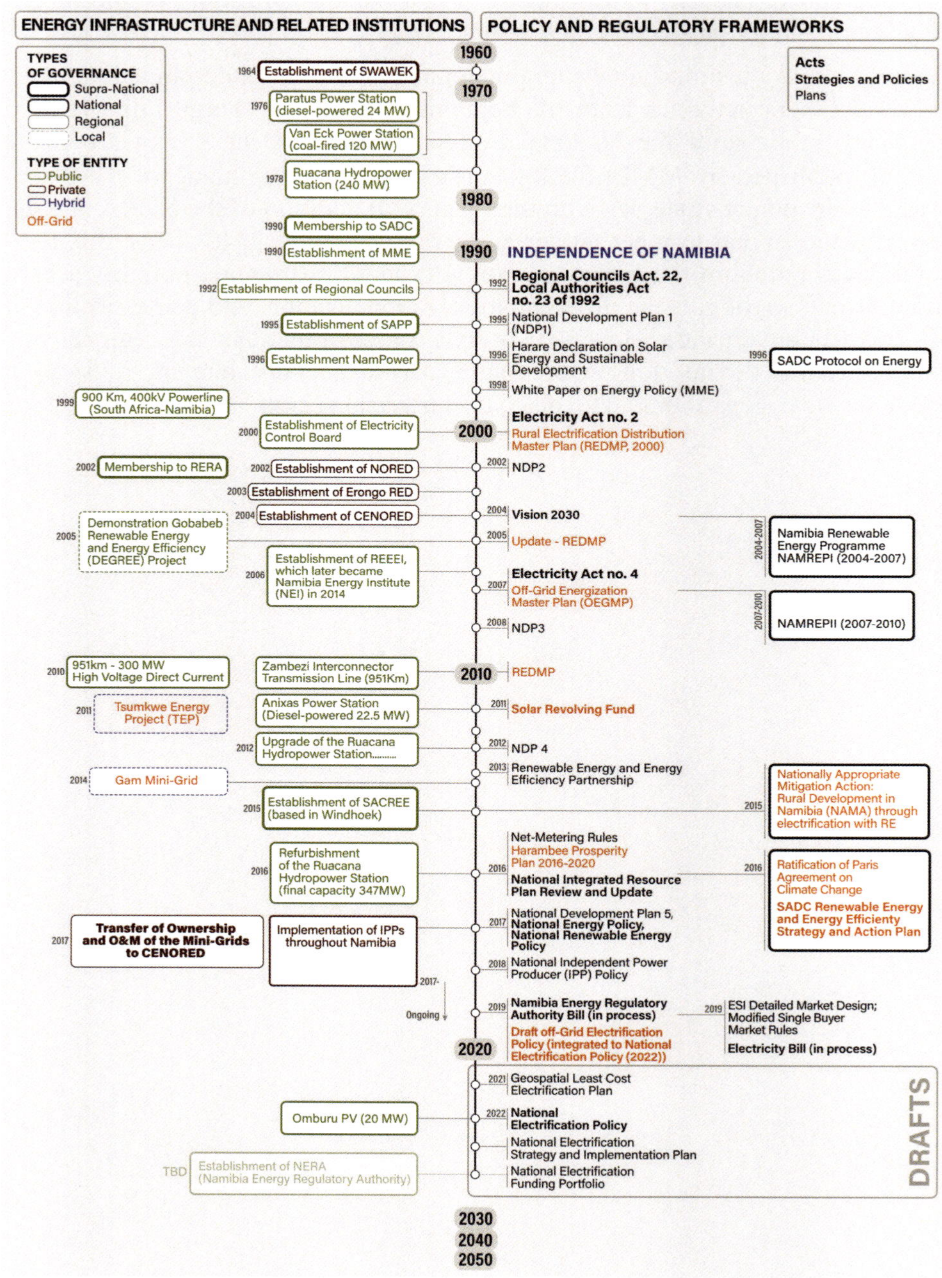


*Figure 6.6. Namibia's Renewable Energy Governance Timeline*
*Source: Author, illustrated by Bentley (2024)*

In parallel with the SADC regional developments, the White Paper on Energy Policy (MME, 1998) was the first policy to recognize the importance of renewable energy and energy efficiency and served as an overarching energy policy for the development of the energy sector. It prepared a solid ground for promoting renewable energies, the necessary institutional and planning framework, public awareness, financing schemes, and human resources. Through a participatory process with experts and officials, 57 issues of concern have been identified. As a result, the objectives of the White Paper were categorized under the security of supply, effective governance, social upliftment, investment and growth, economic efficiency, economic competitiveness, and sustainability. Furthermore, the government considered the (re-)structuring and rationalization of the electricity sector by formulating the Electricity Act, tariff, and electrification targets, as well as harmonizing with the other SADC countries in the region regarding new investments. The situation of the rural areas was recognized; special attention was provided to "the poor urban and rural households that were neglected historically", including rural electrification, rural water supply, solar housing, and water heating with the commitment to "building economy, and improving the quality of life for all its citizens" (MME, 1998). The specific policy on energy, outlining the main challenges and the commitments towards Namibia's development goals, including existing and future energy/electricity actors, highlighted the strong differentiation and rationalization processes that the GRN decided to adopt.

The Electricity Act 2 of 2000 (Republic of Namibia, 2000) concretized the aims of the GRN in restructuring the energy sector. It enabled the establishment of the Electricity Control Board in the same year, which became the regulatory authority in overseeing the Electricity Supply Industry (ESI); this was followed by the formation of REDs within the organizational framework of ESI. The Electricity Act 2 of 2000 was repealed with the Electricity Act 4 of 2007 (Republic of Namibia, 2007), which further detailed the regulations and standards in the provision of electricity.

Furthermore, since energy/electricity was and is considered the most vital in achieving the set objectives in the socio-economic development of the country and improving the quality of life of all citizens, the subsequent Vision 2030 (MME, 2004) was supported as being a long-term policy framework for the national development of GRN. Meanwhile, the National Development Plans and the Harambee Prosperity Plan I (2016-2020) ([HPP], MME, 2016) presented five-year short-term plans provided by the National Planning Commission. The Vision 2030 (MME, 2004) was intended to be a "perception of the future" and to create policy synergies that connect the long-term perspective with the short-term planning for improving the "quality of life of the people of Namibia to the level of their counterparts in the developed world in 2030." It applied sustainable development as a core principle in attaining its objectives in inequality and social welfare, human resources development and institutional capacity building, macroeconomics, population, health and development, natural resources, knowledge, information and technology, and external environment factors (Vision 2030/MME, 2004). It also recognized partnerships among different actors within society as a prerequisite for achieving efficient, dynamic, and sustainable development (Dolunay, 2021).

### Development Plans

After the publication of National Development Plan 1 ([NDP]1, published by National Planning Commission ([NPC], 1995), which was followed by Vision 2030 (MME, 2004), NDP2 (NPC, 2002), NDP3 (NPC, 2008), NDP4 (NPC, 2012) until NDP5 (NPC, 2017). The review of each NDP contributed to designing more detailed plans and objectives. In this sense, NDP2 (NPC, 2002) included a wider perspective of governmental and non-governmental actors. Furthermore, it introduced the outlines of the macroeconomics, sectoral, and cross-sectoral policies, with regional development issues, and elaborated the Public Sector Investment Programme for funding (NDP2/NPC, 2002). NDP3 (NPC, 2008) strengthened the previous plans and visions of an industrialized nation, considered the strategies for increasing local energy generation capacities and distribution networks, improving the regulatory framework, establishing a commercial trading center and enforcing regionally harmonized tariffs, implementing the Rural Electricity Master Plan for remote areas with off-grid renewable energy, extending the urban electricity network, promoting RE and efficient use of energy by introducing public awareness campaigns and technical programs, and increasing local capacity (NDP3/NPC, 2008).

In contrast to the previous three plans, NDP4 (NPC, 2012) adopted a higher-level perspective with overarching goals of high and sustained economic growth, employment creation, and increased income equality to be achieved by 2017 (NDP4/NPC, 2012). The details on how to achieve these goals were left/delegated to the respective ministries and institutions. Logistics, tourism, manufacturing, and agriculture were selected as the key areas to achieve the goals (NDP4/NPC, 2012). NDP5 (NPC, 2017) was published after the Harambee Prosperity Plan. It determined four pillars: economic progression, social transformation, environmental sustainability, and good governance. Energy and electrification were considered in the expansion and modernization of physical infrastructure, which would be prioritized to support industrialization and export development (NDP5, 2017). As a result, the five published NDPs demonstrated the goal, research, and economic-oriented development as a united vision of the country by placing it in the rationalized strategies, plans, and actions toward a prosperous Namibia.

Furthermore, shortly after marking the 25th anniversary of the Republic of Namibia, the Harambee Prosperity Plan (HPP/MME, 2016) served as a complementary plan for the long-term national goal of prosperity with the motto of "One Namibia, One Nation, Peace, Reconciliation, Security, and Stability" (HPP/MME, 2016). It consisted of attaining the goals of effective governance, economic advancement, social progression, infrastructure development, and institutional relations and cooperation. HPP goal 11 aimed to increase the electricity generation capacity from 400MW to 600MW, provide electricity in all schools and health facilities by 2020, and increase rural electrification from 34% to 50% by 2020 (HPP/MME, 2016). In addition, RE would be emphasized through the National Integrated Resource Plan (NRIP) that was in preparation, thus suggesting very quantified targets compared to plans published in earlier years; this allowed more specialization and emphasis not only on renewable energy and its infrastructure within the energy mix but also on how energy infrastructure formed the backbone of economic development and industrialization of the country and the potential cross-sectoral actions that might take place in achieving these goals. Meanwhile, the ECB and MME developed new policies for the expansion of the energy/electricity sector.

**Resource and Energy**

The National Integrated Resource Plan Review and Update (NIRP/MME, 2016) was completed in 2016, updated the NIRP (2012), and further detailed focused action (plans). Thus, the reviewed NIRP (MME, 2016) elaborated explicitly on the electricity sector development and projections for the twenty years between 2015 and 2035, including electricity demand, supply, and costs from various sources at the lowest rate (NIRP/MME, 2016). In addition, the impact of developing other energy resources and the demand side measure for reducing electricity demand was also considered in line with the increase in demand and new generation and transmission prospects (NIRP/MME, 2016). An implementation plan covering short and medium-term plans was also included in the NRIP itself. This is summarized in three key activities for the ESI: securing access to short-term rental generation by 2018, installing fossil-fuel base-load generation by 2021, and continuing programs to install solar PV and wind generation while investigating other RE technologies.

Following the NIRP, the National Energy Policy ([NEP], published by MME, 2017a) was adopted as an update to the first energy policy, "White Paper on Energy Policy," that served for twenty years. NEP (MME, 2017a) aimed to review, update, and create a future-oriented energy policy to accommodate the rapidly changing country and the world. Once more, "energy" was recognized as "an essential ingredient of modern life" and a "critical pre-requisite for Namibia´s socio-economic development" and "energy sector" of "strategic importance to yield maximum development benefits," which would align with the sustainable development and industrialization vision of the country (MME, 2017a). Moreover, specific policy objectives were outlined for rural and urban household access to energy (P29), rural institutional and commercial access to energy (P30), energy as an integral part of national development planning (P32), transparency and good governance (P34), capacity development (P37) and energy-related research and development (P39) among other policy statements in electricity and energy sector. In addition, NEP (MME, 2017a) also framed the envisaged future legal and regulatory framework with the introduction of the National Energy Regulatory Authority (NERA) and NERA Act as an overarching framework for electricity, downstream gas, thermal energy, energy efficiency, and other energy bills, while excluding downstream liquid fuels and upstream oil and gas. As a result, NEP detailed and foresaw more specialized sector and sub-sector activities and included the perspectives of their regulatory framework.

Subsequently, the National Renewable Energy Policy ([NREP], MME, 2017b) was adopted following the path of specialization within the energy field, complementing the NEP (MME, 2017a) and serving as a more explicit commitment, and boost, to the RE implementation. It aimed "to make Renewable Energy a powerful tool for the Government of Namibia to meet its short-term and long-term national development goals, and to assist Namibians to climb the development ladder, empowered by access to energy at levels that facilitate engagement in productive activity" with a vision for Namibia "to become a regional leader in the development and deployment of RE within southern Africa (NREP/MME, 2017b, p. 7). The main objectives were making RE a vehicle for expanded access to affordable electricity in Namibia, confirming Namibia´s commitment to RE, boosting investor confidence, creating an enabling environment for RE development, accelerating the RE sector, and enabling greater participation (NREP/MME, 2017b, p. 7). Nine specific goals and targets and twenty-five core policy statements were also stated to achieve the objectives for both on and off-grid urban and rural areas. It differentiated the regulations and procurement based on their system sizes and considered a harmonization need for the regulatory framework. Thus, for all RE technology installations <500 kW, Net Metering Rules (2016) would apply, a RE feed-in tariff (REFiT) would apply for projects greater than 500 kW and less than 5 MW, and a competitive auction approach for projects greater than 5 MW would apply (NREP/MME, 2017b, p. 27). Thus, NREP (MME, 2017b) served as an enabler for further detailed policies, regulations, frameworks, and rules by recognizing the need to revise the electricity sector market framework, off-grid solutions, and the needs of (private) market actors.

### Private Sector Involvement and Modified Single Buyer Model

As a result, the National Independent Power Producer Policy ([IPPP], MME, 2018) was adopted by including private sector participation in developing, implementing, operating, and owning power generation plants based on a Power Purchase Agreement with an off-taker of the negated power (IPPP/MME, 2018, p. vii). It also classified the IPP Market as small-scale (<500MW under REFIT scheme), medium-sized (5MW-100MW), large power plants (>100MW) connected to the grid, and then off-grid power plants; this also legitimized GRN´s recognition that the investment requirements reaching up to $N97 billion, as identified in NIRP (MME, 2016), could be achieved with private investors rather than GRN alone (von Oertzen, 2019). Consequently, IPPs contributed 699.35 MW (2022) to Namibia's energy mix, equivalent to 9% of the overall mix (NamPower, 2022).

Another component of change over the following years is and will be the shift from the Single Buyer Model (SB) to the Modified Single Buyer Model (MSB), which will enable the IPPs generating electricity to sell directly to REDs, other large customers with the acknowledgment of bilateral agreements, instead of selling solely to NamPower (IPPP/MME, 2018, p. 23). Thus, this will allow the electricity market to have more trading potential among different generators and buyers. The MSB´s new market structure, "Contestable Customers" and "Eligible Sellers," are defined and are required to be registered at the ECB and NamPower and comply with the ECB licensing rules. The transition from SB to MSB consists of three stages until its full implementation (ECB, 2022). In the first phase, 1a, from 1 September 2019 until the end of June 2021, the Contestable Customers may purchase up to 30% of their energy demand from Eligible Generators (ECB, 2022). In the second phase, 1b, from 1 July 2021 to the end of June 2026, Phase 1a activities will continue: traders will be allowed to participate in the MSB market, and distribution-connected customers that are connected to 1MVA and above may purchase up to 30% of their energy demand from Eligible Generators or Traders (ECB, 2022) while NamPower will remain the single importer (ECB, 2022). After July 2026, Phase 2 will take action by keeping the conditions of Phase 1a and 1b, but allowing imports to Namibia between an "Eligible Seller at the border" and the contestable Customer (ECB, 2022). According to ECB, this second phase of allowing the imports is based on the key criteria that Namibia has met 80% or higher of the self-sufficiency in energy supply target (GWh) from local sources (ECB, 2022). As of May 2021, ECB informed that 15 registered Contestable Customers and 5 Eligible Sellers, and more applications, were under review (ECB, 2022).

Thus, MSB will be in complete operation by 2026. The Electricity Bill (2019) and the NERA Bill (2019) are in the review process by the legal drafters, the Net Metering Rules that are in place are under review, and the RE Grid Code will be integrated into the Transmission Code that is operational, but also under review (ECB, 2023a). State-owned companies like NamPower will lose their "single buyer" seat and total control over purchases in the electricity market with the introduction of private actors. This conditional unbundling is expected to bring competition into the sector while contributing to energy generation and self-sufficiency in supply. This illustrates not only an institutional and sectoral differentiation *per se* but also a change and a transition from the centralized power of the state toward a decentralized and shared power with the involvement of other market actors. This transition contributes to achieving the rationalized visions of the country as well as the individualized visions and goals of each actor involved within this new structure. The independence that emerged for each actor within their specialized field has also created new dependencies.

In conclusion, from a temporal hermeneutical perspective, rural electrification and/or access to electricity in the rural unelectrified areas have been considered in the legal and regulatory framework of Namibia since its independence. From one side, GRN invested in restructuring the energy market and, more precisely, the electricity sector to reach the country's goals and visions. These changes over this period, including adopting more specialized policies and plans, also aligned Namibia as an energy market actor in the SAPP in the SADC region. The integration of, or interest in, new technologies continues to be part of the PoM (from its differentiation, rationalization, and domestication). The domestication process uses different sources for energy generation and, more precisely, RE. As a result, they all contribute to the country's sustainable development as a self-sustaining country.

### 6.2.2 Off-Grid Electrification-Related Policies, Plans and Energy Access

As the previous sub-section suggested, Namibia´s policy and regulatory market have been very active for thirty years. From these outlined energy legislation and policies, RE has been considered an important source of energy within the Namibian Energy Mix as well as a practical tool for electrifying inaccessible areas within the vast geography of Namibia. Thus, towards the goal of universal energy access for all residents in Namibia, the following key policies and plans have been active in the field of renewable off-grid electrification (Table 6.1).

*Table 6.1. Key Policies and Plans for Off-Grid Electrification*

| Period | The Beginning (2000 – 2009) | Action Continues (2010-2019) | Moving Forward (2020 onwards)[16] |
|---|---|---|---|
| Development of Policies and Plans | • REDMP (2000)<br>• NAMREP I<br>• NAMREP II<br>• OGEMP (2007) | • REDMP 2010<br>• RE Programs undertaken at the MME<br>• Solar Revolving Fund (SRF)<br>• Draft Off-Grid Energy Policy (2019) | • GLCEP (draft, 2021)<br>• NELP (draft, 2022)<br>• NELFP (draft, 2022)<br>• NES (2022). |

*Source: Author*

#### *6.2.2.1 "The Beginning" (2000 – 2009)*

The beginning of off-grid electrification planning took place, starting with the White Paper on Energy Policy (MME, 1998) and followed by the elaboration of the Rural Electrification Master Plan ([REDMP], MME, 2000b). The national rural electrification in Namibia had been in place since independence and had connected 9,700 households in that year (Emcon Consulting, 1999). Later on,

[16] During the preparation and writing of this study, the new policies were in process of approval.

REDMP (MME, 2000a; MME, 2000b; MME, 2000c) detailed the planning of all regions with the identification of each locality un/electrified in 16 volumes, and it was planned to be updated every five years in 2005 and 2010. Spatial technology, like aerial surveys, digital videos, and GPS (Global Positioning System) for each locality were used to determine its coordinates, and Geographic Information Systems (GIS) were used to digitize the status quo of rural electrification in Namibia in 2000 (REDMP/MME, 2000a, p. 5). Accordingly, 2,865 rural localities in Namibia, 131 in the off-grid areas, and 2,365 in the "future grid-connected" areas were identified (REDMP/MME, 2000b, p. 25). Of the 131 rural off-grid localities, 51 were identified in the Otjozondjupa Region, including Tsumkwe and Gam. Although REDMP (MME, 2000) focused on grid extension while prioritizing the government buildings (Ileka et al., 2017), it consisted of a thorough analysis of the country in every region, including the technical analysis and financial cost for funding. It also used a "point scoring" methodology to identify priority areas (REDMP/MME, 2000c, p. 19). It addressed key issues of the REDMP implementation as well. Thus, this REDMP not only *differentiated* the needs in rural electrification in terms of soon-to-be-electrified and unelectrified areas but also differentiated the priority of government buildings while using digitalization as a tool as a part of the process of domestication in having more control over rural electrification. It also used quantification as a method in the prioritization process, strengthening the rationalization process.

The first edition of REDMP was continued with the NAMREP I that was conducted as a Baseline Study: Barrier Removal to Namibian RE Programme between 2004 and 2007 in order to determine the existing status of solar technologies in Namibia, to improve the solar technology industry and the promotion of cost-effective, and environmentally friendly technologies (Consulting Service Africa, 2005). It also aimed to provide accurate, measurable quantitative data for future interventions and be implemented in the NAMREP – Phase II (NAMREP, 2011). NAMREP aimed to improve rural people's livelihood opportunities through access to off-grid solar technologies, reduce dependency on imported fuels, and remove barriers in capacity, public awareness, institutions, social, financial, and technical areas (Deenapanray, 2011). Phase II focused on promoting the delivery of solar energy services that are commercially, institutionally, and technically sustainable to rural and off-grid communities. The programs were co-funded with the Global Energy Facility (GEF), supported by the UNDP, and implemented by MME.

In the meantime, the Off-Grid Energization Master Plan ([OGEMP], Schultz & Schumann, 2007) for Namibia was issued in consistency with the REDMP (MME, 2005). The OGEMP (Schultz & Schumann, 2007) aimed to "provide access to appropriate energy technologies to everyone living or working in the off-grid, pre-grid, and grey areas" who will not have access to electricity within 20 years (Schultz & Schumann, 2007). The "off-grid" areas, as defined in the REDMP (MME, 2000), would not have access to electricity within 20 years; the "pre-grid areas," according to the updated REDMP GIS Database, would not have access to electricity within ten years (Schultz & Schumann, 2007). The grey areas were not clearly defined in the REDMP 2005 but included informal settlements and farm worker settlements on commercial farms, in which OGEMP (Schultz & Schumann, 2007) would focus on informal settlements first. An "energy shop" approach was proposed, including energy technologies and consumer credit finance revolving funds. In a 20-year plan, 156 energy shops were expected to be established in all 13 regions (Schultz & Schumann, 2007). The shops would be located within a reasonable distance of the targeted communities, supplying energy products and compatible appliances; they were also intended to be payment collection points for revolving funds and customer consultation point (Schultz & Schumann, 2007) s. The annual funding allocated in REDMP (MME, 2000) was increased from N$5 million to N$6.8 million in the OGEMP for support in the off-grid areas. Thus, further specialization within the off-grid areas was completed by laying out funding plans and expanding approaches to their implementation based on an economic unfolding expansion strategy with the energy shops.

#### *6.2.2.2 Action Continues (2010 – 2019)*

In pursuit of the OGEMP, the updated REDMP 2010 (MME, 2010) used GIS to further define rural areas as "those communal areas which are outside of the municipal areas and commercial farms" (REDMP/MME, 2010). The locality was defined as "any rural location with government buildings, including homesteads within a 500-meter radius of a prospective transformer" (REDMP/MME, 2010a). A rural homestead was also defined as "having access to electricity either with the grid connection or within the 500-meter of low voltage grid connection (REDMP/MME, 2010a). Through the REDMP GIS, 13,566 localities were identified and mapped, categorizing 8,270 localities without electricity access (REDMP/MME, 2010a). Of these, 2,879 localities were planned to be electrified systematically within the next twenty years (until 2031) (REDMP/MME, 2010a, p. XI). Since the REDMP (2010) prioritized government buildings, 740 government buildings, 642 schools, and 59,774 homesteads were included in this plan with a total budget of N$ 1,473,361,699, which is equivalent to an annual budget of N$ 73,668,085 (REDMP/MME, 2010a, p. XI). Furthermore, a total of 27 off-grid localities with 17 schools, 17 government buildings, and 592 homesteads were defined in the Kunene, Otjozondjupa, and Omaheke regions, which were located in the remote rural locations and techno-economic electrification to the grid was not possible (REDMP/MME, 2010a). Thus, the updated REDMP in 2010 not only further detailed the following grid electrification plans of the rural localities prioritizing the government buildings but also included the off-grid areas within the off-grid electrification plans. Therefore, MME´s future forecasts for the 20$^{th}$ anniversary of its Independence were further rationalized and domesticated using more digital instruments and digitalization while differentiating the electrification process between on- and off-grid.

In the meantime, MME´s RE Division has been in charge of implementing RE and EE programs, especially with the involvement of supra-national actors such as UNDP, GEF, the Austrian Development Agency, UNIDO, SACREEE, and GIZ, to name a few (MME, 2023c). The division's key functions were providing energy access to off-grid public institutions, implementing and monitoring policies, providing information for investment purposes, creating awareness on energy-related issues, and facilitating research and capacity building in the energy industry (MME, 2023c). The Namibia Energy Efficiency Programme in Buildings (NEEP) (2010-2014), the Concentrated Solar Power Technology Transfer for Power Generation in Namibia Programme (CSP TT NAM) (2014-2017), Southern African Solar Thermal Training and Demonstration Initiative (SOLTRAIN) (2009-2026), and Green People Energy (GPE) are some of the programs they have been involved in (MME, 2023c).

In addition, MME has also been in charge of the Solar Revolving Fund (SRF), which was initially established as the Home Power Revolving Fund to make SHS more affordable for low-income, unelectrified households (MME, 2023b). SRF was integrated within the OGEMP to stimulate the use of off-grid technologies in off-grid areas, but not limited to, with a subsidized loan service to Namibian citizens at a fixed interest rate of 5% per annum for five years (Dolunay, 2021). Solar Home systems (SHS), photovoltaic pumps (PVP), solar water heaters (SWH), and energy-efficient stoves (financed as part of SHS) have been financed through the SRF since 2011. According to the figures of MME, 3,840 SHSs with a total budget of N$ 101 million were financed between 2011 and 2020 (MME, 2021).

All these policies and master plans were also supported through the Nationally Appropriate Mitigation Action: Rural Development in Namibia through Electrification with Renewable Energies (Stockmayer et al., 2015). This action document was developed with the Ministry of Environment and Tourism and UNDP MDG Carbon to support Namibia with the OGEMP goals of providing appropriate energy technologies to everyone living in the off-grid areas. The NAMA was a new tool initiated by the UNDP and first introduced in the Kyoto Protocol in Bali (2007) and the United Nations Framework Convention on Climate Change (UNFCCC) by being voluntary, non-binding policy instruments that provide a framework for pursuing a country's socio-economic and development goals, while participating in the global climate change mitigation agenda (Dolunay, 2021). On the other hand, this would also contribute to the SDGs and enable policymakers to track the impact of the NAMA on environmental conservation, economic growth, poverty reduction, and public welfare (UNDP, 2015). The NAMA aimed to improve private sector involvement, reduce GHG emissions by replacing fossil fuels with RE, increase RE share, and provide income generation and new business opportunities (UNDP, 2015). NAMA´s key instruments for Namibia were interventions A and B. Intervention A focused on establishing mini-grids in the vicinity of schools and potential tourism areas, which would provide electricity for Rural Productivity Zones (RPZs). Intervention B focused on the Energy Zones (EZs) with the scope of supporting the Energy Shops approach proposed by the OGEMP (UNDP, 2015; Stockmayer et al., 2015). The capacity building was a key component in the implementation of the NAMA for the next 15 years, with a total cost estimation of US$ 14 million; it aimed 10 mini-grids and 13 EZs, providing electricity for 1,400 households and 80 new enterprises, serving approximately 8,500 people (Stockmayer et al., 2015). With NAMA in Namibia, the environmental component or impact of using RE in rural development has been considered. However, no further action within this framework has been encountered during the research.

In 2019, the draft Off-Grid Electrification Policy (MME, 2019) was on the table for adoption to stimulate off-grid electrification targets and to reach universal access to electricity goals defined internationally and nationally. However, it was postponed to the next decade. Nevertheless, although REDMP did not explicitly focus on off-grid electrification, it still constituted an essential database for off-grid localities to roll out the next steps. The NAMAs, on the other hand, have paid attention to GHG emissions and low carbon technologies and brought a different component for measuring the impacts of the quantification of the SDGs, being an international tool to oversee the activities on a supra-national level. However, reflecting on the policy/plan development over the past two decades, from 2000 until 2020, and the increase of rural off-grid electrification, Namibia remained behind the envisioned scenarios.

Moreover, the introduction of NAMA also introduced a different tool into the RE deployment system, thus pluralizing the already complex network of RE. The invaluable human resources mentioned in every policy and plan introduced new definitions for implementation rather than pursuing their set targets for action. As a result, the challenge of the national actors within the RE deployment has occurred: while the efforts have been made to reach the supra-national levels with their involvement in continental and global activities, the execution of national-regional-local plans and activities remained behind as a consequence, due to the lack of human resources in different levels of national and sub-national governance presenting a challenging aspect amidst modernization.

#### *6.2.2.3 Moving Forward in Off-Grid Electrification (from 2020 onwards )*

The beginning of the 2020s was challenged by the global COVID-19 outbreak. The Draft Off-Grid Electrification Policy prepared by ECB and approved by the Board remained as a draft, which was later integrated into the draft National Electrification Policy (NELP/MME, 2022b) developed by the MME that is pending Cabinet approval (ECB, 2023a, p. 22). Given the challenges of electrification in Namibia, in 2021, the GRN commissioned a Geospatial Least Cost

Electrification Plan ([GLCEP], MME, ) to evaluate the technologies, costs, and means to reach universal access. A report was also prepared with the support of the World Bank Group. Accordingly, GLCEP´s intention as a technical and economic assistance tool was to provide complementary data-driven tools, identify the least investment-cost (capex) pathways, and conduct a strategic analysis of on- and off-grid solutions for electrification (GLCEP, Perret et al., 2021). Thus, with this plan, an up-to-date GIS-based spatial analysis of investment costs was completed after the publication of REDMP, which only considered rural areas. As a result, GLCEP (Perret et al., 2021) provided eight strategic messages and new definitions for the electricity sector to be further deployed in the electrification policies.

Furthermore, GLCEP's main outputs/strategic messages revealed that the "connectivity rate"—the percentage of households connected to either a grid or off-grid connection—would be 50% by the end of 2019, which is approximately equivalent to 300,000 households, leaving the rest of the country unelectrified (Perret et al., 2021, p. 10). Including the population growth until 2030, 433,000 households would still require electricity, of which 77% (324,000 households) are located within 2 km of the grid (Perret et al., 2021). Thus, according to the spatial distribution, five types of electrification segments were studied: grid densification, grid intensification, grid extension, isolated mini/micro grids, and standalone systems (Perret et al., 2021, p. 11). The total capital investment cost was calculated as N$ 6.1 billion for grid-connected investments for 347,000 connections with a share of 80% and N$ 2.3 billion for off-grid connections of 85,000 households with a share of 20% of the overall connectivity (Perret et al., 2021, p. 13). Accordingly, 17,000 connections, equivalent to 20% of all off-grid connections, were found to be micro and mini-grids (exclusively for solar PV) with a total investment cost of N$ 774,514,841 and 80% of stand-alone systems with N$ 1.5 billion investment cost with 68,437 SHS in total (Perret et al., 2021. p. 15-16). The mini-grids were identified in CENORED and NORED zones, while the micro-grids covered all zones, with the majority being in NORED, including the SHSs. From an installed capacity perspective, this investment translated to 480 kWp mini-grids (720 connections), 9,747 kWp micro-grids (16,245 connections), and 20,531 kWp SHS (68,438 connections) (Perret et al., 2021, p. 70).

GLCEP also used the ESMAP Multi-Tier Framework to define electricity access and accepted a minimum Tier 3 planning standard (Box 3) for mini and micro grids, including pre-accession scenarios of Tier 1 and Tier 2 (Perret et al., 2021, p. 19). GLCEP provided Manifold for GIS Platform and GEOSIM electrification planning software in addition to the report (Perret et al., 2021, p. 9). In this way, moving forward in the third decade of the universal access efforts and universal electricity access for all, MME and ECB mainly updated and revised the electrification planning based on the RED zones, which narrowed down the earlier geographic-region-based plans of 14 regions. This new arrangement in the study illustrates a strategy based on the REDs operational zones, adopting a more target-oriented, rational implementation plan with forecasted installed capacity and investment costs, and perhaps also including services and responsibilities in the actual rollout of the plan in the future. In this way, REDs position have been strengthened and supported.

Moreover, this GLCEP created the basis for the new policy framework, in which the MME, with the support of the ECB, developed the "Electrification Support Mechanism to Improve Electrification of Households in Urban and Rural Namibia (Support Mechanism)" in order to design a program for the expansion of electrification (draft NES/MME, 2021). It is grounded on three main pillars: the National Electrification Policy (NELP), the National Electrification Fund, and the National Electrification Strategy (NES) or Master Plan (draft NES/MME, 2021, p. iv).

Thus, rolling onto the next decade in electrification in Namibia, this support mechanism, which consists of NES (MME, 2021), NELP (MME, 2022b), and NELFP (MME, 2022a) , provides a practical guide and pragmatic framework for leading and guiding the main energy market actors such as the MME, the ECB, the NamPower, the REDs, and other stakeholders in Namibia within the universal access targets by 2040. Therefore, this new mechanism-supported framework introduces a prominent shift in the electrification perspectives of the GRN from the beginning of the 2020s.

*Box 3. Multi-Tier Framework for Measuring Access to Electricity*

| ATTRIBUTES | | TIER 0 | TIER 1 | TIER 2 | TIER 3 | TIER 4 | TIER 5 |
|---|---|---|---|---|---|---|---|
| Capacity | Power Capacity (W) | Less than 3W | At least 3 W | At least 50 W | At least 200 W | At least 800 W | At least 2 kW |
| | Daily Wh | Less than 12 Wh | At least 12 Wh | At least 200 Wh | At least 1.0 kWh | At least 3.4 kWh | At least 8.2 k |
| | Typical Technology | | Solar lanterns | Standalone solar systems | Generator or mini-grid | Generator or grid | Grid |
| | Services | | Lighting of 1,000 lmhr per day | Electric lighting, air circulation, television, phone charging | | | |
| Availabiliry | Daily Availability | Less than 4 hours | At least 4 hours | | At least 8 hours | At least 16 hours | At least 23 ho |
| | Evening Availability | Less than 1 hour | At least 1 hour | At least 2 hours | At least 3 hours | At least 4 hours | |
| Reliability | | More than 14 disruptions a week | | | At most 14 distruptions a week or at most 3 disruptions per week with total duration of more than 2 hours | more than 3 up to 14 disruptions per week or less than 3 disruptions per week with more than 2 hours of outage | At most 3 disruptions pe week with tota duration of less than 2 hou |
| Quality | | Household experiences voltage problems that damage appliances | | | | Voltage problems do not affect th use of desired appliances | |

*Source: Adapted from Bhatia and Angelou(2015, p. 77) and ESMAP(2023)*

From specialized services and priorities provided for governmental buildings like schools, health centers, and other GRN buildings, the new framework provides an inclusive strategy to the county by bringing urban and rural geographic areas and governmental buildings and households by not differentiating them anymore. However, it creates a holistic picture of the electrification strategy for the entire country under SDG7. Although this seems a unifying strategic action plan, the strategy itself re-defines the electricity access terminology; it shapes a more targeted and implementation-oriented framework by including REDs jurisdiction areas, including the ones that are expected to be established, and prepares the basis for a better (renewable) energy governance on a regional and local level. In this, a re-division of labor for the specified electrification targets occurs while the GRN, with its stakeholders, incorporates the domestication process for the same purpose, involving a partial digitalization of the planning, execution, and monitoring activities.

On the other hand, the draft NES (MME, 2021) also integrates three pillars as strategic initiatives: (1) institutional and program management, (2) technical and planning, (3) policy and regulatory, and (4) financial that will run in two parallel tracks, with one emphasizing on the adjustments in the electrification management, regulatory necessities, and capacity building activities, the second will focus on the implementation of the grid and off-grid investment plans highlighted in the GLCEP (draft NES/MME, 2021, p. v). It also identifies the gaps in implementing the strategy under institutional roles and responsibilities, planning and technical, regulations and affordability analyses, and financing. While the MME has been foreseen as the proactive lead in electrification, it also engages with the coordination of all planning and allocation of resources. Thus, it includes the expansion of the Electricity Division to undertake additional roles and responsibilities outlined in the successful implementation of the new support mechanism. Although the MME is projected to be the lead and coordinator in this mechanism having the central power, it also shares the responsibilities according to the expertise of their stakeholders (ECB, NamPower, and REDs) in the use of the geospatial database from the GLCEP (Perret et al., 2021) for the technical and planning purposes as well as for monitoring development. The support of additional regulatory guidelines, standards, tariff guidelines, and requirements is also recognized. The roles and responsibilities of each stakeholder, the new organizational structures, and the primary training and capacity-building needs to guide the electrification toward 2040 have also been considered within the NES (MME, 2021). Private sector involvement is also regarded as essential in reaching off-grid electrification targets. Nevertheless, this would require adjustments to the implementation period of the SDG7 target from 2030 toward a more realistic target of 2040.

Forming part of the NES (MME, 2021), the NELP (MME, 2022b) incorporates a vision "to provide universal access to electricity in Namibia" with the primary goal of "increasing access to electricity services using innovative technological, public and private sector participation and funding approaches, for the benefit of all end-users and Namibia´s development, thereby reaching universal access to electricity by 2040" (NELP/MME, 2022b, p. xi). It adopts nine main objectives, including (P1) optimization of institutional anchoring and increase in private sector participation, (P2) creating a holistic electrification planning and prioritization framework, (P3) expansion and optimization of electrification funding, (P4) guiding the electrification cost recovery, charges, and subsidies, P(5) development and implementation of appropriate minimum standards, (P6) expansion of off-grid electrification, (P7) ensuring awareness, communication, and expectation management, (P8) ensuring adequate and timely capacity development, and (P9) establishing reliable data collection and reporting mechanisms (NELP/MME, 2022b). A clear policy alignment is also presented with the industrialized Namibia and "social upliftment" as per the Vision 2030 (Republic of Namibia, 2004), the NDPs, the Harambee Prosperity Plans (Republic of Namibia, 2016 and 2021), the Paris Climate Agreement, the NEP (MME, 2017a), the NREP (MME, 2017b), the IPPP (MME, 2018) and the NIRP (NELP/MME, 2022b). These objectives highlight a calculable, controllable, goal-oriented, inclusive framework that creates a dependency on the geospatial platform during its implementation phase. At the same time, the GRN holds the lead with shared responsibilities with other stakeholders.

In addition, the guiding principles are framed by placing the government´s role and priorities as a national custodian and coordinator that provides core funding but also other measures to attract additional funding for increasing electricity access; the economic principles explicitly state that "the electricity services will not be provided for free," productive uses will be promoted, socio-economic developments will be prioritized, and subsidies and cross-subsidies will be implemented (NELP/MME, 2022b). The planning and implementation principles align with the NDPs, grid, and off-grid approaches on a "technology-neutral" basis based on the least cost. The institutional principles include harmonization of existing and new electrification implementors, private sector involvement, and "realistic, tangible, and affordable institutional and governance arrangements" (NELP/MME, 2022b). Lastly, electricity service principles underline Multi-Tier Framework (MTF) Tier 3 as the minimum service level. At the same time, the GRN remains "cognizant of the needs of end-users" with a "uniform approach to the determination of connection charges for household electrification" (NELP/MME, 2022b, p. 15). Thus, in addition to the objectives, the guiding principles present the boundaries of the goals to be achieved more clearly. Overall, they constitute electrification's institutional, social, economic, and technical dimensions.

Finally, NELP (MME, 2022b) also integrates an implementation framework in which the institutional actors and their primary roles and responsibilities are described in this new setting; the implementation plan, resource mobilization, reporting, evaluation, monitoring, and dissemination are detailed. It also guides how the resources will be sourced, mobilized through NELFP (MME, 2022a), and applied in tandem with the NELP (MME, 2022b). NELP (ME, 2022b) provides different funding types such as distribution licensee, government, Development Finance Institution (DFI) debt, commercial debt, quasi-equity, equity, and grants and considers projects as public projects, commercial projects, or intermediate projects (NELFP/MME, 2022a).

Moreover, establishing an "Electrification Fund" has been foreseen to channel electrification projects mainly by collecting regular electrification funding contributions from the GRN, SOEs, end-user levies, attracting private investment, and other further investments. Although the ownership of the Electrification Fund has been foreseen by the GRN, with the MME holding its custodianship along with a legal entity, the type of entity (a unit trust, a trust, an en-commandite partnership, a private or public limited company) remains under the research process (NELFP/MME, 2022a, p.48). The governance and management of the Electrification Fund will follow "the highest corporate governance standards to ensure its credibility and attract third-party funding" by appointing a "Management Company" that will be managed by the Board of Directors of the Electrification Fund and governed. As a result, this new legal entity is expected to manage and operate a secure funding portfolio to reach the electrification goals (NELFP/MME, 2022a), which will further contribute to the pluralization of actors to engage in specific purposes and adopt a shared governance structure with the possibility of private management.

In moving toward the electrification goals in Namibia by 2040, new energy actors such as the NERA and the Electrification Fund have been projected. These entities contribute to the differentiation of the (institutional) energy market actors by the pluralization to undertake specific duties for a unified goal of electrification mentioned in all national plans and with the off-grid electrification utilizing solar energy. However, the GRN remains in control without giving autonomy to the new actors, but rather a shared responsibility for the rationalized actions to take place. As is also foreseen in the NELP (MME, 2022b), the ECB is preparing an off-grid electrification regulation aligned with the projected off-grid electrification (PROCEED Conference, 2023). Thus, the modernization process continues by adopting more specific policies and procedures, rationalized actions, specialized entities, and the involvement of the private sector by using digital tools and RE technologies.

As a result, the sum of the acts, policies, master plans, and development plans adopted and forecasted by the GRN over approximately the past 30 years accounted for (n=34). The drafts awaiting the Cabinet's approval have also been added to the count. The rules and standards have been excluded. Accordingly, a process of differentiation and rationalization shall be noticed, with the specialization on energy, followed by RE, electricity being one form of energy, and, very specifically, off-grid-related electrification goals. On the other hand, the process of differentiation within the institutional governance allowed the formation of ECB, REDs, and the NERA in the future for the development of the specific policy and regulatory framework that indicates more specialization and planning. Thus, the off-grid-related policies and plans accounted for 9% of the total, providing the smallest ratio. The efforts in setting up energy policy framework account for 38%, specifically for RE 24%, and electrification 29%, which provides a distributed interest.

In addition, on the vertical hierarchy, 15% was allocated to the supra-national level and 85% to the national level. 20% was dedicated to the appropriate policies, and 43% to the development, resource, and master plans. Additionally, special funding for off-grid RE deployment represented only 6%. Namibia´s electrification demonstrates a process of modernization in the energy and electricity sectors with a pace aligned with its own while taking into consideration the soon-to-be-adopted support mechanism for electrification from a holistic perspective that includes on-grid and off-grid connections, emerging entities, and follow-up regulations, rules, and standards.

#### *6.2.2.4 Off-Grid Electrification in Namibia*

In this sense, in 2004, the pioneer off-grid installation with a capacity of 26k$W_p$ was realized by the Gobabeb Research and Training Centre (GRTC) in the Central Namib Desert (Dolunay, 2021). The mini-grid system supplying electricity to the offices, houses, and other buildings in the GRTC ran under the DEGREEE[17] Project, funded by the Danish Development Cooperation, DANIDA (Dolunay, 2021). Following this project, in 2011, the Tsumkwe Energy Project (TEP), which was funded by the European Union, NamPower, and the Otjozondjupa Regional Council, was put into operation in Tsumkwe, with an installed capacity of 202 k$W_p$ and later upgraded to 303 k$W_p$ in the coming years. As described in detail earlier, TEP was followed by the Gam mini-grid system in 2014, with a capacity of 292 k$W_p$, supplying electricity to the Gam settlement. Furthermore, the B2Gold mining company has its own fuel oil generating power plant as well as a 7 MW solar PV plant (von Oertzen, 2019, p. 44) as an off-grid power supply.

Besides the mini-grids, the stand-alone systems that provide electricity to farms, households, and public buildings are operated privately or by the government. Solar home systems (SHS), photovoltaic pumps (PVP), solar water heaters (SWH), and energy-efficient stoves have been funded under the Solar Revolving Fund (SRF), which provides subsidized loans to Namibian citizens since 2011 (Dolunay, 2021). Accordingly, 3840 SHSs have been financed through SRF from the beginning of the scheme in 2011 until December 2020 (Dolunay, 2021). Moreover, the Off-Grid Renewable Energy Statistics 2019 (IRENA, 2021) stated that solar light and solar home systems (up to 50W) accounted for 115 kW in Namibia. The Off-Grid Renewable Energy Statistics 2019 (IRENA, 2021, p. 14) also stated that solar water pumps for 953 kW, of which 840 kW was for agricultural use and the rest for public water supply.

---

[17] Demonstration Gobabeb Renewable Energy and Energy Efficiency (DEGREEE)

MME aimed to electrify government buildings in remote localities via containerized solar PV systems (Figure 6.7) as a part of the Off-Grid Energization Master Plan (OGEMP). Within this framework, sixteen off-grid systems were funded and installed in the Omusati, Kunene, and Otjozondjupa regions, with a total capacity of 60.28 kW, and NEI continued the dissemination/training activities using the Energy Trailer (Figure 6.8). In addition to these stand-alone systems, there were other initiatives such as "Solar 4 Health" funded by the UNDP, "Sonnenkinderprojekt" funded by an NGO, or other private initiatives (Dolunay, 2021). Lastly, the Sorris Sorris Conservancy and the Sheya Shuushona Conservancy were funded for the installation of each 150kW solar power plant in their respective conservancy areas through the Environmental Investment Fund of Namibia under the project "Empower to Adapt: Creating Climate Change Resilient Livelihoods through CBNRM in Namibia" (Author Correspondence, 2023).

*Figure 6.7. Containerized PV Systems*
*Source: Author*

*Figure 6.8. NEI Energy Trailer*
*Source: Author*

As a result, besides the policies and plans, the off-grid electrification projects indicate multi-stakeholder activities in Namibia, which have remained slow compared to the off-grid electrification endeavors of the government. On the other hand, the historical development of the necessary policy and regulatory framework activities on the national level with the support of supra-national actors illustrates that the process of modernization simultaneously occurs with the deployment of the (RE off-grid) electrification. However, the development of policy-to-action regulatory steps remained missing before the IPPP and the Net-metering Rules that enabled RE integration into the grid for the private sector. Thus, without the necessary governing social, technical, and economic rules, standards, and guidelines that would set the basis for following the policies, the RE off-grid electrification could not reach planned localities. Consequently, Namibia's universal electrification target was postponed from 2030 to 2040.

In this respect, regarding modernization, the differentiation, rationalization, and domestication that has been occurring at the national and supra-national levels have not continued at the same pace at the sub-national level in the past twenty years to implement the planned activities. Although the leading role of the MME in this process creates an expectation by all national and sub-national actors to advance the implementation and reach the desired goals, human resources and their integration into the planning activities at all MLG levels have been found missing or have not been clearly communicated. This situation suggests that the policy, technical, and economic plans have not considered the necessary human resources at all MLG levels that would undertake the expansion of (off-grid) electrification in Namibia. As a result, this creates an added complexity for the already sparsely populated country, unless specific measures are taken.

## 6.3 Complexities and Opportunities for Off-Grid Electrification

Based on the conditions of the un-and electrified areas in Namibia, the interviews were conducted online in 2020 with the national and regional actors (NRA) who have been involved in the works related to off-grid electrification.

### 6.3.1 Complexities and Limitations

The challenges of off-grid electrification in rural areas were summarized as policy, licensing, tariff setting, financial, social, and technical challenges (Hoeck et al., 2022). The intertwined structure of RE governance in terms of its actors and the policies connect them all for the implementation of technically sound, economically viable, socially acceptable, and overall sustainable off-grid projects. In this sense, the NRAs were interviewed, and the following complexities and limitations were identified and presented for off-grid electrification in Namibia (Table 6.2) according to the social, socio-techno-economic, technical and management, socio-economic, and institutional aspects.

*Table 6.2. Complexities for Off-Grid Electrification*

| Type | Complexities and Limitations |
|---|---|
| Social | • Awareness and failed expectations<br>• Expectations of the people in off-grid areas and Ownership<br>• Accessing people |
| Socio-techno-economic | • Grid extension versus solar energy and social acceptance<br>• Energy shops and lack of private sector support |
| Technical and Management | • Operational and maintenance issues |
| Socio-economic | • Inaccessibility of funding and subsidies in rural areas |
| Institutional | • Licensing and/or certification of installers according to the standards that don´t exist. |

*Source: Author*

### *6.3.1.1 Social Complexities and Limitations*

**Awareness and Failed Expectations/Disappointments**

> "So, the **whole issue of awareness** and because it´s just not that, fixing the problem **that are associated to solar**. We **unfortunately have too many solar systems that have disappointed people**. So, **that places** into that **whole mindset** that solar is really not the best because there are so many examples of systems that were put up. And **so much promise and did not deliver"** (NRA2, Pos. 36).

Awareness has been defined as "knowledge that something exists or understanding of a situation or subject at present based on information or experience" (Cambridge Dictionary, 2023). Based on the earlier unsuccessful experiences with solar systems, the NRAs claimed that too many solar systems disappointed people, creating a negative connotation or mindset toward solar energy. In addition, the promises were also not met. Thus, this suggests that the earlier implementations of solar systems involved some promises regarding the system and its benefits. However, after the installations, those promises were not kept or malfunctioned, or perhaps, simply the operation and maintenance did not take place, and the systems failed after a while, leaving the solar systems behind dysfunctional and the beneficiaries disappointed. Two main challenges are the management of expectations and the appropriate public awareness aligned with the action undertaken, which go hand in hand. In the context described by the NRA, the beneficiaries got their awareness through their gained experiences. In another context, with the parallel public awareness or beneficiary/user awareness raising, a different experience might have been obtained.

**Expectations of People in the Off-Grid Areas & Ownership**

> "I think from my side, it´s, you know **perception of the beneficiaries** sometimes, so once you get the electricity, **they expect to be able to do a lot of things**, you know. And then, you **have to bring in the elements** of telling people what are the **limitations**, what are **the demand side measures**. So, there is apart from bringing them, this **wonderful thing such as electricity**, people wanna do also, **they wanna do the same things as if they´re going to town and visit their friends in a grid-connected city**. So, perception is one thing. And I think the second one is to some extent is **the ownership**. You know, if, **if you pay for something you feel that cost, you would be better taking care of somethings** for example. This is now based on experience from smaller system not big mini-grid systems as in Tsumkwe and Gam" (NRA4, Pos. 33).

In continuation of the awareness and the expectations of people, the NRA4 shares the experiences over the perceptions of the beneficiaries that were obtained in the small system off-grid projects, whereby two key issues have been highlighted: (1) beneficiaries who expect to use electricity as like in the towns, where the grid is connected to, (2) ownership of the beneficiaries. In this quote, the limitations of off-grid electricity and, consequently, its "ideal" limited use were mentioned, which contradicts the perceptions and expectations of the beneficiaries. In addition, a conflict of understanding also appears between a "project developer/implementor" and the beneficiary of the solar system, which is partially related to the beneficiary/user awareness and partially related to what the beneficiaries desire in using electricity by all means, like a user in town. The user awareness, in the case of stand-alone systems that way implied in this interview, prior to the installation of a system becomes prominent in order to come to a mutual understanding of different parties involved that the off-grid energy, depending on the techno-economic characteristic provides electricity with some conditions, different than the grid-connected electricity provision in towns.

Secondly, the interviewee explains the connection between the sense of ownership and the care one commits to keeping it: "If you pay for something you feel that cost, you would be better taking care of some things." A sense of ownership has also been associated with taking the responsibility of maintaining what belongs to the owner. Thus, it also opposes the "solar energy for free" mindset found in the fieldwork. Eventually, the solar energy is free until it reaches the surface of the Earth, at least for the moment. Subsequently, it becomes the country's resource where the sun's rays touch the soil. The technology that is required (solar PV) to convert solar irradiation into direct current (DC) electricity, which is then converted to alternating current (AC) power, is necessary to use it in households. Therefore, necessary investment is required to own and use these systems. As a result, the clear and transparent design of public and/or user awareness of such systems identifies the type of ownership, and the expectations attached to its functions.

**Accessing People**

> "... it was **also difficult to access everyone** in the rural area with **no internet, no telephone, communication**"(NRA1, Pos. 12).

Furthermore, aligned with public awareness, NRA1 stated the difficulties in accessing people in rural areas to keep them updated and also raise awareness, when there is no internet connection, telephone, or any other communication tools during the development of the off-grid projects. These challenges partly exist today depending on the rural areas, since telecommunication and internet coverage expand quite fast in Namibia, which was also explained in the previous chapters.

#### *6.3.1.2 Socio-techno-economic Limitations*

**Grid Extension versus Off-Grid Solar Energy and Social Acceptance**

> "Hmm, for us specifically in Namibia, a country being as large as it is with population that we have, population density of about 3% per $km^2$, solar should, and I use the word carefully **"should"**, get more important. Because it is really going to become **much much more difficult** to **extend** the **grid in the future**. However, **there is an awareness that needs to be done**. **I think because the idea that the grid has been forward as electricity and solar is thought like a temporary solution until the grid comes.** That makes the **role of solar very difficult** sometimes. It has got **political connotations** with it so that politicians are probably going to want to preach that, bring solar rather bring grid. Probably, under pressure **to bring grid** or **to bring development** in the form in which **people accept them to** be. If you bring solar, you probably seen it, **you´re undermining some**, you know" (NRA2, Pos. 36).

Without a clear regulatory framework that supports both mini-grids and grid connections, the provision of electricity might be involved in the political debates, depending on the circumstances. In this quote, the NRA 2 emphasizes the importance of utilizing solar energy "should get more important" because the grid connection will not reach every location in Namibia due to its vast geography and low population density, which creates uneconomic transmission network expansion. Reflecting on the earlier experiences by proposing an awareness raising, there is or was an understanding in the earlier times that "solar energy was a temporary solution" until the grid comes. Whether this was thought to be in this way or not, this has been the experience so far, which, as a result, created an image that "solar is temporary" partly because of the lack of its inception/introduction as well as the lack of a necessary regulatory framework that would show the possibilities of "when the grid comes." In this sense, since (bringing) electricity is linked with the social and socio-economic development of people as well as the country as a whole, it also brings a political dimension into the debate, which the interviewee comments, "pressure to bring the grid or development" within the political framework.

However, he also mentions another point behind the perceptions of "solar is temporary" versus "grid-connected," that is, how people feel about the off-grid connections or how solar energy has been seen or perceived by those people using them under the "collective social acceptance of decentralized mini-grids" (PROCEED Recommendations for Action, 2023). The views from which solar has been perceived as an "inferior technology" or off-grid energy users are "second-class citizens" were recommended for action under public engagement and capacity building (PROCEED Recommendations for Action, 2023). Contrary to this terminology, the author´s experience did not involve any quotes that would fall within this category of social acceptance as a "second-class citizen." Instead of this, as it is already framed and explained in Chapter 4 and Chapter 5, the communities in the settlements were more focused on the 24/7 uninterrupted availability of electricity with affordable tariffs. As a result, the overall "social acceptance" experiences in Namibia from the earlier installations until now provide different social acceptance experiences that enable the development of holistic, inclusive, detailed, customized awareness-raising and beneficiary capacity-building programs and activities for future implementations.

## Energy Shops and Lack of Private Sector Support

> "What the government has tried already is that they came up with a concept of an **energy shop to bring the shops to the rural area**. So they can **market the funds** and also **source the technicians**. So, the main objective was to make sure that in each village, for example, 185 shops were supposed to be established. However, due to **limited funding,** this was not realized. And also, it´s a little bit very difficult for Namibia. There is not really a **support mechanism for the private sector** in Namibia to **directly support the private sector**. So, this **remains a challenge even for the shops**. We only assisted them at the **beginning, and we did not then continue with the subsidy** because **they are private** and there is **no direct channel of supporting** them. (…) After there was the solar system, they wanted **to have a champion who was trained by the community with continuous training and support from the government to operate and manage the mini-grid**. But after the government took over, it was **always not clear who took ownership** and all this. As I said, there is a really big challenge, a **big gap to support the private sector directly**. (…) The support is sufficient, but **what we are advocating is that the government should not operate and run the systems directly. Because they work with the budget**, with the lumpsum budget, and they work only with the devel-

> opment budget, and they **do not have an operation budget**. So, there is **a need for government to support the private sector**" (NRA1, Pos. 18-24).

The quote from the interview NRA1 explains multiple technical and economic challenges faced when the *Energy Shops approach* linked with the SRF was introduced with the OGEMP and why it did not function as it was supposed to. One of the reasons for this, according to the NRA1, is that the funding to establish Energy Shops was insufficient. Secondly, the private sector was not directly or explicitly supported by a support mechanism. These two suggest that the private sector involvement was not directly quoted and clarified in the regulatory framework. Moreover, what NRA4 brings into this issue is also linked with the "technical capacity" linked with the financial barriers that refer to the supply of quality products for its implementation by trained people within the Energy Shops approach.

> "It mostly was technical barriers. The **technical capacity** was **not there** to bring this **technology to the installations** even the **quality of products that was supplied**, that was a concern. And then also, **financing barriers**, I think that´s where Solar Revolving Fund was created to address the financing aspect of it" (NRA4, Pos. 8).

With the lack of human resources with "technical capacity" and necessary funding the energy shops could not be implemented as planned, and government support remained limited. The GRN continued listing trained solar energy installers for the installations within the SRF projects but did not provide direct funding for the private sector. NRA1 also refers to the mini-grid within the same framework that the GRN had planned to have a continuously trained "champion" from the community who would operate and maintain it after its installation. However, according to the NRA1, after the GRN took over, it was not clear whose ownership it was, which created a big challenge. These issues underline a lack of procedures in terms of its operation and maintenance that is directly linked with its ownership. Ownership is usually defined within the contractual terms of the projects, and the name of the person to whom it is commissioned and how it will proceed at the end of the project is clearly stated. Thus, in this case, a lack of regulatory framework and the lack of human resources with "technical capacity" appear after the commissioning of the mini-grid. Furthermore, NRA1 provides another reason for the cause of such issues, by being the GRN working with lumpsum budgets and they do not have an O&M budget. Consequently, there is no budget for human resources with "technical capacity" who could also run the O&M. Therefore, NRA1 suggests the GRN not be directly involved in the O&M, since their expertise is in the planning but not in operation. However, the government should support explicitly through certain mechanisms the private sector.

On the other hand, NRA2 highlights the importance of building beneficiary/user awareness about the limitations of using off-grid solar energy and the use of energy-efficient electric appliances. In this sense, using electric cookers or stoves is not supported within the off-grid energy systems due to their high load consumption, which also results in technical failures that might be costly. Therefore, NRA2 proposes off-grid solar energy from a pragmatic perspective in bringing cost-effective electricity provision for off-grid areas that should be expanded on a program basis. This proposal suggests a more structured electrification strategy supported by capacity building and public awareness to expand off-grid electrification.

> "Then, there is the whole issue of just now when is, **what is it that solar can do** and **what is it that it can´t do**. So, you heard people that "ok, now, I have solar system, I can now cook". And then, you hear that burn out. Hmm...Your regulator or **some important something doesn´t work**. So, lots of work to be done going forward, but it is **solar "should" from pragmatic economic perspective** be more important and be rolled out at the **much program based than** it is currently" (NRA2, Pos. 38).

#### *6.3.1.3 Socio-Economic Limitations*

**Inaccessibility of funding and subsidies in rural areas**

> "Since now, because till now the funds is, is existing and it´s growing for the number of installation. It just that before it was outsourced to private sector, and the subsidy did not really benefit**, not really benefit directly the community,** that's why the government decided to bring it back to the ministry **to directly have impact. I see** to **the community with some difficulties**. Then, the government **subsidies but then**, all **operational and cost goes to the company which manages and not directly to the consumer**" (NRA1, Pos. 14).

> "Yes, the government took over **but still it has limited capacity, people, few people and it´s only still concentrated in Windhoek**, is **not well marketed in the rural area**. It is **not easily accessible to people** who are in rural area to still sometimes you go in the rural area and this people did not yet have?" (NRA1, Pos. 16).

NRA1 informs that the (SRF) funding was initially outsourced to the private sector, which was then taken back to the MME to create a direct impact. However, the concern arose at the community level, where the beneficiaries in the rural areas did not directly benefit. It was primarily beneficial for the installers and O&M companies, who could use the funding from an economic perspective, but the community beneficiaries did not receive an economic benefit. Then, after MME took over, they had limited human resources capacity to conduct and continue the program. In addition, the concentration of skilled installers remains in the Windhoek area, and the rural areas that need them cannot benefit from it. According to the 2021 update from the MME, there is a total of 169 certified PV-installers (42%), solar thermal installers (15%) and RE suppliers (43%) in Namibia certified by the MME (Dolunay, 2021). However, as the NRA1 also expresses, the regional distribution remains unproportioned; while the majority of the installers are in the Khomas Region (54%) and Oshana (20%) regions, the SRF-financed installations took place mainly in the Omusati (31%), Ohangwena (14%), Oshikoto (13%), and Oshana (11%) regions in 2021 (Dolunay, 2021). Therefore, the lack of human resources with the required technical and management capacity also becomes prominent for expanding installations in rural areas that still do not have access to such funding.

### *6.3.1.4 Technical and Management Limitations*

**Operational and Maintenance Issues**

> "But challenges were more on, now, that you have more **community in the rural area deep** and the **technicians did not stock up for also renewable energy technology** if the … Is broken, if batteries broken and so also resulted that some **replacements were necessary** and **no evaluation** was really done to find out if this whole home power programme batteries are functioning. I mean the systems are functioning out there because **no one do evaluate**. **People are left on their own**. And some but some still communicated to the technicians and managed to have.. so the main challenges was just that the **technicians were mainly concentrated in Windhoek** and not in the rural area. And sometimes you have communities out there who needs services but they **don't get the service they need** for some years. I mean **for some time**" (NRA1, Pos. 10).

The accessibility of the off-grid areas provides the challenge itself, not only for electricity connection. After the installation, the expectation is that the RE technology will be operated and maintained properly. However, the experience of NRA1 is that mostly the technicians were not prepared to carry out the O&M, did not keep the necessary stock of the spare parts at their premises, and were located in Windhoek. Since they did not have any spare parts on-site or on their base in Windhoek, they were unable to attend to the needs of the communities and commit to their services. Since these installations have been funded through the MME, NRA1 explains that there has not been any evaluation of the systems after the completion of the installation. Thus, there is an assumption that they all function properly despite the O&M challenges. As a result, in relation to the SRF-funded projects, one might claim that there has not been any evaluation to track the functionality of these services, which is also connected with the image of solar system acceptance in the peripheries.

##### *6.3.1.5 Institutional Limitations*

**Licensing and/or certification of installers according to the standards that don´t exist**

> "Yeah, it´s a whole new structure that solar needs to come into place. Because you **have standards**, for example, you don´t go anywhere and do an electrical installation **if you are not licensed, fully qualified. You don´t have that set-up in the solar industry**. Earlier, it is best to register or to acknowledge certain people as qualified, not even qualified just recognize installers. but these are not really people that would be licensed, fully qualified, and able to do the work, so.. You have **standards that are much reduced**, therefore the work sometimes is not the best quality. And then, **you don´t really have a structure** or **an enforcement of what kind of standards are used in doing such systems**. So, it´s a combination of many many things that result in again, unfortunately, many systems not working perfectly" (NRA2).

In this quote, NRA2 informs another dimension of the lack of standards for the installers, which is also linked with the lack of standards in general for the solar industry. NRA2 states that there are standards for electrical installations; if one is not qualified or licensed, it is impossible to do installations. However, this standardization does not exist for solar installations, which is a challenge that results in system failures. NRA2 suggests having standards in place for solar installations and registering or recognizing the installers to start with prior to the standards to prevent malfunctioning systems. In this way, the recognition of the installers would provide a technician pool specialized in solar energy installations, which would later comply with the system standards once they are in place. Therefore, the lack of standards is also a part of the lack of regulatory framework for off-grid electrification that was previously mentioned, which is within the scope of the ECB, in charge of the policies, regulations, rules, and standards development necessary for the electricity market. As a result, with this extended institutional regulatory framework, the process of bureaucratization as a part of rationalization continues evolving within the electricity sector to find a technical conformity that fits best.

### 6.3.2 Opportunities for Off-Grid Electrification

The opportunities for off-grid electrification collected from the NRAs have been summarized in Table 6.3.

*Table 6.3. Opportunities for Off-Grid Electrification*

| Type | Opportunities |
|---|---|
| Social | • Awareness Raising<br>• Continuous Capacity Building (SRF Trainings, PROCEED Training)<br>• Community Engagement and Psychological Ownership of the Mini-Grid |
| Institutional | • Top-Down Approach "The Communication has to come from the top and then, trickles down"<br>• Rural Electrification – "New Customers" |
| Economic | • Cross Subsidization<br>• "State or any philanthropic association or system"<br>• Operation with Profit |
| Technical and Economic | • New Technologies and Pricing<br>• "Utility Model"<br>• Grid Encroachment "What happens when the grid comes?" |

*Source: Author*

#### *6.3.2.1 Social Opportunities*

**Awareness Raising**

> "Then, **the whole big, huge important aspect of awareness**, I think awareness not only from the knowing what you can do and can´t do with solar. But **awareness on the importance of solar why it can be viewed on the same level any electricity whether is from the grid or not**. That´s how to better take care of yourself so that you just help people **changing their mindsets about solar**. Hmmm, the best to take the range of everything. So, what would people **listen** the most? For example, in rural areas where **radio perhaps the best mode of, on bringing information**. So, **people talk about**. Just, most of the time **keeping that message coming**. Especially for the population of West need solar. And I think f**rom Gam respectively should have campaigns** where we closer to those people that need that information more. And **awareness also comes through just doing a better job of projects**" (NRA2, Pos. 40-42).

Regarding the challenges described in the negative connotation of solar energy as a "second-class" energy source by the collective social acceptance, the NRA2 proposed public awareness as "the solution" to overcoming the misconceptions about solar energy. The goal was to improve the community's acceptance of solar energy as being on the same level as energy from the grid. It was also an opportunity to inform the beneficiaries about using it according to the grid's requirements and using energy-efficient appliances. As a public awareness-raising tool, NRA2 mentions radio as a communication device that feeds information continuously and makes people talk about it. Thus, it creates a continuity in reaching more people over time. Another mode is awareness campaigns, which NRA2 thinks necessary in Gam and the West (of Tsumkwe), where people will need solar. Thus, they will need this information. Nonetheless, successful projects realized in off-grid areas are also part of a good awareness-raising tool on-site for delivering the necessary information. In this respect, NRA2 reflects on the awareness as the primary tool in efficiently using the off-grid systems, in which the delivery of this service is expected to be realized by the project developers in the form of campaigns and successful projects, and the information continuously disseminated through radio streaming for this effective implementation. Thus, this contribution aligns with the past, present, and future energy policies and NDPs.

**"Continuous Capacity Building"**

> "Renewable energy cannot operate on its own to be fully balanced. If we are bringing in **energy security**, of course, **storage need** to come but these **technologies are all new in Namibia. We have few people who understand the technologies and who can operate and maintain**. So, while combining financing we also need to combine it with **capacity building and continuous capacity building to train individuals and to also start with the local**. We need the robust, this is what we call, **robust capacity building**. So, CENORED actually, the electrical company, which supposed to, but even them when they are operating Tsumkwe you can see, some, **some knowledge is still lacking**. So **we need continuous training and also we need continuous financing programme, supporting also private sector**, so they can employ more to create jobs and while basically somebodies are making profit" (NRA1, Pos. 28).

Furthermore, NRA1 explained that security of supply is an essential component of RE (solar energy for this case) that needs to be included in delivering a balanced system, which is directly linked with the new technologies, whereby there are not many people who understand them in Namibia. NRA1 proposed a "continuous capacity building" of the individuals by starting with the locals precisely, making it a "robust capacity building" to increase the capacity with the new technologies, their operation, and their maintenance. NRA2 shared the experience with the CENORED, who took over the O&M in Tsumkwe and Gam, and they realized that their training in operating such mini-grids was lacking. NRA2 emphasized the need for more training and continuous financing to create jobs, including in the private sector, that will eventually make a profit.

> "The ones that we used, or installed at schools and other public institutions. uhm, so, it´s **just means that you as a project developer, you need to do two or three extra things to ensure that they totally understand what is it, how they can use it and also, how to feel confident about** it. Because if you get something, and you feel it´s not really

doing what you want, then, it´s like "I don´t really want it", you know. **So you have to also work with people to make them, give the necessary information about what they are getting and so forth**" (NRA4, Pos. 33).

On the other hand, NRA4 expressed the expectations from the project developers with the clarity of the following statement "it´s just means that you, as a project developer, you need to do two or three extra things to ensure that they totally understand what is it, how they can use it and also, how to feel confident about it." In this way, NRA4 interpreted the responsibility of delivering "capacity building service" to the project developer in a more implicit way. Thus, it is assumed that the beneficiaries would take responsibility for their equipment, take care of it, and not throw it away when something goes wrong. NRA4 also provided experiences with the SRF and PROCEED training, in which the installers receive refresher training about the technologies and their operations. CENORED´s training regarding the O&M of the mini-grids (after three years, they took over the systems) realized by the PROCEED project team was highlighted as an important action that belongs to the capacity building programs, which they will need in the future for other REDs, too. However, whether this implicit expectation of training as a capacity building activity by the developers are included within the budget of the project either by the contractor or by the GRN remains unclear.

### SRF and PROCEED Trainings

"Especially from the solar revolving fund, there has been **several update** or **refresher trainings to installers to get them a breath of new technologies**, also the operations of the fund. uhmm but **the NEI has also the task to see or what type of trainings they can bring and especially trainings** like, of partners like the PROCEED Project is one of them where it identified some of the issues on, especially **on the grid side, on the mini-grids**. There was trainings done to the **regional distributor** to their staff, how that **system should operate and operate optimally**. So, **through this kind of activities** there is also **that capacity building element** (…).I think that will really give some or enhance their skills. The mini-grids currently is just in one region, and jurisdiction of that region ... But if we want to go in the future to other regional distributors in their area if there are mini-grids. Then, definitely, that same kind of training needs to happen as well" (NRA4, Pos. 27-29).

In summary, capacity building has been perceived as magical as the "RE" or the "electricity" itself. It has been used quite often in energy policies (NEP (MME, 2017a), NELP (MME, 2022b)), as well as in NDPs, and the documents of national actors. Capacity building is "the process of developing and strengthening the skills, instincts, abilities, processes, and resources that organizations and communities need to survive, adapt, and thrive in a fast-changing world" (UN,

2023). Furthermore, it is accepted as a "core concept of development policy" that peaked in theoretical debates in 1995 and 2005 to strengthen the capacity of individuals and institutions in developing countries (Zamfir, 2017). Capacity building is not only seen as an "increasingly important strategy or international actors in their efforts to promote security and development" but also "seeks to promote a bottom-up approach" (Edmunds & Juncos, 2020, p. 3). From an ethnographic perspective, Douglas-Jones and Shaffner (2017, p. 1) analyzed capacity building, hope, and potential from the perceptions of absence and insufficiency of the present towards a more desirable future. They discovered that capacity building, adding to its "change" component, also describes how it might occur. In addition, they concluded that "capacity building was intended to be transformative"; however, it also transformed itself" (Douglas-Jones & Shaffner, 2017, p. 11). Furthermore, the term "capacity building" has been widely used, an emerging new terminology in development community "capacity development" differentiates the two (Zamfir, 2017) by strengthening the existing skills and knowledge, rather than "building something new from the ground." As a result, the concept of "capacity building" and "capacity development" as it was intended to be from the use of the terminology in the development initiatives, policies, in local, national, and international levels, its complexity of implementation also appears within RE governance and implementation in Namibia.

#### *6.3.2.2 Institutional Opportunities*

**Top-Down Approach "The Communication has to come from the top and then, trickles down"**

> "So, what the **Ministry** is doing, is to use that **at highest level**. So, you want to have **policy that embraces renewable**, policy that gives it as much support as possible, and then, through that **having your leadership, country leadership, your politicians also embrace it**. Because the **communication really has to come from the top and then, it trickles down**. So, once that is, so that´s where you have to start. But then, **on the ground as well, having a team doing a good job of constantly training installers so that they do a better job**, doing also some more **work with the standards**, so that your standards **for what equipment you used**" (NRA2, Pos. 40).

Regarding the top-down approach, NRA2 explains how everything starts under the leadership of MME, supported by the policies embraced by the politicians, and followed down the lower levels of the hierarchy. However, in contrast with the regular hierarchical structure, NRA2 suggested "on the ground as well, having a team doing a good job of constantly training installers so that they do a better job," emphasizing the capacity development of the installers on the ground by their continuous training and the standards that needed to be collected in order to use qualifying equipment for a sound installation. In this respect, capacity development from the bottom to the top has been expected and intended. Thus, it contributes to a relatively small movement within the RE governance.

**Rural Electrification – "New Customers"**

> "We are, we have got a department that does, we call it "**rural electrification**". So we have got a team that **is targeting these rural areas to bring power to them** and we are going to start looking at off-grid. I know that off-grid is differently target or an **area where strategically we need to start looking for new customers**. And then, **new customers are unexplored**. It´s a complete new customer base which we never had. So, we are definitely looking into that. **We also don´t have a solution yet** on the table to say "**ok this is gonna be our.. we need 40 households, we need a 1000 people, this is the payment system, this is the technology we´re gonna use**". We haven´t come up with it yet" (NRA3, Pos. 15).

On the other hand, in a different institutional setting, a department is differentiated for "rural electrification" to achieve goal-oriented work in bringing power to the targeted rural areas. The differentiation of the specific work also translated to the beneficiaries as "new customers," whereby individuals have been "quantified" as income generation points aligned with a business, profit-oriented mindset rather than providing electricity services for people-residents as it is often advocated in the policies. Furthermore, looking into this rural electrification expansion, there was no set strategy to electrify those areas. However, it could be interpreted as a rather strict top-down procedure based on how it was described. As a result, although there is a tendency to expand rural electrification through differentiation and rationalization, it remains under hierarchical control.

### *6.3.2.3 Economic Opportunities*

**"Cross-Subsidization"**

> "The best I've come across is anything that has a **cross-subsidization approach**. And the rationale behind cross-subsidization approach is, is simply that economies of scale existed levels above that the most mini-grids are capable of handling. So, you will find that the breakeven point needs to be like, for example, 10,000 meters or 15,000 meters or somewhere along that sorts of lines or you need a concentration. And you will **only find that in a utility setting**. I'm still to seat really work well in a micro-grid or mini-grid setting. And so, **the ownership is anything that will allow for that cross-subsidization**. So, having said that, most governments are not comfortable, the governments that I've worked with, which would tend to be socialist or left-leaning governments are

> not comfortable, which you can understand in subsidizing something that **allows profitability for private enterprise**. Which means they're basically topping up and making profits for private enterprise. One would rather than do it **as a social enterprise**. And if there are **any losses, you just absorb it**. And so, that would be, so that's why I am saying the **utility model is one that allows the cross-subsidization**; **it allows for an in-built subsidy arrangement that by any call for funds or ad-hoc funds from time to time"** (NRA5, Pos. 59).

Based on his experiences in the field of energy and electricity, NRA5 proposed a "utility model" with cross-subsidization for future off-grid electrification projects. According to his expertise, the only model that would allow the handling of the economies of scale can be achieved by the utility company, which can develop a model at an economic break-even point customized for micro-grids or mini-grids. An (electric) utility is defined as "a corporation, person, agency, authority, or other legal entity or instrumentality aligned with distribution facilities for delivery of electric energy for use primarily by the public, "which shall be investor-owned, municipal, state, federal utilities, or rural electric cooperatives (Energy Information Administration [EIA], 2024). The wide use of cross-subsidization enables governments to provide utility services to people experiencing poverty by charging higher prices to high-income residents to subsidize lower-income residents (Li & Li, 2018; Li et al., 2018). However, Li et al. (2018) stated that in the event of financial deficits in the case of Colombia, the utilities decreased perceived service quality to subsidized consumers; thus, they suggested that funding subsidies, strengthening quality regulation, with increased transparency and reliability, should be ensured by the government. Consequently, similar to the cases in Tsumkwe and Gam, where the illumination of public areas is the responsibility of the operator of mini-grids (CENORED), the community reported that it did not function properly (PROCEED Conference, 2023).

**"State or any philanthropic association or system"**

> "If what I've just explained incorporated and understood by policy makers, ´coz at the end of the day, it´s **gonna be the policy makers that enforce it**, it´**s not going to be the private sector**. Private sector is, business school one-o-one says that. Private sector **is motivated by profit,** it´s not motivated by social needs**. So, therefore, only the state or any philanthropic association or system,** you know, the Gates Fund or something like that, or the Clinton Foundation. They are the only one who can step in and **fill that bridge which still goes** to **cross-subsidization.** The money is coming from one, from the left hand and been placed into the right hand. And they all side dealt with one to support an unsustainable commercial model that does not have state sanction. So, it goes back, the answer goes back to one that has state under-

standing and a **utility model that would allow for encroachment growth, expansion, upgrades, all the other things that will require the system and sustainability**." (NRA5, Pos. 61)

Following up the cross-subsidization in a utility model, NRA5 suggested a social enterprise, State or philanthropic association that would be funding such systems, due to the fact that in the conditions of Namibia whereby mini-grids do not result to be profitable based on the economies of scale. And that international funding would bridge the economic gap together with the government. On the other hand, he also brings attention to the policymakers who are in charge of the enforcement of such activities through the policies, not the private sector. Thus, he proposed a utility model that is fully supported by the GRN, co-financed with international funding, and operated by utility companies to provide cross-subsidized electricity for off-grid rural areas. In this way, through this model, a shared model of public-private-donor actor setting appears in order to ensure affordable tariffs and sustainable mini-grids.

**Operation with Profit**

"**To operate and manage directly with profit**, while making profit. They can employ a number of people. But with the off-grid is till today, since Namibia is not locally manufacturing, no batteries, everything need to be imported. **There is no way one can run an off-grid technology without the government support**. But one need **to operate it profitably**. But **with direct subsidy knowing** that this community also need to **have access to electricity to the development but only if there is a private sector support programme**" (NRA1, Pos. 26).

NRA1 shared views similar to NRA5's about the private sector´s profit-making model and said that a mini-grid without a profitable operation was impossible. GRN support and private sector support seemed a "must" for off-grid. Thus, public-private partnerships appear to balance the conditions of the off-grid implementations and operations in the off-grid areas toward a more controllable and calculable future.

### *6.3.2.4 Technical and Economic Opportunities*

**New Technologies and Pricing**

"I think that **batteries are gonna play a major role in off-grid**. Because we can get PV and batteries there. I think we want **to move away from fuel, fuel off-grids**. That just doesn´t help the problem, taking fuel to the site. So, I think, batteries will definitely make a difference in terms of off-grid. Really looking long-term into the future, whether something **like hydrogen will play a role, I don´t know**. So, I think the batteries **will definitely be the equalizer, I think that is where the new technologies coming available**. Obviously, as you know, ... flow

> batteries and capacitor batteries and liquid-metal batteries, and so. I think once these technologies unlock the price will come down. I think it´s just a price issue. That´s stopping the electrification of off-grid" (NRA3, Pos. 23).

For the future of off-grid electrification, the NRA3 thinks that new technologies will play a key role in their expansion, moving away from fuel use in the off-grids. Listing flow batteries, capacitor batteries, and liquid-metal batteries, NRA3 forecasts that the availability of those will make them cost-competitive for greater use. Thus, they will introduce new pricing for the systems. However, he remains reluctant for the use of hydrogen in the future for off-grid systems.

**"Utility Model: mini-grids should have been coordinated as if they belong to a utility"**

> "Uhm, **mini-grids should have been coordinated as if they belong to a utility**. And what that would do, that would **prevent grid encroachment issues that are fear**. It´s feared from an **investment point of view** and it´s feared from a technical point of view. So, for example, one of the things that I don't stand by is a pure levies mini-grid. Because what it then means is that, if **the grid was to appear tomorrow morning because they found gold**, it´s one of the favourite statements that I´d like to make to people. **If you find gold or some mineral, I can promise you the grid will be there overnight**. They will be no talk about the economics and productive use of energy on whatever.. We need that! So, then, you would find all this infrastructure that was so stressful to put in, and went to all these hoops is just discarded and abandoned" (NRA5, Pos. 55).

With this statement, **"mini-grids should have been coordinated as if they belong to a utility,"** NRA5 suggested that the investment and technical fears regarding the grid encroachment of mini-grids are unnecessary if the utility model has been applied. He highlights that the conditions of an unparticular off-grid area might change drastically overnight if there are multiple mining interests, and all the infrastructure might be discarded if the mini-grids lie on levies only. In this case, he implies that the "utility model" will not fall into this risk.

**"Grid Encroachment: What happens when the grid comes?": Grid-Harmonized Solar PV Systems**

> "But we can go to the within 1 km radius or 2 km radius and we can set up infrastructure that mirrors the grid while not being the grid. If that makes sense. So, like Tsumkwe, the way that Tsumkwe was done was even from the military base which I give them credit, they already had a 415V generator, 3 phase, that fed through a step-up transformer and fed into a 11kV grid network. So, **all we had to do was improve and ex-**

> **tend the grid network and upgrade it to 33kV and get it ready for when the grid came**. So, **the grid encroachment would not be a problem.** And also, **the equipment that was put in in terms of the smart inverters and so on would allow for this interlink**. So, you could be charging from either end and you sending power in the other direction or that could be working in tandem. And that was taken to down" (NRA5, Pos. 56).

In continuation, NRA5 explains the technical conditions in Tsumkwe, how the grid network was improved from the military time and upgraded to 33 kV, and how it was made ready for the "when the grid comes" scenario, including using smart inverters. Moreover, he continues explaining how they made some PV installations taking into account the "if grid comes" scenario; they used grid-compatible equipment, which will only need a net meter to connect to the grid for the systems that are at 400V level, and the battery system can be transported somewhere else. In this way, he claimed that "the choice will be between the utility and the customer." However, he also warned that not all systems are grid-compatible in Namibia, and in the case of grid encroachment, the systems will most likely be abandoned.

> "And what we did is these systems are not MV level grids, there are 400V grids. But they, the way which PV system was put into the system was one that the PV would then become grid-connected PV, if the grid was to appear. So, no inverters needed to be changed, no PV systems need to be taken down, and you basically have a net-meter customer. And the battery system housing would just be lifted up, put on the back of a truck and taken away or it could stay. **The choice would be then to the utility and the customer, at that point in time**. So, we created 3 grids that as I said they are not MV grids, so they are not 11kV or 33kV, there are strictly at 400V level. **But they are ready for the grid**. And, every other site I visited across Namibia is not ready for the grid. So **you'll get abandonment**, you'll get **incompatibility**, because the **inverters are not smart** inverters, they can´t grid capable, they´re not switch capable, that sort of stuff. They're just battery inverters. So these are some of the issues around grid roll-out that I think it´s very very important and is **missed even by some of our top engineers and policy"** (NRA5, Pos. 57).

### Community Engagement and Involvement as a part of Psychological Ownership of the Mini-Grid

As indicated earlier, data collection delays occurred due to travel restrictions. The validation of the findings of this research has been presented during the PROCEED workshop in March 2023 with the national and sub-national level energy stakeholders, including community members from Tsumkwe and Gam. In addition, as a part of the workshop, a *World Café* format group discussion was held, whereby participants swapped tables to answer particular "table questions." Four groups explored the ownership models in one of the four discussion tables. Participant 1 (P1), P2, P3, P4, and P5 are representatives of the energy stakeholders, including community members, and "O" is the discussion table responsible (author). The section from these round table discussions demonstrated and confirmed the gap within the local and regional governance and how the mini-grid operator implemented a solution to bridge this gap. An excerpt from these discussions has been presented below.

> "P2: So, any committee that may be nominated and this time by the community, they will also go into the big learning curve.
> P5: Yes. That´s true.
> P3: Yes.
> P1: Yes.
> O: Would this work? What do you think?
> P3: I think if they are involved, sort of let´s say community funding, I don´t know whatever. I think they also be responsible sort of.
> P4: Exactly.
> P3: They will **feel like it´s "ours", that ownership**. They **will take care of it.**
> P5: Yes, it will work, because **we can learn a lot**, to try to save everything and know how to use it and then, protect that also, have experience and then, generation to generation they will learn from it and it will run smoothly.
> P2: And prevent potential, your fellow community members for example, running that inefficient electrical cooking things, you know this.... your responsibility to make sure that your neighbor does not use it.
> P5: Yes, and **this responsibility** we started already. It´s the things that Mr. XXXXX did not say it in public. So, **we selected a group of members in the area itself**, in Tsumkwe, **for us, to go door to door to see who are using gas, who are using things which are tapping the power**. And I am one of the members. So, **we are monitoring everybody**. So these things of trapping until now, we don´t know from where it is coming.
> P4: That´s part of the operation.
> P1: **Community involvement**, yeah!
> P5: It´s where even, I came up with the idea of getting a stove, to put in the shop, so we are planning. We are even planning to bring a depot in Tsumkwe. If we, yeah, because the population it´s what stops now the sponsor of guests.
> O: **So,** Like this is a monitoring team?
> P4: That´s **demand side management**.
> O: Is it?
> P3: Yes, it is demand side management.
> O: Ok.

> P4: **Community-based demand side management.**
> P3: **That´s consumer education.** Consumer education, that´s part of the demand side management" (*World Café* Ownership Group, Pos. 38-58).

Among these four separate group discussions, the importance and the necessity of "community involvement" have been agreed upon by "all" participants, regardless of their background. Moreover, CENORED´s acceptance of community involvement and engagement has brought another aspect: "community-based demand side management." This has emerged through the increasing consumption of energy based on the high peak consumption and inefficient appliances, as explained in Chapter 4, such as hot plates, blow driers, irons, and water heaters. Thus, CENORED decided to engage the community and formed a team monitoring the use of these appliances in order to ensure mini-grid stability and meet daily energy demands. A (psychological) ownership has been in process, which refers to "the state in which individuals feel as though the target of ownership or a piece of that target is 'theirs'" (Jami et al., 2021, p. 698). This shall be a particular object, idea, place, or relationship that goes beyond the physical ownership and might include emotional attachment and identification with the object; expressions of "mine," "my," and "our" are used to show it.

Research also shows a positive correlation between the material goods owned and the individual´s self-esteem, which depends on how one is valued and considered by others (Jami et al., 2021, p. 700). These possessions are ways in which individuals confirm and communicate their identity to gain self-esteem while affecting their owners' social position and relational value (Jami et al., 2021, p. 699). Furthermore, researchers suggested a direct connection between the individual´s self-esteem and psychological ownership and between self-esteem and prosociality (Jami et al., 2021). In this way, improving self-esteem also increases prosocial behavior, which are behaviors, actions, or attitudes that contribute positively to society for the benefit of others. Thus, Jami et al.´s (2021, p. 712) research confirmed this relationality by presenting self-esteem as a "mechanism linking the two." Psychological ownership may take several forms for diverse purposes, such as business, marketing, management, and other services, fostering strong, healthy, and sustainable relationships.

As a result, these findings also confirm that local and regional knowledge is available to take the necessary steps to overcome the hurdles they face. They serve as a validation for the community's involvement from the early stages of the mini-grid development and their engagement in the lifetime of the mini-operation with the necessary capacity-building and development activities to be designed and included within the mini-grid management.

## 6.4 Future Visions at National, Regional, and Local Levels

During the second field research in Namibia, the visions of the State regarding off-grid electrification were verbalized at the inauguration ceremony of the connection of Mangetti Dune in the West of Tsumkwe Constituency to the national grid. This event gathered the officials of local, regional, and national authorities and local people together. The Deputy Minister (DM) of Mines and Energy, the Governor of Otjozondjupa Region, the Tsumkwe Constituency Councilor, and CENORED representatives participated in the event. Table 6.4 summarizes the key viewpoints of the national actors to understand their future visions.

*Table 6.4. Future Visions at the National Level*

| Governance Levels | Market Actors | GRN Related and Development | Electricity | Support of stakeholders |
|---|---|---|---|---|
| National | MME | • Narratives of the GRN,<br>• Dissemination of National Commitments and Promotion of the Actions | • "Electricity is a modern necessity of life."<br>• "MME remain committed to make sure that we extend these lines to Tsumkwe, to Gam, and to the rest of Otjozondjupa." | Support for CENORED |
| | ECB | "to be the leading electricity regulator in Africa" with the mission to regulate the Namibian Electricity Supply Industry in a sustainable manner, in the interest of all stakeholders with regard to efficiency, affordability, safety, and accessibility." | | |
| | NamPower | "to be the leading electricity service company of choice in SADC." | "to provide the electricity needs of our customers, fulfill the aspirations of our staff, and satisfy the expectations of our stakeholders in a competitive and evolving environment." | |

*Source: Author*

### 6.4.1 National Level – Off-Grid Electrification

#### *6.4.1.1 Ministry of Mines and Energy*

MME´s vision as an institution is to "provide access to Namibia´s geological, mineral and energy resources for sustainable economic growth, equal benefit, and prosperity to all Namibian citizens," is announced with the mission "to formulating required legislations and policies to regulate the activities accordingly and to stimulate investment for sustainable economic development and benefit to all Namibians" (MME, 2023a). The core values have been stated as accountability, transparency, responsiveness, innovation, and integrity, which provide the basis of actions in the work of MME and how the issues have been approached, which are aligned with the GRN´s national plans of good governance.

The speech from the DM of MME during the inauguration ceremony confirms the actions of MME. The works of grid electrification of Mangetti Dune started in 2020 and was finalized in 2022. The 120 km 33 kV medium-voltage overhead line from Marulaboom to Mangetti Dune and the installation of low-voltage networks cost N$ 40 million, funded by the MME and realized by CENORED. Consequently, the operation and maintenance were transferred to CENORED, with the addition of electrifying 183 HH and three teachers´ houses in three localities (Field Notes, 2022). DM also informed the community that what the Minister of MME had promised during his visit to Tsumkwe Constituency in 2019 with the Mangetti Dune grid extension and the 410 connections to the Gam mini-grid was duly realized.

**"Narrative of the Government is not doing anything must stop from today"**

> "The Ministry´s effort and collaboration with Tsumkwe Constituency stretches back in 2011, when the MME, with the assistance of the EU, established the first commercial mini-grid in the country: The Tsumkwe Solar Hybrid System. The MME further funded containerized solar systems. As we have heard, there was no connection to the grid, and the MME and **the government did not keep quiet**. We made sure we looked at the best alternatives we could do by then. And we provided for containerized solar systems in the places such as XX, XX, XX, that was in 2015 and in 2018 we also made sure, we put up containerized solar systems in XX, XX, XX and Noma Roidajeb, we delivered. **The narrative of the government is not doing anything must stop from today. Applauses**. **Together** with the **regional** leadership, CENORED, and other stakeholders, the MME will continue pursuing more opportunities and collaborations to ensure that every HH, business, and institutions in the region and the country is supplied with electricity" (Mangetti Dune Inauguration Speech, DM, Pos. 21).

By listing all the services and actions initiated and provided by the GRN, more precisely, the work undertaken by MME, the DM reflects the complaints received. With the opportunity to address the local communities and the completion of the electrification of Mangetti Dune, the DM ensures that the GRN did not forget about the communities or the overall electrification plans. They continue working on them with the stakeholders to achieve the goals of the Rural Electrification Programme, which has been on the table since 2000, for Namibia to have a "world-class" infrastructure.

> "This development is so necessary. Without it, the prospect of growth of individuals, businesses and the community at large, are greatly hampered whether in the village, in the farm or in early settlement. But the government of the Republic of Namibia and the MME, in particular, through its Rural Electrification Programme, is pleased to witness another addition to **Namibia´s world-class infrastructure portfolio**" (Mangetti Dune Inauguration Speech, DM, Pos.18).

**"Electricity is a Modern Necessity of Life"**

> "Electricity is a modern necessity of life and **should be in every village, in every farm, in every settlement**. However, Namibia´s vast space and low population density make this .... mission a bit challenging and costly. Therefore, you might still find that many households in our country that are still without electricity, that is **about 50% of the HH, which is 300,000** that are still without electricity. **This is a national challenge**. And the government through the MME is **committed to tackle it**. To this end, the MME has **set a target to have all HH in Namibia electrified by 2040**" (Mangetti Dune Inauguration Speech, DM).

Based on the thorough studies that were realized starting from the 2000s, as explained in the previous sub-section, MME has a very accurate database of the unelectrified localities and the number of households. The fact that electricity is considered a modern necessity of life, it is also expected to be on every farm, in every settlement, in every village in order to have light without any health risks, a clean environment, and to study at night all attend online learning activities as it was introduced within the "new normal of the COVID-19 pandemic" conditions. On the other hand, it is expected to boost small and medium businesses and create productive-use opportunities, of which the beneficiaries are called to take this opportunity for socio-economic development activities.

**Energy Practices**

> "I visited the health center here and that was the first time **I have seen a baby delivered by cell phone light because there was no electricity**. Now, that is something **of the past** and **we are not going to go back**

> **there**. (…) But most importantly, you, the beneficiaries of this investment, therefore I plead with you to make the most of this investment by **being efficient** and **using the electricity for productive purpose**. Hon governor, ladies, and gentlemen, I am confident that this project will bring much needed **social benefits** that can be harnessed in two tangible economic activities. **The pure electricity enables** the establishment and extension of **small and medium business** enterprises. It enables **our children to do their schoolwork** allowing them additional hours at night to study and revise for their school exams under **a clean environment** which also **minimizes the burning of the houses** by candle flames. It further enables schools to conduct the **so much required e-learning practices** under **the new normal as imposed by Covid 19 pandemic**. And also facilitates the protection of our **environment by minimizing on deforestation for further fuel**" (Mangetti Dune Inauguration Speech, DM, Pos. 18-22).

This speech confirms the essentialness of electricity in everyday life, from health through social benefits to education and productive uses, environmental friendliness, and safety, by focusing on the identified energy practices in the caste study areas. Furthermore, it also expresses the "lived experience" of witnessing the birth of a baby under the light of a mobile phone and how this experience can foster even more electrification endeavors and create an understanding between the local communities and the decision-makers. On the other hand, it underlines that "that is something **of the past** and **we are not going to go back there"** and the irreversibility of the process when electricity comes, as it is in the modernization process. As a result, this quote also confirms how energy practices have a bridging role within the MLG of off-grid electrification through its lifeworld experience that are mutually shared and experienced by all people.

#### Praise for CENORED and Support for Other Stakeholders

> "Ladies and gentlemen, I am **overjoyed and applause CENORED** for responding to the call and working with the MME to tackle this national challenge. This initiative shows that **CENORED as a company have a broader vision and have brought into the national vision of reaching universal access**. I **urge you as CENORED** to continue with this **good gesture** and work **toward expanding the distribution and connectivity to communities** within the vicinity of the newly installed electricity infrastructure **and beyond**. On the same note, I would like to **congratulate CENORED** Board of Directors and Management **for the completion**. Well done, **CENORED and the Team**! Applauses" (Mangetti Dune Inauguration Speech, DM, Pos. 23).

> "Achieving this goal, we **require a collaborative effort from all stakeholders such as ourselves from the MME, NamPower, our regional electricity distributors, in this case CENORED for**

> **Otjozondjupa Region, Regional and Local Authority Councils, but many more**" (Mangetti Dune Inauguration Speech, DM).

Furthermore, DM emphasizes the involvement of CENORED in completing the connection to the grid and the role they play within the government´s plans in expanding its network, thus electrifying unelectrified off-grid locations. On this occasion, it is clear that although CENORED has been established as a private company, there is a strong relationship with the government. Furthermore, in terms of the off-grid electrification plans, she provides the commitment of MME for the rest of the Tsumkwe Constituency in extending the grid network until Tsumkwe and Gam. At the same time, the techno-economic debate continues on the side of CENORED as the owner of the mini-grid systems in both places.

> "Let me also comment and **congratulate the impacted community** for yet another chance of **economic liberation afforded** to them through this project. Either wish **to encourage the community to take the full of advantage of the usage of this infrastructure to develop, to call in investors and uplift the local economy of the** ........as well as of the region, Otjozondjupa. With this commissioning comes **the call for the community** to work closely with CENORED as an operator **to ensure** that this infrastructure is **well-maintained**. Please take good care of all these assets. Please pay your electricity bill consistently and on time. By doing all this, you will definitely **enable CENORED to continue its great work of ensuring that you have safe, secure, and reliable electricity supply**" (Mangetti Dune Inauguration Speech, DM).

**"Way Forward"**

> "Of course, one might ask, now, we have electricity up to Mangetti Dune. **What is the way forward?** The government and the MME remain committed to **make sure that we extend these lines to Tsumkwe, to Gam, and to the rest of Otjozondjupa.** Applauses... And to make sure, from now on, just like the way we did, with availing of the resources, we will continue to plan, whether it is in the **short term or long-term**, to make sure that **connectivity** has taken place in the **whole of Otjozondjupa region**" (Mangetti Dune Inauguration Speech, DM, Pos. 23).

A way forward is described for electrifying Tsumkwe and Gam through the extension of lines of the national grid. Although the excerpts of this inauguration speech have political connotations, the Mangetti Dune, and surrounding communities did have a chance to get acquainted with the "developments" in Namibia and the plans and visions of the GRN firsthand directly from DM. This illustrates that the MME is not disconnected from the localities; they hold strong ties with the local communities.

#### *6.4.1.2 Future Visions of the ECB and NamPower*

The visions of ECB and NamPower are reflected in their websites, plans, and annual reports. In this sense, the ECB´s vision, according to the Integrated Strategic Business Plan (2021-2026), is "to be the leading electricity regulator in Africa" with the mission to regulate the Namibian Electricity Supply Industry sustainably, in the interest of all stakeholders concerning efficiency, affordability, safety, and accessibility" (ECB, 2021). The ECB has the mandate of "exercising control over the ESI with the main responsibility of regulating electricity generation, transmission, trading, distribution, supply, import and export in Namibia through setting tariffs and issuance of licenses" as per the terms of the Electricity Act 4 of 2007 (ECB, 2021). With this role, the ECB oversees the efficient development and functioning of electricity in Namibia, ensures efficient provision of electricity and its security, with a competitive environment in the electricity industry, and promotes private sector investment in the electricity industry (ECB, 2021).

On the other hand, NamPower envisions itself as very similar to ECB, "to be the leading electricity service company of choice in SADC" (NamPower, 2019, p. 12). In terms of the role of electricity:

> "NamPower **understands the important role that electricity** plays in all aspects of socio-economic development – **not just in driving the machines** and equipment that keep Namibia's factories, mines and farms going, but also in powering our schools, our offices, **and our very lives"** (NamPower, 2023).

Although the role of electricity is firstly associated with machinery, the "lives" are also recognized within the limits of their services. Furthermore, NamPower´s mission is identified as "to provide the electricity needs of our customers, fulfill the aspirations of our staff, and satisfy the expectations of our stakeholders in a competitive and evolving environment," along with the values of "customer focus, integrity, teamwork, accountability, empowerment, health, safety, and environment" (NamPower, 2019, p. 12). In order to achieve the vision and mission, their strategic goals are to ensure the security of supply, unlock the value of electricity sector collaboration, optimize financial sustainability, and drive operational and organizational excellence (NamPower, 2019, p. 12).

ECB and NamPower, as state-owned enterprises (SOE), envision themselves as "being a leading company" at the supra-national level within the SADC region. They admit their responsibilities within the "national" level only, without strengthening, supporting the sub-regional endeavors, and enlarging their visions in the lower levels of the hierarchy. Instead, they opt to become a regional leader at a supra-national level. Thus, reaching people living in the unelectrified localities in Namibia does not explicitly take the form of an objective to be accomplished. On the other hand, the companies work with good governance principles, including capacity building and development activities and corporate social responsibility. However, being a competitive actor in energy trading takes priority; this brings into discussion the role of the SOEs in the country´s development versus the country´s representation as a leading company on a supra-national level.

The SOEs are established according to the State-Owned Enterprises Act of 2006, amended in 2008, and governed by the Ministry of Public Enterprises (MPE) (Weylandt, 2016). A hybrid governance model was established in 2016 due to the inefficiencies of the earlier versions (Weylandt, 2016). Accordingly, there are three types of SOEs: commercial, non-commercial, and extra-budgetary. According to MPE´s website, there are 22 commercial, 38 non-commercial, and 11 extra-budgetary enterprises (MPE, 2023). Under this new system, NamPower is listed as a commercial SOE, reports to the MME, and is regulated by the ECB (NamPower, 2023). The non-commercial SOEs to which the ECB belongs remain under the line ministries, and the Ministry of Finance takes charge of the extra-budgetary SOEs (Weylandt, 2016). The vision of MPE is "to be a catalyst and adaptive institution that promotes sustainable economic prosperity for all" with the mission "to ensure prudent public finance management and setting the direction of Namibia´s economic and fiscal policy for sustainable, inclusive economic growth and development" (MPE, 2023). Thus, all SOEs serve as efficient and effective actors in their field of expertise to implement NDPs aligned with the country's visions at a national level. In the case of ECB and NamPower, as mentioned earlier, the nationally and corporately owned objectives address the national and supra-national levels more without specifically prioritizing the strengthening, supporting, and empowerment of the regional and local levels for the unelectrified areas in the case of electricity provision.

### 6.4.2 Future Visions of Off-Grid Electrification at the Regional Level

At a regional level, the future visions support the national policies while adopting their own goals. Table 6.5 presents the key findings of the future visions at the regional level concerning local, regional and national visions.

*Table 6.5. Future Visions at the Regional Level*

| MLG Visions | Local | Regional | National |
|---|---|---|---|
| Regional Governor - Otjozondjupa | **"Lighting in Mangetti…A joyous moment"** | "CENORED have been **doing magic!"** | "Electricity is not a luxury but an enabler!" for Development |
| | "Our Target and Aim: Grid Electrification of Tsumkwe and Gam Settlements" | | Step by Step in the Rural Electrification Programme – Support of the Government |
| CENORED | "Electricity is one of the essential basics human needs and (the) key driver for economic and social growth of any community" | "Goal – Contribution to Socio-Economic Development" | "We believe that electricity is a basic necessity for equally quality of life and equality." |
| | Call for Safety – Illegal Connections | | "Common Goal with MME: Deliver Sustainable and Effective Services to the people in Namibia." |

*Source: Author*

### *6.4.2.1 View of the Otjozondjupa Region*

Following the national level actors, at a regional level, the governor of the Otjozondjupa Region and CENORED representatives presented their views regarding the off-grid electrification in the region. The support of CENORED was observed not only at a national level but also at a regional level during the ceremony. The governors of the regions are appointed by the President of Namibia, according to the Act 15 of 2010 Special Advisors and Regional Governors Appointment Amendment (Shejavali, 2015). However, their powers have been separated from the Chairperson of the Regional Council (who was previously considered as one role, as Regional Governor) according to the Act 16 of 2010 Second Regional Councils Amendment (Shejavali, 2015). In this sense, Regional Governors act as central government representatives, investigate and report to the President or the Ministry, and are the link between central, regional, and local authorities and Regional Councils. They are appointed for five years (Shejavali, 2015).

**"Electricity is not a luxury but an enabler!" for Development**

> "….Today, we are gathered here to witness the actual commissioning of this electricity. This is to live **by our mantra that "service delivery is**

> **key**". Electricity is **not a luxury** but **an enabler**. It is an enabler and brings about **the necessary development in an area**. Today, we are witnessing the electrification of the 33 kV voltage line that stretches for a 120 km hon DM. **One of the key things that we were busy with this line was that we should make sure that the whole Tsumkwe West is covered and every ... is electrified**. (…) As **we are talking about development**, development comes **with a lot of benefits**. The CEO has put it to us today that **110 local community members were employed** and benefited directly from this project. That´s what we call development because **you are employed locally on your own project** that you have worked on" (Mangetti Dune Inauguration Speech, Governor OTRC, Pos. 8-11).

The statement "electricity is not a luxury but an enabler" has the connotation of a period when electricity used to be a luxury that the governor recalls and refers to by introducing its new position of electricity as an "enabler" that makes something possible (Cambridge Dictionary, 2023). The "mantra of service delivery is key" informs about the political dimensions of bringing the service to the community. Furthermore, the development was stressed in this quote through the electrification project, the employment created for local community members, and the function of electricity as an enabler for making many things possible. The choice of the word *mantra*, a word in Sanskrit that originated in Hinduism and Buddhism, that is used in the form of a "word or sound repeated to aid concentration in meditation" (Oxford Reference, 2023), illustrates their connection to the other geographies in the world. On the other hand, local employment through regional projects has been emphasized, providing regional and local collaborations.

### "Lighting in Mangetti… a joyous moment"

> "….I **know vividly** that when I came here in December, I had a situation whereby I came to visit the hospital for some relatives of mine, **and I found that the whole of Mangetti lighting like nothing.** And to me, that **was a joyous moment** because you started so on, and eventually the result is bigger and everybody is benefiting. That´s what **we call bringing development to the people by the government of the people for the people under the government of the Republic of Namibia**. And we are grateful of that" (Mangetti Dune Inauguration Speech Governor OTRC, Pos. 10).

*Figure 6.9. Grid Connection in Mangetti Dune*
*Source: Author*

The governor´s "vivid" lifeworld experience of *seeing* the Mangetti Dune *lit up* with lights created a "joyous moment" and a profound recent memory of an enabler rather than a luxury of the past. In addition, the accompanying joy of a completed electrification project (Figure 6.9) for the benefit of all communities brought another dimension of contentment and success. Nevertheless, as part of his speech, he linked the GRN and the people according to Namibia's hierarchical structure. Thus, the people "under" the GRN of Namibia received electricity "from" the GRN, a development bringer, a light-bringer. As a result, on a regional level, the governor transferred the power of the hierarchical structure from top to bottom, from the GRN to the region, and then to the localities.

## Step by Step in the Rural Electrification Programme – Support of the Government

> "... These are all part of the government programme brought us in what we call the Rural Electrification Programme to make sure that the rural community is covered and benefits from the mainstream developmental activities" (Mangetti Dune Inauguration Speech, Governor OTRC, Pos. 9).

> "…We are determined fully hon DM and all stakeholders, to make sure that we live up to the expectation of the citizens....... that the government of the people will come to them wherever they are located and **development will reach them wherever they are**. So we are leaving to this one, **a clear testimony**, as regional leadership, **we are committed to live up to** the … and with the good working **relationship that we have amongst ourselves**, with the good support that we get from the **central government** through you DM and others, we are confident that indeed, **everything else is surmountable and service delivery to the people will be realized in the set time frame**" (Mangetti Dune Inauguration Speech, Governor OTRC, Pos. 13)

Following up, the governor continued naming every locality and a number of connections connected to the grid, and the investment was realized within the Otjozondjupa Region. He informed the audience about the earlier stages of Mangetti Dune and Mkata when they were partly electrified with solar energy. Now, they are connected to the grid within the Government's Rural Electrification Programme. The rural communities are part of the development activities by emphasizing the commitment of GRN to reaching every corner of Namibia, the good relations among stakeholders and central government, and the commitment to delivering service according to their plans for development. Thus, he underlined, once again, how electricity is interlinked with development and its political dimensions while delivering the necessary services for people.

### "CENORED have been doing magic! "

> **"CENORED have been doing magic on this project, moving thi**ngs from one place to the other with the **good expertise** they have, with the **good capabilities** that they have. They have **demonstrated once again** that, indeed, every challenge put before them is **surmountable**. And CENORED for that I would like to thank you for the good and hard work that you have dedicated in ... this time"( Mangetti Dune Inauguration Speech, Governor OTRC, Pos. 8)

Reflecting on the very initial meaning of electricity changing worldviews, reenchanting, and being a magical tool, the governor expressed that "CENORED has been doing magic" on the grid electrification by connecting households, businesses, and public institutions into the grid, bringing a magician within this context. The physical state of electricity, being invisible to the human eye but measurable, enhances this meaning and analogy with magic and the magician. On the other hand, the magic also implies CENORED´s work in this project and its successful completion despite the challenges they encountered. Therefore, he contributed by building trust in CENORED and its activities in the Otjozondjupa Region.

### "Our Target and Aim: Grid Electrification of Tsumkwe and Gam Settlements"

> "**Our target and aim** as far as electrification are concerned as we are putting today, we said, Tsumkwe West is covered up to Mangetti. We have two off-grid systems in Tsumkwe Constituency, one in Tsumkwe itself in the Tsumkwe Settlement and the other one in Gam. **The MME also supported strongly, the expansion of the electricity in Gam**, whereby an additional 400 houses were actually added to the already existing grid system, and that was commissioned already. So, hon minister you must also need to know that the Gam community people have benefited, the Tsumkwe community equally benefited. And out of the

> whole country, the only off-grid systems are prevalent only been seen in Tsumkwe Constituency.
>
> That is how government is committed **towards green energy**. It is about **being green** and **going green** as far as renewable energy is concerned. So, we are grateful for that. But the aim that we are sitting here hon DM, the electricity in Tsumkwe and Gam somehow .... **are off-grid systems in a way doesn´t live up to the total expectation in terms of the workload** ... put on the system. So, we were thinking how best to expand this.
>
> **One of the ideas was to take the electricity from Mangetti**, where we are, 80 km, **to Tsumkwe.** Then, this system is fitted into the existing system of Tsumkwe. We have another electricity line that came all the way from the border post in Omaheke Region up to Eiseb. And **Eiseb is only 50 km away from Gam. So, the idea was why don´t we then, tap this electricity, which is a grid system up to Gam** and **tap the Mangetti grid up to Tsumkwe.** In that way, **you have electrified the whole Tsumkwe Constituency,** and you are done with the electrification thereon. So this is a project that we have been thinking of. .... with CENORED expertise are doing a **detailed assessment and detailed design** to give us option to what we can go for. So, hon DM, when we approach you of this, and **seeking support of this project**, be aware that those are the initiatives and intents that we have at our tables and electrify the whole of Tsumkwe" (Mangetti Dune Inauguration Speech, Governor OTRC, Pos. 12).

The dilemma between an up-grade in the mini-grids in Gam and Tsumkwe versus the grid potential of grid connection has been in the debates of the PROCEED Project as well as the stakeholders in Namibia. The recommendations for action in social, economic, and technological aspects were delivered at the PROCEED Conference (2023). In this respect, the governor informed about their regional assessment and stated that the grid-electrification option was also on the table to connect Tsumkwe via Mangetti Dune and Gam via the national grid from Eiseb. He also differentiated between the on-grid and off-grid systems by stating that "… [they] are off-grid systems in a way doesn´t live up to the total expectation in terms of the workload" and delivering his concerns about the reliability off-grid electricity in terms of its technical capacity. He completed his speech by requesting the support of MME for this matter. This suggests that the regional government, in this case, has been more inclined to the grid connection rather than the expansion of mini-grids in Tsumkwe and Gam Settlements, which could mark a rural (grid) electrification target within the Tsumkwe Constituency.

#### *6.4.2.2 The Central and North Electricity Distribution Company (CENORED)*

The CENORED is the third licensed RED in 2004, after NORED (2002) and Erongo RED (2002), with the shareholding of the Regional Councils of Kunene, Oshikoto, and Otjozondjupa Regions, the Local Authorities within its supply areas, and NamPower (CENORED, 2023). Its jurisdiction area is the Otjozondjupa Region, part of the Oshikoto, Omaheke, and Kunene Regions, with approximately 34,000 clients in the main supply area and also serves approximately 9,700 clients in the Omaheke Region and Okahandja (CENORED, 2023). It covers an area of 120,000 km$^2$ with 8,000 km of power line infrastructure to provide electricity for single-phase prepaid clients to large power users (CENORED, 2023). Two hundred one people are employed in CENORED, headquartered in Otjiwarongo. The vision of CENORED is to "Enable sustainable access to affordable electricity to improve lives" with the mission "We distribute and supply the electricity in a customer-responsive manner that stimulates socio-economic development" (CENORED, 2022). They also determined their values as *passion* by being committed with heart and mind, *creative*, curious, *hardworking*, and delivering results beyond expectations. They are also customer-focused by being ambassadors dedicated to the cause (CENORED, 2022, p. 12). Thus, CENORED depicts a corporate, regional company portrait with profit and public shareholding, bringing passion and people into their work and mission.

**"Electricity is one of the essential basics human needs and (the) key driver for economic and social growth of any community"**

> "In Mangetti Dune, Otjozondjupa Region, this project formed part of government of the Republic of Namibia and CENORED Electrification Programmes **to ensure** that **governmental plans are duly executed to accelerate the development** in Namibia. More our ... program planned **to create us to the electricity grid for off-grid areas within our area of operation and the country enlarge**. Under the Programme we are particularly concerned **about improving the living standard of our people** and .... ..**electricity is one of the essential basics human needs and key driver for economic and social growth of any community**" (Mangetti Dune Inauguration Speech, CENORED Representative).

The representative from CENORED confirmed their support for the governmental plans through the rural electrification program in delivering electricity supply for off-grid areas in order to improve the living standards of the people by stating, "*improving the living standard of our people*" in an inclusive way and acknowledging electricity as a "*one of the essential basics human needs and key driver for economic and social growth of any community.*" Furthermore, he also introduced justice and equality by mentioning, "*We believe that electricity is a basic necessity for equally quality of life and equality*". Thus, he also recognized the transcendental meaning of electricity, universal for all.

**Call for Safety – Illegal Connections**

> "However, electricity safety is of outmost important, therefore I am hereby **appealing to the beneficiaries or residents of this locality not to engage into unsafe practice**s, just as electricity through bypasses, rendering in illegal connections" (Mangetti Dune Inauguration Speech, CENORED Representative, Pos. 6).

Furthermore, he called on the beneficiaries or residents not to engage in unsafe practices and illegal connections while CENORED being aware of the undesired use of electricity. Thus, this warning is also meant as a reminder based on the experiences of operating and maintaining the distribution lines and supplying electricity.

**"Common Goal with the MME: Deliver Sustainable and Effective Services to the People in Namibia"**

> "… **CENORED and the MME have a common development objective which is to deliver sustainable and effective services to the people of Namibia**" (Mangetti Dune Inauguration Speech, CENORED Representative, Pos. 2).

> "The successful completion of this project that we are celebrating today is the result **of partnerships** and **a collective learning approach between the two parties** in our efforts to construct **service delivery to the people** and move quicker towards the **achievement of our national development agenda** for Namibia. Development and provision of services to communities **require stakeholder consultation and cooperation.** Thus consultation, engagement continuous and targeted investment for **both public and private sector**. It should be noted that event that it´s during the responsibility of the Regional Councils to identify areas to be electrified, thereafter, CENORED through electrification program will then electrify the identified area for sustainable community development" (Mangetti Dune Inauguration Speech, CENORED Representative, Pos. 3-4).

The quote above strengthens the ties with the development visions and the common goal with the MME in delivering sustainable and effective services to the residents of Namibia. Recognizing stakeholder consultation, cooperation, and public and private sector collaboration, they envision providing services with continuous and targeted investment that inform about the multiplicity of actors with a rationalized vision. However, in terms of electrification, within the regional hierarchical governance structures, they expect Regional Councils to identify the areas to be electrified for CENORED to pursue the electrification program for "sustainable community development." This suggests that CENORED does not develop its programs as a regional electricity distributor but waits for the official request from the regional authorities. Based on the history of rural electrification development in Namibia, this statement does not support the past electrification plans, in which a thorough identification of the unelectrified localities was conducted twenty years ago and updated frequently, which do not need necessarily further identification, but rather a status-quo update for action. Furthermore, it underlines the existing hierarchies at a regional level as well. As a result, it shows that at a regional level, although the public-private partnership has been recognized to bring successful implementations, the authority of the regional authorities subsists simultaneously.

**"Goal – Contribution to Socio-Economic Development"**

> "That goal contributes to **attaining high level national objective and contribute immensely to the industrialization of the country**. Therefore, we need to **strengthen and encourage public private partnership** with the primary goal of better understanding of our respective contributors to our common development objectives. On this very happy occasion, I time off by also reminding all of our **electricity is paramount for social, economic development**" (Mangetti Dune Inauguration Speech, CENORED Representative, Pos. 6).

As a result, as the quote above suggests, "electricity is paramount for social and economic development," and attaining the electrification objective also delineates the rationalization within CENORED´s perspective in aligning with the national objectives by meeting the regional ones.

## 6.5 Conclusion

### 6.5.1 "Going from Government to Governance"[18] in Namibia

This chapter outlined, described, and analyzed the actors, the networks, and the policies regarding the RE governance in Namibia from an MLG perspective; analyzed and interpreted the complexities and opportunities, as well as the visions of the national, regional, and local actors. The government can be found within this shifting process from "government to governance:" the modernization of the RE sector and its governance, especially for (off-grid) electrification. This shift occurs with its differentiating actors for more specific responsibilities to be undertaken in the field of RE by dispersing the authority to more specialized actors, involving institutions, public and private actors, and

[18] Lo (2018).

national and supranational perspectives. Although the communities are involved in the earlier phases of RE projects, their engagement and involvement in a more comprehensive way have not yet been established. Thus, the government and state institutions still have the power to conduct a shifting process or "transformation" (Lo, 2018, p. 1) from the hierarchical governance model with limited actors into a more pluralistic form regarding the actors and networks. Consequently, this shift presents a meta-governance, also known as "governance of governance," whereby the government keeps its crucial position while exercising it (Lo, 2018, p. 4). Furthermore, it creates a dialectical coexistence of two governances by transforming the hierarchical governance and the meta-governance (Lo, 2018, p. 4).

Figure 6.6 illustrates the (off-grid) RE governance structure in Namibia. The triangular pyramid has been chosen as an illustration advancing the two-dimensional triangular form of hierarchical (and meta) governance to the three-dimensional approach, interpreting it within the context of MLG. Thus, the triangle on the left represents the supra-national level, with the state forming its base and active in the supra-national activities with the SADC-related, EU-related, and global actors and networks. Then, the triangle on the right (Figure 6.10) represents the sub-national level similarly, with the state constituting the base with the national, regional, and local actors. Thus, while modernization has been occurring on each triangular side, meta-governance has been occurring in Namibia at the center of each triangular pyramid, bringing two governances together.

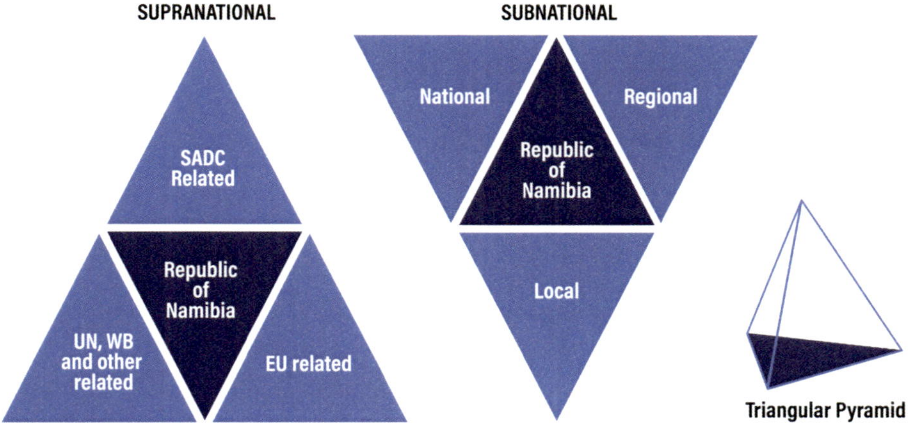


*Figure 6.10. Supranational and Subnational Governance in Namibia*

*Source: Author*

In this respect, the meta-governance demonstrates a space for actors and networks led by the government that allows the modernization process to take place in the (off-grid) RE governance field, particularly for electrification. When modernization has been investigated based on differentiation, rationalization, domestication, and individualization processes, the differentiation in the governmental structures and inclusion of the private sector and other related actors enable actor and network processes for the goals that are intended to be attained within the plans and policies that are steered by the government in Namibia. From a very specialized lens, NamPower, as a SOE, seeks to secure the country's energy supply while taking its role within the SAPP as its energy trading institution. Besides handling the transmission network and its development, the inclusion of IPPs within the energy mix supports Namibia´s energy demand, and the rest is supplied by NamPower and SAPP. Adopting policies and regulations enhancing the participation of private actors and using renewable energy sources contribute to the government´s role in meta-governance as a steering wheel within these multiple intertwined processes. For the state, energy is associated with socio-economic development and industrialization. Therefore, the responsibility of securing the energy supply has been shared with NamPower, with the allocated power to increase the generation capacity and the transmission network. In this sense, from the point of domestication in the energy industry, more energy resources, such as wind, biomass, and solar, have been considered in addition to their existing energy infrastructure. Thus, it brings more technology that will enable the control of multiple technical and technological processes that will enable the necessary electricity generation, which will also create more dependencies on technology as a part of domestication.

Furthermore, the fast-forward development of green hydrogen production in Namibia introduced additional actors in the field of energy as well as new public-private partnerships. In addition to the electrification targets, Namibia is now considering benefiting from the income generation of green hydrogen for the electrification projects. Although the existing business plans and economic contribution paint a positive portrait of the economic future of Namibia, the real action plans for the off-grid electrification execution depend on the human resources and the capacities for execution, follow-up, and monitoring while ensuring the overall sustainability of the mentioned systems, especially in the off-grid areas.

### 6.5.2 Policies, Complexities, and Opportunities

Namibia´s ESI dependency on the imports from SAPP (with 65% still representing approximately 50% of the population connected to the grid and the remaining 50% mainly living in rural areas) has been neglected in the key actors' reports. In this sense, the National Electrification Strategy and Implementation Plan, which is in the process of approval from the Cabinet and includes the unelectrified areas, illustrates a holistic and inclusive approach to the lack of electrification in Namibia that has been recognized since its independence.

On the other hand, green hydrogen suddenly appeared in Namibia's energy landscape in the modernization of the energy mix and future scenarios. Green hydrogen has been the biggest hope for creating the necessary funding for electrification in Namibia, which was not even in the picture in 2020. The changes the world has been going through with COVID-19, the solutions with fewer carbon footprints, green energy options, and the impacts of the subsequent war between Russia and Ukraine have enabled different solutions for the energy supply globally. On the one hand, green hydrogen solution came into the energy mix to create jobs and green fuel. On the other side, paradoxically, it has shifted the focus from electrification (as a plan of the past towards the universal access goal) to generating fuel. They are two sides of the same coin regarding supply and demand. However, reflecting on the history of energy/electricity in Namibia since 1990, as discussed in this chapter, the unelectrified areas remain neglected in this course due to the sparsely populated geographical structure of Namibia, which also introduces limited human resources and the socio-techno-economic bottlenecks in electrifying those households or businesses.

The findings of the complexities suggested interdisciplinary categories such as socio-economic, socio-techno-economic, and socio-technical alongside social and institutional complexities by justifying "social" as the common denominator and proving that "energy is a social issue with a technical component, not the other way around," (Pasqualetti, 2021, p. 3). The failed expectations of the past, the lack of awareness raising, and difficulties in accessing people have been some of the main social limitations. The social acceptance of the grid, the lack of private sector support in implementing energy shops, operational and maintenance issues, and the lack of standards have not enabled off-grid electrification to be implemented as planned. Furthermore, although funding and subsidies existed, accessing them has not been easy.

In return to the mentioned complexities, national and regional actors presented more (mono)disciplinary opportunities for successful implementations, such as social, economic, institutional, and techno-economic. These opportunities presented a top-down hierarchical approach acknowledging community engagement and ownership of the mini-grid with the hybrid partnerships as a sustainable model for mini-grids, allowing cross-subsidization or grid encroachment techno-economically viable solutions. In this way, the actors contributed to the ongoing MLG meta-governance in the Namibian energy sector based on their experiences.

In conclusion, as underlined earlier, capacity building and capacity development have a broader definition than its connotation to the economic and technical disciplines, which has been utilized when designing productive use business cases, LCOE calculations, or technical optimization of these systems at the utility scale. At a local level, the introduction of electricity into everyday life conveys a transcendent meaning beyond its physical properties, as explained in Chapters 4 and 5. Therefore, the capacity building and development necessities, contents, methodologies, and characteristics for the communities that are the beneficiaries of these systems would differ from the needs of the institutional levels often mentioned in complexities and opportunities.

### 6.5.3 Visions of Off-Grid Electrification at the National, Regional, and Local Level

Reflecting on Namibia's policy and regulatory framework development, MME, along with NPC and ECB, made many efforts to cover the unelectrified areas for electrification. The recent history of off-grid electrification policy demonstrates many policies, development plans, and action plans and the strong connection of electricity to socioeconomic development. On the other hand, although the policies and plans recognized the crucial importance of building and developing capacity, they did not quantify neither the necessary human resources nor its required budget to undertake the implementation of electrification plans, thus risking their realistic applicability under the conditions in Namibia. In fact, quantifying the necessary human resources would contribute to rationalization. Hence, it would also depict the new employment fields for the county for the envisioned socioeconomic development and industrialization.

Furthermore, on a supra-national level, Namibia has been pursuing the commitment to the Paris Agreement and universal energy access 2030, which was revised to 2040 in the NES, and to being actively engaged in the SAPP in the SADC region. The policy and regulatory framework have been aligned to meet the nexus of supra-national and national needs to successfully implement the elaborated policies in the country to reach the development and industrialization goals. However, as the complexities of RE deployment are outlined with an accompanying modernization process, the translation of these policy and development efforts into implementation actions does not occur; hence, electrification does not reach the unelectrified rural areas, where the communities expect to have employment opportunities, training centers, reliable and affordable electricity in order to have a better life. The differences between the activities of supranational levels and subnational levels and how the expansion of electrification does not occur as planned may have multiple reasons.

Firstly, the socio-techno-economic RE deployment has different dynamics at all MLG, and the actors involved have not been equally distributed at all levels. Secondly, the RE off-grid implementations are closely connected to the development of communities in various layers, such as necessities like water and food, health, education, income generation, self-sustaining, reducing poverty, and contributing. Thus, the local level is also closely linked with the country's socio-economic development, and many more are interlinked to these areas. Third, decision-makers consider the off-grid RE a magical tool to bring wealth, health, prosperity, development, and industrialization. However, for these multiple level impacts, multiple disciplinary efforts have not been met. Furthermore, this magical tool view, in return, has also been shared by the local communities, which interpret electricity as "magical," which cannot be denied. However, within the modernization process of governance, the different dynamics within the social, technical, economic, and institutional processes define the rhythm of realizing off-grid electrification and its related goals.

# 7 Discussion

Chapters 4, 5, and 6 presented the research findings of off-grid communities and the MLG in the RE field for electrification through a life-world analysis with a hermeneutical interpretation. This Chapter discusses the key findings and compares the similarities and differences between two local communities at local, regional, national, and supranational levels.

## 7.1 MLG and Meta Governance with the Process of Modernization

The previous chapter's findings indicated that energy governance operates at supranational and subnational levels, emphasizing a meta-governance approach. This involves transitioning from traditional government roles to a more complex governance processes, where the government remains the principal authority in overseeing actors, networks, and policies in the modernization process. In this context, Figure 7.1 aims to depict this process from a three-dimensional perspective, advancing beyond the two-dimensional interpretation (Lo, 2018) and illustrating it within an MLG framework. The state of Namibia is the leading authority at the apex of each pyramid, sharing a common base structure. The sides of the pyramids represent national, regional, and local levels at the subnational level. In contrast, the supranational level encompasses regional (SADC-related) and EU-related levels, as well as the global level, which includes organizations such as the WB, or the UN.

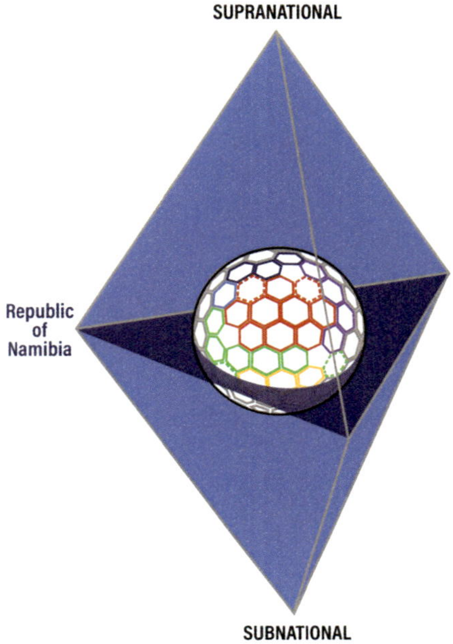


*Figure 7.1. Multi-level Meta Governance of Renewable Energy*

*Source: Author*

The hexagonal sphere of actor-world(s) in the center illustrates the differentiating actors to carry out different tasks within RE governance, which is not fixed but in motion, and active, as in the figurational process described by Elias (1976).

### Differentiation of Actors and Pluralization of Energy Services

On a local level, the findings suggest a pluralization of actors in Tsumkwe and Gam represented in the settlements, such as public institutions, ministerial offices, local private shops, NGOs, CBOs, churches, parastatals, residents, and one actor for mini-grid electricity related matters. In Gam, compared with Tsumkwe, there is a lower representation of ministries and electrified shops, no parastatal, and fewer NGOs, and CENORED is not represented in the settlement. On the national level, the actors have been specialized in terms of their tasks, such as ECB in the regulation of policy and regulatory frameworks, and NamPower in the generation of power, its transmission, and partial distribution depending on the region; NEI in the research and implementation; IPPs in the generation of electricity and feeding into the grid, are some of the key players. The distribution of electricity has been planned under five regions, of which only three are active and operational, and the Regional Councils, NamPower, are also shareholders of this privately established company. The pluralization of actors demonstrates the specialization of energy generation, transmission, distribution, and consumption components.

The specialization in energy generation includes renewable energy sources, with electricity as one of the specialized energy types. Furthermore, specialization involves private companies, public-private establishments, public actors, community actors, and international actors involved in the generation. The more specialized the actors, the more involved they are in differentiating the individual actors. Hence, the more differentiation that emerges with specific tasks, the more goal-oriented and cost-benefit-oriented actors appear in renewable energy governance. These goal-oriented and techno-economic considerations have been taking place at every level. On a national level, the state seeks to provide energy security to its citizens through its renewable energy mix. When national financial resources have been limited to attain this goal within a given period, including private actors with defined conditions, such as the IPPs, occurred. However, the off-grid rural regions, even with their more specialized conditions, still await further specialization in the pursuit of electrification. This suggests that while differentiation in the localities has been taking place in the off-grid rural areas, there is also differentiation occurring first at a national and then at a regional level while allowing the share of tasks and specialized work without giving away the state's authority.

However, on a supranational level, the state of Namibia signed the SADC Energy Protocol and has been taking part in SAPP and SACREEE at a supraregional level while also being involved in global agreements (i.e., Paris

Agreement) and EU agreements in advancing and enhancing the implementations in Namibia by aligning them at the national level. However, when the events and related policies of the past 30 years are taken into consideration, there is a tendency to work on a national level by working on the REMDPs or NDPs, which include rural electrification and related rural development, and initial policies, while also working and being involved in the supra-regional and supra-national networks. The goal-orientedness within the rationalization process, the support received from global actors, and the scarcity of human resources in Namibia have resulted in the delay of the "trickle down" effect of the government at regional and local levels.

Another factor delaying the implementation of already existing rural electrification plans within the MLG framework is that supra-national actors have already been specializing in executing specific tasks, while in Namibia, this specialization has not been present or not with the same authority as in the supra-national counterparts that would enable mutual cooperation in the case of electrification, which also demonstrates the meta-governance in action. As a paradox of differentiation, the boundaries tend to blur among the actors, providing space for multiple actors to cooperate and collaborate in realizing different tasks. In this sense, although global initiatives support the countries at the national level, the country's human resources constitute the limiting factor in spreading and enhancing these support mechanisms. In Namibia´s case, the human resources align with the supra-national actors. However, executing electrification at regional and local levels still needs support from human resources.

Therefore, Namibia´s sparse population, rich renewable resources, especially solar energy, and vast geography, place the country in a particular position for the development of off-grid electrification. This situation also makes Namibian residents invaluable for every development step; however, they remain overlooked under these specific conditions. Thus, one might say that the residents of Namibia are a treasure in Namibia valued even above the mines beneath the earth. As a result, only with people-centered policies, regulatory frameworks, and capacity building, can the development program be reached in local settlements.

**Intertwining Processes of Modernization in Off-Grid Energy**

Figure 7.2 illustrates the intertwined and irreversible processes of differentiation, individualization, rationalization, and domestication within the modernization process. As the findings suggest, these four processes have been taking place at different paces at all levels of the RE MLG. This incorporates additional challenges to the existing complexities of RE deployment, especially in off-grid electrification.

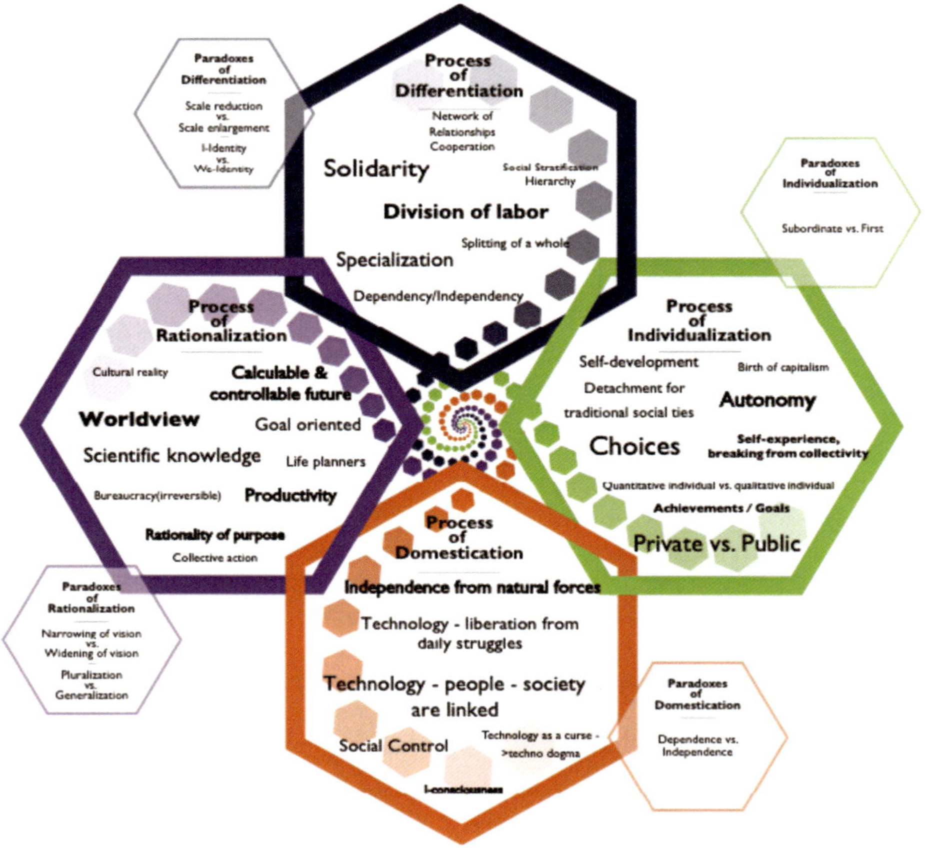


*Figure 7.2. Process of Modernization - Revisited for Off-Grid RE Governance*
*Source: Design by Author (based on van der Loo and van Reijen, 1992), illustrated by Bentley (2024)*

Descartes´ ambition to overcome the human eye´s limitations centuries ago has not ceased; it continues evolving, moving further into a world equipped with more digital tools and processes to overcome human-related limitations. The tools and processes have been developed to overcome a specific issue within a certain boundary. Thus, not only has machinery been developed in the technical sense, but technology and technological processes have also been renewed with digital tools and processes in our daily lives and work. Elias (1995) discusses machinery and its processes as a part of the technization process without the digitalization component (at the time of his analysis, digitalization was not yet included within the scope of what constituted machinery). However, in the first quarter of the 21$^{st}$ century, the author considers digitalization a part of the domestication process in modernization. She agrees with Elias´ description of processes as "unplanned long-term processes that travel in an observable direction without particular long-term objectives while intertwining, intersecting, or entangling with many planned activities" (Elias, 1995, p. 7). In this sense, digitization and digitalization also contribute to *a better life* idea for some in the short term without actually considering its long-term impacts, as they also contribute to people´s everyday lives by utilizing different appliances for different purposes and exploring their different characteristics in quest of *a better life*. As a result, following in the steps of computer technologies, from palm devices to laptops, from the first mobile phones to smartphones, and from analog to digital cameras, technology has progressed in many directions; these have expanded to specialize in particular disciplines, but also, significantly, to introduce all-in-one concepts such as merging multiple services into a tablet or a smartphone that can take a picture, make a movie, use it as a computer, or a phone, a video call. Through these technologies, distances have been effectively dissolved. However, the limiting factors of all this techno-progress have been electricity, the internet, and the accompanying paradoxes.

Namibia has also been transitioning to digitization and digitalization in line with global trends. At the national level, policies include using digital tools, such as new software, to plan and monitor the operationalization of Namibia's electrification. Electricity distributors have been adopting digital payment collection methods while using websites as a platform to communicate their activities and stay in touch with customers. At the local level, in both settlements, the findings suggest that the communities have adopted digital payment systems, where digital devices and internet services are available, and have acquired the competence to use them. Community members are also engaging in social media and internet activities, which link them to their families, acquaintances, and the rest of the world. Furthermore, in Tsumkwe, several projects digitizing and archiving the oral San culture were realized. As a result, digitization and digitalization have occurred at the local and national levels.

## Individualization and Energy Practices

The individualization process is intertwined with differentiation, rationalization, and domestication. From a phenomenological perspective, the "light" that enlightens everyday life and energy practices also allows each individual to live with some personal items, expanding the range of energy practices or converting its use to electric appliances. One important energy practice achieved through artificial "light" was that individuals had the lighting in their individual/family houses for the children to study at night or for them to watch their own television without going to the neighbors´ houses while extending the "living" time of the day into the night hours. In this sense, the meaning attached to it as a source of light, as a *sun of the night*, would prolong their personal lived experiences. Nonetheless, television or radio also connected them to other places in the world and keep them informed while they were in their individual surroundings.

Mobile phone use in analog and digital form is also part of resident's energy practices. Digital smartphones serve as multi-skilled devices for communicating, as general phones but also for following up on news and events, managing and executing digital payments or other digital processes, and lastly following social media. However, local findings suggested that not only was the ownership of a mobile phone counted as significant for the residents, but also the possibility to charge them on their own, in their own houses made the main difference compared to earlier times, when they needed to go to the neighbors′ houses to charge their phone. Similarly, the ownership of appliances within the home, like a refrigerator for keeping food and drinks fresh or irons for ironing clothing, also significantly impacted the individual's everyday life. These energy practices are in continuous interaction with the life-world lived through all the senses. Thus, through the "lit" and multiplied energy practices (through electrification), the community members can have access to a broader scope of everyday activities in the settlements, which makes the appliances not only light bulbs or devices a physical component of an energy practice but provides a transcendent meaning in everyday lives.

One of the primary changes in energy utilization within the context of domestication has been the shift from cooking with firewood to cooking with electricity, gas, or combinations of these resources. Although solar cooking is known and has been introduced to the settlements, several discussions in several places suggest that its use did not result in its adoption as a cooking practice, a fact later confirmed by community members. Contrary to what was found within local practices and supranational intentions, solar cooking has been widely promoted on a supra-national level as a clean cooking device.

Considering the temporal component, as within the temporal hermeneutics that keeps past and future in the present loop of understanding, the communities have been experiencing electricity access in everyday life in tandem with digitalization. Moreover, even though electricity is limited to those community members with the appropriate devices and is mainly limited within the space of their individual homes, all community members are nevertheless faced with domestication and digitalization, even those who reside in (un-)electrified households. Additionally, since computers and the internet are also used in schools, this suggests that most community members are already familiar with digital devices. Similarly, at the regional and national levels, as indicated, digital devices are widely used, making their use inevitable in work and private life, but they also create an irrevocable process.

As discussed in preceding chapters, modernization's paradoxes, which occur at all levels of governance, accompany the process of modernization. At a local level, through the domestication of energy practices, although individuals maintain "autonomy," they also become dependent on electricity and electronic appliances, potentially followed by dependency on the internet. At the regional and national levels, digitalizing certain services provides autonomy for monitoring a particular process at any given time, such as monitoring a power plant or mini-grid with remote service providers and the internet. Government forecasting, implemented using digital tools, also makes the user immediately dependent on digital devices, software, and the internet.

Furthermore, as part of the differentiation, scale enlargement and reduction have occurred simultaneously through the differentiating actors or services. This situation creates a focus on reducing the scale of a particular work or service. An example of this at the local level is the work realized by the conservancies as a part of the CBNRM or electricity distribution managed from the office in Tsumkwe for both Tsumkwe and Gam. This paradox is also closely linked with widening and narrowing the vision. At the national level, the government´s vision is to become an industrialized Namibia and to extend socioeconomic development to its residents. However, it also aims to reach all electrification, following SDG7 on supranational and subnational levels. In this way, widening the vision occurs for a particular vision, while narrowing the vision to reach a greater vision becomes inevitable.

Lastly, through individualization, the subordinate-first paradox appears. This suggests that the individuals accustomed to being group members are instead elevated to central importance as individuals; they are viewed at the individual level, in contrast to a group member. Within this context, individuals become the end-users, the consumers, or clients. This paradox is difficult to address in its current stage since individualization is currently in its initial phase, and shared community still the dominant model. From an energy practice´s perspective, it is easier to argue but does not include all aspects of life. However, at the regional and national level speeches and discussions, there is the debate about a "first-class citizen," which might have different implications. However, it supports extending public services to all Namibian residents, recognizing them individually and not only as communities.

## 7.2 Visions with Limitations and Opportunities

"Electricity is like a Source of Life," "Electricity means Light," "Electricity is Life"

Electricity is embedded in life-worlds. Findings of the local level suggest that "electricity is like a source of life," "electricity means light," and "electricity is life." It is an "enabler" for socio-economic development at regional and national levels and a necessity of life. Electricity is also the driver of Namibia's ambition to become industrialized, prosperous, and self-sustaining under the guidelines of Vision 2030. The "light" enlightens the "life" at night and all the senses of the life-worlds to live, to experience everyday life. As a result, it is an indispensable, irrevocable component of life; it is essential. Hence, this is how it has been envisioned by all actors within the MLG.

Through these meanings, a re-enchantment with worldviews takes place, which brings a contrasting perspective of the disenchantment that would be brought with the controllability and predictability of life according to Weber's modernization worldview. Electricity is a "magical thing" at a local level; at the national level, it brings "magic" in solving development issues such as poverty, hunger, health, education, water, employment, and many more related to electricity, as highlighted in the national visions and development plans, along with the individual SDGs. This contributes to the overarching vision of Tsumkwe and Gam's settlements becoming towns with shops, malls, banks, (vocational) training centers, roads, posts, and entertainment centers. In Tsumkwe, certain voices explicitly mentioned the wish for a "San" town, and Gam voices considered these infrastructural developments part of a town-like life and services. Only a few students from the 7th grade in Tsumkwe Primary School envisioned a Tsumkwe with a metro line (mentioned in one of the chats), further enhancing the collective vision of a town.

The wishes and expectations surrounding employment as a means of generating income and living a better life are also present in Tsumkwe and Gam. In some cases, it is expected that employment will be provided by other people, companies, projects, or the government; in other cases, individuals desire to start their own business as a means to self-employed and independence. The reliability of electricity, uninterrupted throughout the day and week, is considered a "must" for businesses (big shops or banks) to open and continue growth. Reliable electricity also provides a form of predictability and controllability for businesses as well as security from theft. The unavailability of reliable electricity was found to be one of the key limitations to establishing new businesses, or allowing existing businesses to grow, at the local level. At the national level, Vision 2030, adhering to NDPs and energy policies, emphasizes repeatedly the prominence of energy in creating jobs, generating income, productive use of energy, and contributing to socio-economic development.

Local communities also consider having a training center for building and further developing their capacities, particularly for the youth, who comprise nearly one-third of the population in both settlements, significant at a local scale. They stated their willingness to engage in community activities related to the mini-grid and electricity expansion. Furthermore, this willingness was confirmed within the CENORED´s demand side management activities in Tsumkwe and Gam, which involved community members in raising awareness about not using high-consumption appliances such as hot plates, electric heaters, or irons. Community involvement and engagement are considered prominent themes/topics in the sustainability of the mini-grids.

Although community involvement and engagement, along with public or community awareness, are known, the findings suggest that the details of these core components need to be clearly defined to address the broader MLG framework. Furthermore, the terminology used in multiple disciplines (social, technical, economic) to address the issues of capacity building, capacity development, community involvement, community engagement, and public awareness have multiple meanings when referring to off-grid generated electricity. Therefore, the introduction of electricity in a previously unelectrified locality also requires not only community involvement at a particular stage of the project development or execution, but also the engagement of the community in the operation and maintenance of such systems; this also requires continuous learning and awareness-raising activities that are linked directly to the community. This suggests that the beneficiaries of these capacity building, capacity development, and public awareness programs should be communities directly, and that they should be elaborated and executed with participatory learning and teaching methods to ensure sustainability and continuity. As a result, the social aspect of socio-techno-economic off-grid electrification requires customized approaches catered to individual locations, which need to be duly prepared and considered.

## 7.3 From Mini-Grids to National (Off-Grid) Energy Infrastructure

The transcendent meaning of light is intertwined with the everyday lives and energy practices of local experience; however, at the regional and national levels, these local meanings are often overlooked in favor of national-level techno-economically focused plans and programs. These meanings have been encapsulated within the numbers, technologies, and economics, which are in line with rationalization and domestication first but also contribute to differentiation and individualization. This situation also calls attention to the overlooked contribution of the social sciences in off-grid energy, electrification, and policy. However, energy practices provide an interlinking approach, connecting the meanings to materials through competences. Thus, socio-techno-economic findings contribute to the scalability of off-grid electrification in widening access to energy. However, it is important to remember the past and reflect on the present before continuing with energy practices. Thus, the case study findings have been triangulated and discussed within the context of the off-grid MLG, including their differences and similarities.

### 7.3.1 Differences between Case Studies

*Table 7.1. Differences between Tsumkwe and Gam*

- Historical background and population
- Local actors
- Electrification rate
- Electric appliances
- Electricity expenses
- Electrified wishes

*Source: Author*

The key differences between energy practices and governance in the historical background, population growth, local actors, electrification rate, electric appliances, and electricity expenses are shown in Table 7.1.

#### *7.3.1.1 Historical Background and Population*

Following Namibian independence, the Tsumkwe settlement inherited generators and a limited electricity infrastructure from the SADF. Therefore, though small, the settlement was already partially electrified (primarily public institutions and facilities). In 2011-12, with the installation of a mini-grid, around 100 households initially had mini-grid connections. The Gam settlement was established in 1993 by repatriated Herero people returning from Botswana. They, in contrast, had no previous electricity infrastructure. The mini-grid was installed in 2014. In the meantime, the population of Tsumkwe had grown from 1,404 inhabitants in 2011 to an estimated 3,215 inhabitants in 2021, while Gam had grown from 1,229 to an estimated 3,195 inhabitants respectively (Table 7.2). As a result, both settlements had grown in population only at different rates, and under different circumstances, and were electrified accordingly, but not entirely.

*Table 7.2. Comparison in the Population and Electricity Connection*

| | Population 2011 | Population 2021 (estimated) | Total Power Connection | Connected Households (2022) |
|---|---|---|---|---|
| Tsumkwe | 1,404 | 3,215 | 393 | 330 |
| Gam | 1,229 | 3,195 | 611 | 556 |

*Source: Author*

#### *7.3.1.2 Electrification Rate*

Similar growth occurred in mini-grid connections for both settlements. In Tsumkwe, the total number of connections reached 393, of which 330 were households in 2022. Meanwhile, Gam settlement initially had 200 connections, which increased to 611, of which 556 households, by 2022. According to these findings, in 2022, the estimated electrification rate for households in Gam was 73% and 50% in Tsumkwe, placing both above the Namibian rural electrification average (33.2%) and Gam well above the overall average for Namibia (Table 7.3).

*Table 7.3. Comparison of Electrification Rate*

| | Tsumkwe | Gam | Rural Namibia | Namibia Average |
|---|---|---|---|---|
| Electrification Rate | Approx. 50% | Approx. 73% | 33.2% | Approx. 50% |

*Source: Author*

These estimates represent a trend in the steadily progressing electrification of houses and businesses not previously experienced by the locals. This data, therefore, proved valuable and useful output for this research. In 2022, due to the mini-grid's operation and maintenance issues, it was already not sufficient to meet the consumption needs of the electrified households and businesses; this highlights the question of what the future will bring for the unelectrified households (those awaiting applications), and whether the potential additional demand in the electrified places will remain within the capacity of the existing mini-grid. Therefore, population growth and the dynamics of its energy practices linked with the PoM reveal a critical point in planning micro or macro systems to meet present and future demands. As a result of this process, the following energy practices and findings address the practices developed and practiced over the ten year period in Tsumkwe and seven year period in Gam, based on the data collected in 2021.

#### *7.3.1.3 Local Actors*

Among the many local actors from the settlements' public, private, and community sectors, Tsumkwe Settlement had more public institutions, ministerial representations, local and regional authorities, regional/national private companies, NGOs, and parastatals. In contrast, Gam Settlement had more Traditional Authorities and churches and no CENORED office. The presence of public institutions and regional and ministerial representations suggests that the government, LA, and RA are present in the settlements (Table 7.4). However, questions of these public actors' productivity and actions will need to remain for research in a future study. This type of research would enable understanding local governance in general in various fields and reaching the visions of the communities and the government without leaving the communities feeling left behind.

*Table 7.4. Comparison of Local Actors*

| Number | Public Institutions | Traditional Authority | Ministerial Representation | Churches | Parastatals |
|---|---|---|---|---|---|
| Tsumkwe | 6 | 2 | 5 | 8 | 3 |
| Gam | 5 | -4 | 2 | 11 | - |

*Source: Author*

#### *7.3.1.4 Electric Appliances*

The field data demonstrate that a maximum of twelve different appliances (with a mean of 6.33 used in everyday energy practices) were used in the households of Tsumkwe, with a distribution of 21.6% (1-3 appliances), 25.5% (4-6 appliances), 33.3% (6-9 appliances), and 19.6% (10-12 appliances). Gam's figures show nine appliances used, with a mean of 5.32 used in everyday energy practices, and a distribution of 28%, 36%, and 36%, respectively. These findings show both similarities and differences in terms of the different appliances in use. Tsumkwe, has a greater diversity of items, a 19.6% difference, than Gam (Table 7.5). The majority of different appliances are centered between 4-9 appliances in each settlement. Additionally, the 1-3 appliances that constituted the backbone of the Multi-Tier Framework (Box 3) were limited to 21.6% and 28%; this does not meet the visions of the policies with a minimum Tier 3 scenario, which presents an unsustainable goal for electrification. These figures suggest that residents in both settlements lead lives beyond the Tier Framework or at the edges of Tier 4 and Tier 5, and that, remarkably, it has taken them only ten to twelve years to reach this consumption level. As a result, this situation highlights the community's understanding of "electricity is life," and electricity can be adapted into multiple practices of everyday life when it is provided reliably and uninterruptedly.

*Table 7.5. Comparison of Electric Appliances*

| Number/ Household | Different Type Appliances | Mean Appliances | (1-3) Appliances | (4-6) Appliances | (7-9) Appliances |
|---|---|---|---|---|---|
| Tsumkwe | 12 | 6.33 | 21.6% | 25.5% | 33.3% |
| Gam | 9 | 5.32 | 28% | 36% | 36%- |

*Source: Author*

The mean household room size was 2.24 rooms and 6.08 people in Tsumkwe and 1.82 rooms and 5.52 people in Gam. Concerning the appliance groups, a wide distribution was noted in single, double, and triple-room households, with 62.8% of 1-9 appliances corresponding to single and double-room households in Tsumkwe. In contrast, in Gam, the sum of single- and double-room households corresponded to 82.1% with the ownership of 1-9 appliances. This difference demonstrates that the energy practices in Tsumkwe are more diversified than those in Gam. Furthermore, in Tsumkwe, the higher the education, the more the tendency to have more appliances in the household and vice versa. In contrast, in Gam, the number of appliances per household is more distributed across education levels.

#### *7.3.1.5 Electricity Expenses*

In contrast to the appliance use findings, the mean monthly electricity expenses in Gam were 253.06 NAD (with 50% spending between 1 and 300 NAD), and in Tsumkwe 178.55 NAD (with 80% spending between 1 and 300 NAD). These findings suggest that more high-consumption appliances were used in Gam than in Tsumkwe. Furthermore, considering the regional monthly electricity consumption calculated in Chapter 4 at 346.67 NAD/month/customer and the Namibian average of 1,083.33 NAD/month/customer, excluding the fixed charges, the households in Tsumkwe and Gam consume less than the regional and national averages, which are grid-connected (Table 7.6). This places the settlement's electricity consumption above the average mini-grid customers worldwide while they remain below the regional and national averages.

*Table 7.6. Comparison of Electricity Expenses*

| Expenses/ Household | Monthly Mean | Regional Average | National Average |
|---|---|---|---|
| Tsumkwe | 178.55 | 346.67 | 1,083.33 |
| Gam | 253.06 | | |

*Source: Author*

#### *7.3.1.6 Electrified Wishes*

Residents of Tsumkwe and Gam most desired a television (24% and 16%, respectively), followed by a fridge (14% and 9%), with the average of the three priorities that were asked also ranking the first two appliances (Table 7.7). However, the electrified wishes differed in Gam, with the deep freezer (16%) and electric stove (8%). In contrast, music players and fans were preferred in Tsumkwe at a rate of 9%. In summary, the (food) cooling practices in Gam, with a total of 30%, support their animal husbandry-based livelihoods as well as their dietary traditions. The shift occurs when comparing Tsumkwe and Gam in appliances related to cooking. As a result, the top four preferred appliances in Gam are more energy intensive that account for 62% of the respondents, compared to Tsumkwe´s preferred appliances (43%).

*Table 7.7. Comparison of Electrified Wishes*

| | Tsumkwe | | Gam | |
|---|---|---|---|---|
| | Priority no.1 | Average of Priorities | Priority no.1 | Average of Priorities |
| Television | **16%** | 10.7% | **24%** | 17% |
| Fridge | **9%** | 10.3% | **14%** | 11% |
| Deep-freezer | - | 2% | **16%** | 8.7% |
| Electric stove | 4% | 4.3% | **8%** | 8.7% |
| Music player | **9%** | 8% | - | 1% |
| Fan | **9%** | 7% | - | 2.7% |
| Nothing | 20% | 35.3% | 5% | 13.3 |

*Source: Author*

In contrast, the desired list of appliances in Tsumkwe was extensive and more diversified. Music players and fans each ranked at 9%. The results also show that communication and media practices had a higher priority in Tsumkwe, at a rate of 26%, followed by food and air cooling practices (18%), totaling 44%. Twenty percent of the respondents did not wish to have any further appliances, which differed from Gam, where five percent of the respondents did not wish to have any further appliances. These findings suggest that the residents' wish to have more appliances, and a considerable proportion of these are energy-intensive cooking and cooling items, requiring a 24-hour energy supply. Furthermore, the residents' dietary habits and livelihoods impacted the type of appliance they acquired.

### 7.3.2 Similarities between the Settlements

The similarities between the settlements concerning electricity have been categorized under energy practices, electricity supply satisfaction, lack of mini-grid capacity for universal access, and energy use beyond the Tier 3 framework (Table 7.8).

Table 7.8. *Similarities between Tsumkwe and Gam*

- Electricity supply satisfaction
- Lack of mini-grid capacity for universal access
- Energy use beyond Tier 3 consumption
- Energy practices and time

*Source: Author*

#### *7.3.2.1 Electricity Supply (Dis)Satisfaction*

The survey findings illustrate that the community in Tsumkwe was 72.55% satisfied with the mini-grid electricity supply, and their satisfaction was related to both being able to use their appliances and having 24/7 uninterrupted electricity. The primary source of dissatisfaction was electricity prices and interruptions, especially on rainy days. In Gam, 82.1% of the community was satisfied with the supply and showed the basis for satisfaction and dissatisfaction as residents of Tsumkwe. Therefore, the continuous availability of electricity plays a crucial role in everyday life, especially after it becomes available, which was emphasized in both settlements. Thus, it demonstrates that the communities live above the Tier Framework and average electricity consumption for off-grid areas.

#### *7.3.2.2 Lack of Mini-Grid Capacity for Universal Access*

From a local perspective, Tsumkwe and Gam do not yet have 100% electrification rate, and the mini-grid capacity is not currently meeting the electrified households' and businesses' demands. Therefore, generators that consume non-renewable and polluting diesel fuel are used during the evenings. Additionally, differentiation has also been created between electrified and unelectrified households and businesses at the local level. The findings also suggest that in Tsumkwe, some households were not yet connected to the grid due to the costly connection fee or the additional fee due to the distance of the house to the connection point. However, this research shows that once households have electricity, consumption is inevitable and irreversible. Although the affordability of the initial connection is the primary limiting factor in electricity access, the lack of actual planning for electricity distribution appears to be another prominent factor. Thus, two different levels of disparity have been created through partial electrification, even though the intention is present for reaching universal access. The first disparity is related to the affordability of the essential connection fee at the household level; the second is related to the infrastructural disparity of the mini-grid electricity distribution lines that have not been planned yet to cover or reach all households in the settlement. To achieve ubiquitous electrification, it is evident that a comprehensive settlement *and* distribution plan should be considered. In the meantime, the discrepancy between community members who live in the electrified households and interpret electricity as life, and those who live in the unelectrified households and do not interpret as such, still remain.

### *7.3.2.3 Energy Use Beyond Tier 3 Consumption*

At the national level, off-grid electrification strategies and plans adopt a Tier Framework: A minimum of Tier 3 services with at least 1 kWh/ day, at least 8 hours of daily electricity, and 3 hours of evening electricity are provided, with a minimum of 14 disruptions per week or a minimum of three disruptions with a total duration of more than 2 hours (Bhatia & Angelou, 2015). The residents of Tsumkwe and Gam have been using electricity for ten to twelve years at consumption rates ranging from 4 kWh to 330 kWh/month, continuously, compelling AMDA´s and Tier 3´s consumptions (Table 7.9). Thus, the findings of this study demonstrate that Tier 3 presents neither a reliable nor a realistic starting point, nor a strategy that can be maintained to either envision or plan universal electrification in Namibia; in this case, Tier 3 could only be a temporary solution for the localities or settlements like Tsumkwe and Gam, and, furthermore, over the short or long-term this would also not satisfy the current or future consumption needs of the residents. It is also unlikely that the mini-grids will remain sustainable throughout the system's techno-economic lifetime. Therefore, the social, socio-economic and socio-technical implications should be considered during the planning and operation of each location that is to be developed.

*Table 7.9. Comparison of Electricity Consumption*

| Consumption (kWh)/ Household | Monthly Min. | Monthly Max. | AMDA Mini-Grid Average | Tier 3 (Monthly Minimum) | Tier 5 (Monthly Minimum) |
|---|---|---|---|---|---|
| Tsumkwe | 4 | 330 | 6.2-10 | 30 | 246 |
| Gam | 10.5 | 300 | | | |

*Source: Author*

### *7.3.2.4 Energy Practices and Time*

The core household energy practices explored during this research include cooking, lighting, cooling, heating, communication, media, leisure activities, sewing, ironing, and washing. Compared with pre-electrification, electric appliances and access to electricity have had a life-changing impact on the residents of both settlements. These changes in energy practices have brought independence, autonomy, convenience, (self-)esteem, self-development, solidarity, and respect. They have enabled residents to envision their settlement becoming a town with all the desired facilities available in one place. The energy practices in both settlements were enabled through the electricity supply from the mini-grids and the electric appliances community members acquired. Appliances use was eagerly adopted into the daily lives of residents who appreciated the reliable power supply that served them without damaging their appliances. In contrast, the lack of connection to the mini-grid and the unreliability of the current electricity supply are also recognized as drawbacks by both residents and business owners. Residents are aware that not all households have electricity access, and business owners cite unreliable electricity supply as a disadvantage to business activities and expansion.

The temporal parameter is also important to consider when planning the longevity of a mini-grid. The cases of Tsumkwe and Gam illustrate that mini-grid capacities can already be exceeded within the approximate economic and operational half-lifetime (approximately 10 years) of a mini-grid. Thus, the electricity demand by the end-users is place-based and place-sensitive, especially in settlements with 2,000-3,000 inhabitants. Considering the process of modernization alongside the techno-economic planning reveals that the adoption of electricity within the communities prompts the development of energy-related practices amongst individuals, which are, in turn, accompanied by the wishes and visions of the future following in a relatively progressive and relatively irrevocable direction. Furthermore, electrification is accompanied by digitalization that connects the communities, both local (with Namibia) and global connections. As a result, energy´s overarching meanings, discussed in section 7.2, enable the mini-grid's electricity to be more than simply the light, water, and income sources but also include the processes of individualization, domestication, rationalization, differentiation, and their paradoxes that span through time and across space. Energy becomes a technical component of everyday life.

## 7.4 "Electricity is Light" versus "Electricity is Life"

The findings of the life-worlds and energy practices conclude that the overarching meanings of "electricity is light" and "electricity is life" stretch beyond electricity's physical and material aspects. These meanings highlight how individuals live this experience from the light as a cognitive act of "*Video ergo sum*[19]" and/or "*Scio, ergo sum*[20]" of the everyday life energy practices that introduce enhanced everyday experiences filled with practices that improve life and make it more convenient. Moreover, Box 1 provides a set of literature for its multilayered understanding of "light," and Box 2 presents some background about "life" to seek and understand how the findings of the case studies relate to these broader meanings. It was found that both themes have been widely discussed across disciplines, such as astrophysics, physics, biology, chemistry, philosophy, humanities, and the social sciences.

The lived experiences of community members demonstrate the modernization process through household energy practices made possible by partial electrification within the settlements. However, the results suggest limitations for the residents of the settlements in undertaking the energy practices of everyday life. These limitations are primarily due to (but not limited to) the economic and techno-economic affordability of connections, the affordability of the tariffs,

---

[19] I see, therefore I am.

[20] I know, therefore I am.

and the affordability of electric appliances. However, partial electrification extends electricity to only a portion of the community while the others remain without, resulting in different social and socioeconomic development paths.

Furthermore, households and businesses connected to mini-grids have had the opportunity, albeit limited, to connect to the internet for digital practices. As a result, throughout their 10-12-year electricity connection, some individuals or households have experienced energy and digital practices simultaneously. Thus, they do not live *off-the-grid*; instead, they are connected, particularly through social media, to the rest of Namibia, the continent, and the globe.

Internet use, an initial and rapidly spreading digital practice across most geographies, has been more gradual, in contrast to the nearly simultaneous experience of electricity and the internet in settlements. Global technological developments have begun incorporating AI into our everyday lives, and new digital devices provide these AI-integrated services. Based on this study's findings, incorporating AI into residents' daily lives is not far off, considering fast developments in science and applied sciences fields. These findings suggest a quicker transition to this energy practice and foresee positive and negative consequences. Therefore, as van der Loo and van Reijen (1922) explain, the PoM also has paradoxes: Two sides of the same coin create a whole, and modernization takes place with its associated paradoxes. However, the author also suggests that, in this frequently used analogy, the “whole” should include the often-neglected thickness, which can be interpreted as the linking, binding, or bridging section of two different sides of a three-dimensional coin. The character of energy practices can also be seen as part of the same coin; energy practices have material and immaterial characteristics where the subjects execute the action. Within this concept, the practice also presents the full spectrum of energy practices in the 3D-coin.

Further to this analogy, at the national level, the material aspect of the coin in energy practices creates the basis for identifying the consumption or demand of electricity that defines the need for necessary energy infrastructure for the country. Thus, this material aspect is linked to social and socio-economic development at a national scale, forming the other side of the coin. Additionally, through modernization and its paradoxes, MLG is shaped by the interconnections and intertwindedness of off-grid RE´s social, economic, and technical aspects. However, the current NES in Namibia foresees a minimum of Tier 3 electricity access, which, when considering the PoM and its paradoxes, creates risks in the long-term planning, implementation, operation, and sustainability of the mini-grids.

In conclusion, this study revealed unexpected areas for further research exploring the relationship between light and life through electrical practices.

## 7.5 Human Geography in the Interdisciplinary Research of Off-Grid RE Governance

Silvast and Fouds (2022, p. 92) suggest that "interdisciplinary research should be studied as a social activity and the scientific ideas that it generates explained by sociological dynamics." In line with their argument, a macro-sociological approach through the findings (but not limited to) of the PoM to off-grid energy governance brings together, intersects, and intertwines multiple theories and methodologies for innovative, pragmatic implementation at the broader scale and at different levels of governance as well as in the research and implementation of interdisciplinary projects.

Firstly, this work contributes to research in energy geography through its links and intersections with macro sociology, human geography, and social geography. Secondly, using a phenomenological epistemology to explore phenomena of electricity from a bottom-up, multilevel framework enabled an alternative perspective to approach the place-based issues of everyday life. A theoretical contribution was made by applying and using phenomenological approaches to human geography topics: Schütz's lifeworlds methodology was used to identify and explore the energy practices and their spatial and bodily experiences (Rothfuß & Dörfler, 2023). Third, this research contributed to phenomenological practice, often absent in energy governance or human geography research. Accordingly, a phenomenological approach to the policy design and regulatory framework, including the necessary guidelines for planning and implementation and the utilization of energy practices in everyday life, revealed how energy governance may or may not lead to the desired visions of a community, region, or country. Additionally, from a macro perspective, the processes occur at all levels of off-grid energy governance in different intensities, priorities, or combinations.

This research also explained life-worlds and the processes of off-grid RE governance through the PoM defined by van der Loo and van Reijen (1992). It also contributed to modernization theory with new findings in the research field, such as the digitization of the archives and cultural heritage and digitalization of the payment processes in rural communities, being the processes which have accelerated exponentially over the past 40 years and can also be found in the rural areas. It also demonstrated the ongoing processes of the new era, including the emerging fields and use of AI, and explored MLG using a life-worlds approach.

In summary, reflecting on Elias´ extensive research about the "processes" in general, his contribution to disciplinary and interdisciplinary research and the PoM (van der Loo & van Reijen, 1992), interdisciplinary (socio-techno-economic) energy research could benefit from these social perspectives and, thus, bridge the gap of an overlooked area of the social sciences in the techno-economic policy research and implementation. In a 1970 interview, Elias (1970s, as cited in Norbert Elias Foundation, 2019, 00:00) emphasizes the prominence of sociology in producing more knowledge for practical implementation. Elias' point can be interpreted for social scientists or the social sciences, not just for sociologists, in that social sciences have the potential to make significant contributions to the interdisciplinarity of socio-techno-economic research.

> "We are not yet clear that our human world is to a large extent an undiscovered world. We do not really understand ourselves. And that is the central task as I understand it of sociology to gain more certain knowledge of the human level of the universe which we form with each other. Let us try, my feeling is that we will be able to produce more knowledge which held a practical application than we do today (...) Indeed, what I personally think that sociologists could produce more knowledge which is practically useful for vital problems of our living together" (Elias, 1970s, as cited in Norbert Elias Foundation, 2019, 00:00).

Finally, to understand the issues within the (intersections of) multilevel off-grid RE governance at various levels, it is important to look first at the "human level of the universe which we form with each other" (Elias, 1970s, as cited in Norbert Elias Foundation, 2019, 00:00). This view will produce the necessary knowledge crucial for practical application to address critical problems. Thus, the interconnectedness of sociology, human geography, social geography, and energy geographies, benefit new understandings and potential solutions that could contribute to the socio-techno-economic discussions of RE, especially in off-grid implementations. In addition, the issues concerning energy and society that have cross-cutting themes in interdisciplinary, multidisciplinary, and transdisciplinary research and implementation, might gain a different momentum by linking micro and macro approaches on a broader scale, especially for country-specific implementations; these would elevate and help improve the visibility of social science research compared to the economic and technical disciplines within the (interdisciplinary) energy and policy studies (Royston & Foulds, 2021).

## 7.6 Limitations

The intertwined nature of the concepts and theories explored and investigated in this research made explaining them linearly over consecutive chapters challenging. Therefore, to support the linear explanation of intertwined concepts and theories, several visual interpretations and illustrations were utilized to demonstrate the social phenomena and the modernization experienced by the MLG while shifting from the earlier versions of energy governance. Furthermore, as indicated earlier, the COVID-19 pandemic hindered the original research plan from the outset. On-site field research was limited for two whole years; the survey process was interrupted, two factors that resulted in a smaller sample size being used than initially intended. However, these challenges allowed more reflection and helped focus the research on collected data. Digital research methods, like digital calls, were also introduced due to the restrictions imposed. Unfortunately, a digital survey method with digital mapping would also have brought valuable insights into this research and provided practical implementations at the settlements, which was not possible in the initial years due to internet connectivity limitations to pursue such a study in the settlements.

# 8 Synthesis

This chapter concludes with two sections that synthesize the findings. The first summarizes the overall research conclusions, while the second presents recommendations for the policy and regulatory framework based on multilevel off-grid governance research, which aims to enhance ongoing policy and implementation efforts for off-grid electrification in Namibia.

## 8.1 General Synthesis

"Electricity is life," and "energy is a social issue with a technical component, not the other way around," a component interwoven in all activities of life (Pasqualetti, 2021, p. 3). This research investigated off-grid electrification in Namibia using a social sciences perspective by addressing the MLG of renewable energy from local to national and supranational levels using a bottom-up approach. First, this research explored off-grid RE governance through the visions of the local communities of Gam and Tsumkwe. Second, the investigation continued at the subregional, national, and supranational levels, where RE and off-grid policies were explored from the perspectives of local, regional, and national actors. Further, an inductive, exploratory mixed-methods approach, guided by three broad research questions, was used to learn about the research topic. A life-world approach was implemented using hermeneutical interpretation within the methodology. The findings suggest that a process of modernization has been occurring at all levels.

The PoM (van der Loo & van Reijen, 1992) is defined by four main processes: differentiation, rationalization, domestication, and individualization; these processes are characterized as having neither beginning nor end and proceeding in an irreversible direction through time. The findings of this study were explained within the context of these four processes. Lighting, cooking, cooling, heating, sewing, laundry, welding, communication, media, and leisure practices constitute the most prominent energy practices present in the everyday lives of the settlement residents. Among community members, the primary points of satisfaction surrounding electricity access at the household were the use of power and appliances and their 24/7 reliability and availability, and the primary points of dissatisfaction were the interruptions due to technical reasons, weather conditions, and tariffs. Although the communities have different historical backgrounds, the settlements expressed a collective future vision of becoming a "town" with "town-like" services, for example, shops, malls, streetlights, banks, hospitals, tar roads, or vocational training centers. Thus, they highlighted *sine qua non* characteristic of "*electricity is light*," "*electricity is life*," and "*electricity is like a source of life*."

At the national and regional levels, the differentiation of actors in terms of electricity supply, transmission, and distribution services addresses goal-oriented solutions to electricity supply. The government, beginning with the independence of Namibia, has continuously adopted the required acts, policies, strategies, plans, and regulatory frameworks. Furthermore, electricity has been accepted in all plans and policies as a crucial component of socio-economic development within the context of Namibia's Vision 2030 goal of becoming an industrialized, prosperous country that also meets the visions of the communities. However, as of 2022, Namibia´s electricity access is/was estimated to be approximately 50%, with 33.2% access in rural areas. Even with this access rate, Namibia depended on the energy supply from the SAPP, prompting the government to include IPPs within the energy market actors (2018) in adding RE-based power plants feeding into the grid and RE into the energy mix. However, although numerous policies and plans have been adopted since 2000, rural electricity access has not met the plans.

Between 2021 and 2023, national energy actors prepared new NES, GLCEP, and NELFP policies to tackle existing shortcomings and electrify rural areas, depending on their locations, with on-grid or off-grid energy solutions. This new strategy, with its policy and regulatory components, was still in process during this study; it includes new institutions such as NERA and the Electrification Fund in the RE governance for (off-grid) electrification. The expansion of actors for specific tasks also demonstrates the process of differentiation and rationalization in the process with the use of digital tools such as geospatial planning and new software tools in planning, managing, and monitoring the future energy systems, suggesting a domestication process. Furthermore, the analysis also revealed a link between the Namibian state and the energy market dynamics in the SADC region with SAPP, RERA, and SACREEE and with global actors, for example, the EU, the WB, and the UN. Thus, there has been a shift in the governance of (renewable) energy through modernization within a multi-level meta-governance, which allows the government to be the leading authority while still including other actors and networks in the process.

Although several off-grid electrification plans were adopted and made available, the electricity access rate illustrates that these policies and plans have yet to be realized. Additionally, the techno-economic requirements for reaching these off-grid areas in Namibia's vast and sparsely populated geography remain in the deliberation process. The human resource component that could be utilized to address the challenges facing the electrification process has been largely overlooked, incorporating another set of complexities. Therefore, similarly to the neglect of social sciences within socio-techno-economic RE implementations, including policy development, a similar situation is occurring in Namibia, which has resulted in the prioritization of technical and economic aspects. The lack of consideration given to the human and social aspects in planning and implementation has hindered the progress of electricity expansion in rural off-grid areas. Furthermore, this situation also foregrounded the prominence of community involvement and engagement, capacity building, capacity development, and public awareness, which thus far have only been explored and interpreted through a technical or economic lens. In contrast, a social science perspective, which explores individual experience with energy through everyday life, offers more inclusive, comprehensive, customized, or customizable programs tailored explicitly to the needs of the communities in off-grid RE and EE-related areas.

In conclusion, this research addresses a gap in the interdisciplinary sciences, bringing a phenomenological approach to MLG of RE and, more precisely, to off-grid areas, including the perspectives and visions of the communities. Two case study areas within an MLG framework have intensified the pages and lengthened the study by elaborating on policies and the perspectives of regional and national actors. Due to the two-year pandemic restrictions, the sample size of surveys remained below the intended average. However, the author is confident that the sample was sufficient to reveal the most crucial attributes of the datasets to demonstrate the temporal dimension of the connection growth with the future visions of the communities in guiding the future visions for grid connection or mini-grid upgrade scenarios. As a last note, future research in the interdisciplinary RE with its social, technical, and economic aspects could be extended and applied in areas to enhance the sustainability and success of the research and the projects with the inclusive integration of human geography and methodologies.

## 8.2 Policy and Regulatory Framework Recommendations for Namibia

To extend the recommendations as a part of the conclusion for Namibia, this section presents the policy and regulatory framework recommendations for policymakers and practitioners in Namibia who (could potentially) work with the policymakers to implement more human-centered approaches based on the findings of this research.

**Recommendation 1:** Participatory Action (Research) as a Methodology

Participatory Action (Research)- PAR is the recommended methodological approach for addressing the multi-layered energy issues. Within this context, PAR involves multi-level actors from diverse backgrounds, particularly from within communities, to participate in the research and decision-making processes, from the initial policy and regulatory framework development to all stages of the off-grid system, solar heating, or energy efficiency implementations. This would allow international actors to get involved at different stages. This methodology has already been implemented in Tsumkwe, where it was used to incorporate community empowerment, enabling research and action simultaneously by involving and engaging the community, thus contributing to capacity building and development with tacit knowledge transfer (Juujärvi & Lund, 2015). Furthermore, using digital methods and tools to reach off-grid areas for this purpose would tackle the issue of human resources and geographical considerations. As a result, this would introduce a bottom-up approach to empower communities by integrating customized programs.

**Recommendation 2: Inter-institutional cooperation and collaboration in Namibia**

- Action 2.1: Integration of Social Sciences in RE Research and Implementation
- Action 2.2: Participation of Local Actors (Traditional Authorities, Local Councils, Conservancies, etc.)

Inter-institutional cooperation and collaboration are also recommended to bridge the gap between the social and techno-economic aspects of RE implementation. In this regard, NSA, UNAM, and other relevant institutions could be involved in the future phases of electrification and planning in support of national objectives. Furthermore, it is also suggested that the NEI, the leading research institution in RE, should integrate social scientists, social geographers, and human geographers and link local participation when developing new projects and expanding to new areas. Through integrations and/or community involvement, adequate and suitable system designs and off-grid system sustainability will be ensured.

This research reveals the modernization process occurring at the national, regional, and local levels; considering this finding, it is recommended that the community members be involved and engaged at the local level and be supported throughout the planning, dissemination, capacity building/development, and operational steps of mini-grid implementation. In addition, Traditional Authorities, Local Councils, and Conservancies should also be integrated into each stage of local action undertaken; these should be defined and identified by the collectivity and work towards capacity building, development, and public awareness. Finally, local-level actors and implementations should also be integrated at all levels of the regulatory framework promoting further electrification expansion.

**Recommendation 3: Capacity Building and Development for RE, Revisited with Social Sciences and Digital Tools**

- Action 3.1: Elaboration of comprehensive capacity building, capacity development, and public awareness programs for Communities (NEI, NUST, UNAM)
- Action 3.2: Youth-focused employment strategies to be developed for communities (Research institutions and productive-use-centered actors)
- Action 3.3: Establishment of *Research and Training Center* in Tsumkwe and Gam (NEI, NUST, UNAM, and all stakeholders that conducted research in the settlements)

Through inclusive and interdisciplinary consortiums, it is recommended that a comprehensive and holistic framework for capacity building, capacity development, and public awareness activities be designed and developed specifically for place-based implementations. Capacity building and development ensure the continuous empowering of communities to utilize renewable solar energy efficiently. At a local level, this would promote a continuous cycle of general programs addressing the entire community and specialized programs for children and youth. Finally, relevant productive-use skills development programs and building technical capacity for local technical assistance are also recommended.

**Recommendation 4: Inclusion of Off-Grid Electrification Infrastructure Developments in the ESI Reports**

- Action 4.1: REDs to include off-grid developments in the annual planning and reports,
- Action 4.2: NamPower to include off-grid developments in the annual planning and reports.

If necessary, the key institutional actors leading these programs should include a similar budget for their expansion at all levels. Since the new NES was designed to combine off-grid and on-grid electrification, it is recommended that the institutions involved, such as NamPower and REDs, should include the off-grid infrastructural requirements in their annual reports. This way, multi-level implementation would be ensured as part of the MLG.

**Recommendation 5: Transparency and Clarity in the Off-Grid Electrification**

- Action 5.1: List of the off-grid localities and the schedule for off-grid electrification to be made publicly available (online) (MME and other involved institutions)
- Action 5.2: Plan and allocate budgets for the local human resources necessary to undertake off-grid electrification in MLG, which will support the techno-economic plans.

In conclusion, it is recommended that the MME and the future Electrification Fund have a transparent electrification strategy by announcing publicly the sites targeted to be electrified in the years to come as a part of the public awareness process and be the bridge in bringing clarity to the following steps, budgets, and requirements, which could also attract the interest of more donors than expected. Lastly, to meet the envisioned goals of future electrification by 2040, the budget for the required human resources to execute the forecasted techno-economic planning should be prepared and included in the main budget.

Universal access can only be achieved if locally produced meanings and resources are integrated at all levels of the techno-economic infrastructure plan of off-grid electrification, thus transforming it into socio-techno-economic planning to achieve multilevel goals and visions.

## References


Adriaansens, H. P. M. (1980). *Talcott Parsons and the conceptual dilemma*. Routledge & Kegan Paul.

*African Development Bank (AdDB) (2023)*. https://www.afdb.org/en/about/mission-and-strategy

Africa Minigrid Developers Association. (2022). *Benchmarking Africa's MiniGrids Report, 2022*. Africa Minigrid Developers Association (AMDA). https://africamda.org/wp-content/uploads/2023/02/AMDAs-Benchmarking-Africas-Minigrids-Report_2022.pdf.

Al-Amri, M. D., El-Gomati, M., & Zubairy, M. S. (Eds.). (2016). *Optics in our time* (1st ed. 2016). Springer International Publishing : Imprint: Springer. https://doi.org/10.1007/978-3-319-31903-2

Alvesson, M., & Sköldberg, K. (2018). *Reflexive methodology: New vistas for qualitative research* (Third edition). SAGE.

Ansell, C. K., & Torfing, J. (2016). Introduction: Theories of governance. In J. Torfing & C. K. Ansell (Eds.), *Handbook on theories of governance* (pp. 1–18). Edward Elgar Publishing.

Asemota, G. N. Osarumwense. (2013). *Electricity use in Namibia. Developing algorithms to encourage more efficient consumer behaviour and motivate more environmentally friendly utility practises*. iUniverse, Incorporated. https://www.barnesandnoble.com/w/electricity-use-in-namibia-dr-godwin-norense-osarumwense-asemota/1114479912

Bache, I., & Flinders, M. (2004). *Multi-level governance*. Oxford University Press. https://doi.org/10.1093/0199259259.001.0001

Bachmann, K. (2018). *Genocidal empires*. Peter Lang D. https://doi.org/10.3726/b13834

Bhambra, G. K. (2011). Historical sociology, modernity, and postcolonial Critique. *The American Historical Review*, *116*(3), 653–662. JSTOR.

Bhatia, M., & Angelou, N. (2015). *Beyond Connections: Energy Access Redefined* (ESMAP Technical Report 008/15; ESMAP Papers). © World Bank. http://hdl.handle.net/10986/24368, License: CC BY 3.0 IGO

Bhattacharyya, S. (Ed.). (2013). *Rural electrification through decentralised off-grid systems in developing countries*. Springer London. https://doi.org/10.1007/978-1-4471-4673-5

Bhattacharyya, S. C. (2013). Introduction. In *Rural electrification through decentralised off-grid systems in developing countries* (pp. 3–12). Springer London. https://doi.org/10.1007/978-1-4471-4673-5

Bhattacharyya, S. C., & Palit, D. (2013). Summary and conclusions. In *Rural electrification through decentralised off-grid systems in developing countries* (pp. 285–293). Springer London. https://doi.org/10.1007/978-1-4471-4673-5

Biesele, M., Gordon, R., & Lee, R. (Eds.). (1986). *The past and future of !Kung ethnography: Critical reflections and symbolic perspectives: Essays in honour of Lorna Marshall*. Helmut Buske Verlag.

Biesele, M., & Hitchcock, R. K. (2011). *The Ju/'hoan San of Nyae Nyae and Namibian Independence* (1st ed.). Berghahn Books. https://doi.org/10.3167/9781845457549

Boamah, F. (2020). Desirable or debatable? Putting Africa's decentralised solar energy futures in context. *Energy Research & Social Science*, *62*, 101390. https://doi.org/10.1016/j.erss.2019.101390

Boeree, C. G. (1998). *Qualitative methods: Part one*. Shippensburg University. http://webspace.ship.edu/cgboer/qualmethone.html

Bogner, A., & Mennell, S. (Eds.). (2022). *Civilisations, civilising processes and modernity – A debate: Documents from the Conference at Bielefeld, 1984*. Springer International Publishing. https://doi.org/10.1007/978-3-030-80379-7

Borah, A. (2015, November 25). *Dispesh Chakrabarty's provincializing of Europe*. https://notevenpast.org/dipesh-chakrabartys-provincializing-of-europe/

Botelle, A., & Rohde, R. (1995). *Those who live on the land: A socio-economic baseline survey for land use planning in the communal areas of Eastern Otjozondjupa*. Ministry of Lands, Resettlement and Rehabilitation.

Bowen, G. A. (2009). Document analysis as a qualitative research method. *Qualitative Research Journal*, *9*(2), 27–40. https://doi.org/ 10.3316/QRJ0902027

Boyd, D. J., Ashton, B. T., Bisacky, L. H., & Lopez, J. C. (2012). *Lights on the horizon: A socioeconomic impact evaluation of rural electrification in Tsumkwe, Namibia*. Worcester Polytechnic Institute; Digital WPI. https://digital.wpi.edu/show/5q47rp237

Bridge, G., Barr, S., Bouzarovski, S., Bradshaw, M., Brown, E., Bulkeley, H., & Walker, G. P. (2018). *Energy and society: A critical perspective*. Routledge, Taylor and Francis Group.

Bristow, W. (2023). Enlightenment. In E. N. Zalta & U. Nodelman (Eds.), *The Stanford Encyclopedia of Philosophy (Fall 2023 Edition)* (Fall, 2023). Metaphysics Research Lab, Stanford University. https://plato.stanford.edu/entries/enlightenment

Brown, R. P., Baker, A. N., Wang, Z., & Morlath, R. C. (2011). *Developing energy and economic initiatives for rural Namibia*. Worcester Polytechnic Institute.https://digital.wpi.edu/concern/student_works/vd66w032w?locale=en

Cal, A., & Tehmarn, A. (2016). *Phenomenological epistemology approaches and Implications for HRD research and practice*. National Institute of Development Administration. https://www.ufhrd.co.uk/wp-content/uploads/2016/10/paper_128.pdf

Calvert, K. (2016). From 'energy geography' to 'energy geographies': Perspectives on a fertile academic borderland. *Progress in Human Geography*, *40*(1), 105–125. https://doi.org/10.1177/0309132514566343

Calvino, I. (1985). *Invisible cities* (W. Weaver, Trans.). Harcourt Brace Jovanovich.

Cambridge Dictionary. (2023a). *Awareness*. Cambridge Dictionary. https://dictionary.cambridge.org/dictionary/english/awareness

Cambridge Dictionary. (2023b). Enabler. In *Cambridge Dictionary*. Cambridge Press and Assessment. https://dictionary.cambridge.org/tr/s%C3%B6zl%C3%BCk/ingilizce/enabler

Cameron, R. (2015, July 2). *Mixed methods research*. Mixed Methods Research Workshop, Melbourne, Australia. https://www.deakin.edu.au/__data/assets/pdf_file/0020/681023/Dr-r-cameron_mixed-methodology.pdf

Cattivelli, V. (2021). Institutional methods for the identification of urban and rural areas. A review for Italy. In A. Bisello, D. Vettorato, D. Ludlow, & C. Baranzelli (Eds.), *Smart and sustainable planning for cities and regions: Results of SSPCR 2019—Open Access Contributions* (pp. 187–207). Springer International Publishing. https://doi.org/10.1007/978-3-030-57764-3

Central North Regional Electricity Distributor (CENORED). (2022). *2022 Annual Report. Towards a dynamic and sustainable electricity supply industry with the MSB, and innovative institutional adaptability* [Annual Report]. https://cenored.com.na/wp-content/uploads/2022/12/Annual-Report2022.pdf

Central North Regional Electricity Distributor (CENORED). (2023). *CENORED - About*. CENORED - About Us.

Chakrabarty, D. (2000). *Provincializing Europe*. Princeton University Press; JSTOR. http://www.jstor.org/stable/j.ctt7rsx9

Chakrabarty, D. (2011). The muddle of modernity. *The American Historical Review*, *116*(3), 663–675. JSTOR.

Chang, K.-S. (2017, December 4). *Reflexive modernization*. Wiley Online Library.https://onlinelibrary.wiley.com/doi/abs/10.1002/9781118430873.est0835

Charmaz, K. (2006). *Constructing grounded theory*. Sage Publications.

Chiguvare, T. M. (2019). *An evaluation of human development through renewable energy provision in an offgrid Tsumkwe Settlement Area of Otjozondjupa Region in Namibia* [Master of Arts, University of Namibia].

https://repository.unam.edu.na/items/55de98ff-c326-4d83-9288-73044782a0b8

Chiguvare, Z., & Chiguvare, T. M. (2016). The state of renewable energy technologies and applications in Sub-Saharan Africa. Where do we stand, and where do we go? In O. C. Ruppel & B. Althusmann (Eds.), *Perspectives on energy security and renewable energies in Sub-Saharan Africa: Practical opportunities and regulatory challenges* (Second revised and expanded edition, pp. 9–20). Macmillan Education Namibia.

Chiguvare, Z., & Ileka, H. (2016). Challenges and opportunities for increased energy access in Sub-Saharan Africa, with special reference to Namibia. In O. C. Ruppel & B. Althusmann (Eds.), *Perspectives on energy security and renewable energies in Sub-Saharan Africa: Practical opportunities and regulatory challenges* (pp. 21–40). Macmillan Education Namibia. http://old.agora-parl.org/sites/default/files/kas_44837-1522-1-30.pdf

Commonwealth Local Government Forum (CLGF). (2023). *Namibia key facts*. Commonwealth Local Government Forum (CLGF). https://www.clgf.org.uk/regions/clgf-southern-africa/namibia/

Consulting Service Africa. (2005). *Baseline Study: Barrier Removal to Namibian Renewable Energy Programme (NAMREP) Final Report* (p. 131). Ministry of Mines and Energy. Directorate of Energy.

Costelloe, T. M. (1996). Between the subject and sociology: Alfred Schutz's phenomenology of the life-world. *Human Studies*, *19*(3), 247–266.

Council of Traditional Leaders Act 13 of 1997, Pub. L. No. Council of Traditional Leaders Act 13 of 1997 (GG 1706) brought into force on 31 March 1998 by GN 64/1998 (GG 1828) as amended by Council of Traditional Leaders Amendment Act 31 of 2000 (GG 2462) came into force on date of publication: 28 December 2000, 7 (1997). https://www.lac.org.na/laws/annoSTAT/Council%20of%20Traditional%20Leaders%20Act%2013%20of%201997.pdf

Creswell, J. W. (2007). *Qualitative inquiry & research design: Choosing among five approaches* (2nd ed). Sage publ.

Creswell, J. W., & Plano Clark, V. L. (2007). *The mixed methods reader*. Sage Publications.

Curtis, A. (2015). *Why studying literature is important*. Unpublished. https://doi.org/10.13140/RG.2.1.1580.4324

Dabaren, A., & Mensah, J. T. (2023, July 23). *Ten facts about digital technology adoption in Ghana*. World Bank Blogs. https://blogs.worldbank.org/ africacan/ten-facts-about-digital-technology-adoption-ghana

Declaration of an Area as a Conservancy: Nature Conservation Ordinance, 1975. (2006). https://www.lac.org.na/laws/2006/3726.pdf

Deenapanray, P. (Sanju). (2011). *Terminal Evaluation Report (Final version) MME/UNDP/GEF Barrier Removal to Namibian Renewable Energy Programme (NAMREP) Phase II.*

DeLellis, A. J. (2000). Clarifying the concept of respect: Implications for leadership. *Journal of Leadership Studies*, *7*(2), 35–49. https://doi.org/10.1177/107179190000700203

Department of Public Enterprises. (2024). *Public Enterprises Commercial*. Department of Public Enterprises. https://mpe.gov.na/public-enterprises-commercial

Dieckmann, U., Thiem, M., Dirkx, E., & Hays, J. (Eds.). (2014). Otjozondjupa Region. In *Scraping the pot: San in Namibia two decades after independence*. Land, Environment and Development Project of the Legal Assistance Centre and Desert Research Foundation of Namibia.

Dolunay, Ö. (2021). Reaching People. An overview of the electricity access with a focus on the renewable off-grid electrification in Namibia. *Asian Institute of Technology Mastering Energy Supply for Isolated Areas (MESFIA) 2021 International Conference*, 1–9.

Dörfler, T., & Rothfuß, E. (2017). Place, life-world and the Leib: A reconstructive perspective on spatial experiences for human geography. In B. B. Janz (Ed.), *Place, space and hermeneutics* (Vol. 5, pp. 413–425). Springer International Publishing. https://doi.org/10.1007/978-3-319-52214-2_29

Dörfler, T., & Rothfuß, E. (2018). Lebenswelt, Leiblichkeit und Resonanz: Eine raumphänomenologisch-rekonstruktive Perspektive auf Geographien der Alltäglichkeit. *Geographica Helvetica*, *73*(1), 95–107. https://doi.org/10.5194/gh-73-95-2018

Douglas-Jones, R., & Shaffner, J. (2017). Capacity building in ethnographic comparison. *The Cambridge Journal of Anthropology*, *35*(1), 1–16. https://doi.org/10.3167/cja.2017.350102

Driessens, O., & Nærland, T. U. (2022). Mediated recognition: Identity, respect, and social justice in a changing media environment. *Communications*, *47*(4), 505–515. https://doi.org/10.1515/commun-2022-0068

Dugard, J. (2019). *Confronting apartheid: A personal history of South Africa, Namibia and Palestine*. Jacana.

Dumont, L. (1967). *Homo hierarchicus: Essaie sur le systeme des castes.* Gallimard.

Dumont, L. (1977). *Homo aequalis 1: Genèse et épanouissement de l'idéologie économique* (Nachdr.). Gallimard.

Edmunds, T., & Juncos, A. E. (2020). Constructing the capable state: Contested discourses and practices in EU capacity building. *Cooperation and Conflict*, *55*(1), 3–21. https://doi.org/10.1177/0010836719860885

Eisenbach, B. J., Hanlon, C. H., & Ingalls, E. (2009). *Tsumkwe energy skills assessment* (p. 131). Worcester Polytechnic Institute. https://digital.wpi.edu/concern/student_works/hx11xf972?locale=en

Eisenhardt, K. M. (1989). Building theories from case study research. *The Academy of Management Review*, *14*(4), 532–550. JSTOR. https://doi.org/10.2307/258557

Eisenstadt, S. N. (2000). Multiple modernities. *The MIT Press - JSTOR*, *129*(Winter 2000), 1–29.

Eisenstadt, S. N., & Schluchter, W. (1998). Introduction: Paths to early modernities: A comparative view. *Daedalus*, *127*(3), 1–18. JSTOR.

Electricity Act, 2000 (Act 2 of 2000) (2000). https://www.lac.org.na/laws/2000/2270.pdf

Electricity Act, 2007 (Act 4 of 2007) (2007). https://media.namiblii.org/files/legislation/akn-na-act-2007-4-eng-2007-10-18.pdf

Electricity Control Board (ECB). (2007a). *Information on Regional Electricity Distributors (REDs) 1. Overview of REDs-frequently asked questions*. Electricity Control Board (ECB).

Electricity Control Board (ECB). (2007b). *Information on Regional Electricity Distributors (REDs) 2. History of REDs*. Electricity Control Board (ECB).

Electricity Control Board (ECB). (2007c). *Information on Regional Electricity Distributors (REDs) 6. The Electricity Bill, 2007*. Electricity Control Board (ECB).

Electricity Control Board (ECB). (2021). *Integrated Strategic Business Plan 2021-2026*. Electricity Control Board - Integratd Strategic Business Plan. https://www.ecb.org.na/about-us/integrated-strategic-business-plan/

Electricity Control Board (ECB). (2022). *General notice: Update on the implementation of the Modified Single Buyer Market Model and the promulgation of the market rules*. Electricity Control Board (ECB). https://www.ecb.org.na/wp-content/uploads/2022/07/2106-ECB-GENERAL-NOTICE-MSB.pdf

Electricity Control Board (ECB). (2023a). *Annual Report-2023. Securing Namibia's future through a sustainable electricity supply industry* (p. 128). Electricity Control Board (ECB). https://www.ecb.org.na/wp-content/uploads/2023/11/ECB-Annual-Report-2023-web.pdf

Electricity Control Board (ECB). (2023b). *Regional Electricity Distributors*. https://www.ecb.org.na/regional-electricity-distributors/

Elias, N. (1987). *Über den Prozeß der Zivilisation: Soziogenetische und psychogenetische Untersuchungen*. Suhrkamp.

Elias, N. (1995). Technization and civilization. *Theory, Culture & Society*, *12*(3), 7–42. https://doi.org/10.1177/026327695012003002

Elias, N. (2000). *Uygarlaşma süreci 1: Batılı dünyevi üst tabakaların davranışlarındaki değişmeler* (E. Atesman, Trans.; 1st edition). lletişim Yayınlari.

Elias, N. (2004). *Uygarlaşma süreci 2: Toplumun değişimleri bir uygarlaşma teorisi için taslak* (E. Özbek, Trans.; 2nd edition). İletişim.

Emcon Consulting Engineers. (1999). *State of the Environment Report on Namibia's Industrialisation Environment—Draft Final Report*. Ministry of Environment and Tourism. http://the-eis.com/elibrary/sites/default/files/downloads/literature/State%20of%20the%20environment%20report%20on%20namibia%20industrialisation%20environment.pdf

Energy Information Administration (EIA). (2024). *Electric utility*. U.S. Energy Information Administration (EIA) Glossary. https://www.eia.gov/tools/ glossary/index.php?id=Electric%20utility

Erfani, G. (2022). Reconceptualizing sense of place: Towards a conceptual framework for investigating individual-community-place interrelationships. *Journal of Planning Literature*, *37*, 452–466.

ESMAP (Energy Sector Management Assistance Program). (2023). *Multi-Tier Framework (MTF) for measuring access to electricity by ESMAP (Energy Sector Management Assistance Program)*. Multi-Tier Framework (MTF) for Measuring Access to Electricity by ESMAP. https://mtfenergyaccess.esmap.org/methodology/electricity

Evers, H. D. (2022). The civilising process: World figuration or world system? In A. Bogner & S. Mennell (Eds.), *Civilisations, civilising processes and modernity – A debate: Documents from the Conference at Bielefeld, 1984* (pp. 17–24). Springer International Publishing. https://doi.org/10.1007/978-3-030-80379-7

Fanon, F. (2002a). *Black skin, white masks* (Repr.). Pluto Press.

Fanon, F. (2002b). *The wretched of the earth* (C. Farrington, Trans.). Grove Press.

FSG Remaining Social Change. (2021). *Guide to the World Café method*. FSG Social Change. https://www.fsg.org/wp-content/uploads/2021/08/World-Cafe-Method_0.pdf

Fusaro, F. (2023, March 10). *Digitization, digitalization, and digital transformation explained – understanding differences incl. Examples*. More Than Digital. https://morethandigital.info/en/digitization-digitalization-and-digital-transformation-explained-understanding-differences-incl-examples/#Definition_of_Digitization

Gewald, J.-B. (1999). *Herero heroes: A socio-political history of the Herero of Namibia, 1890-1923*. James Currey ; Ohio University Press.

Gewald, J.-B. (2000). *"We thought we would be free...": Socio-cultural aspects of Herero history in Namibia 1915-1940*. Köppe.

Gimpel, H., & Schmied, F. (2019). Risks and effects of digitalization: A multi-level taxonomy of the adverse effects of using digital technologies and media. *Proceedings of the 27th European Conference on Information Systems (ECIS),*. European Conference on Information Systems (ECIS), Stockholm&Uppsala. https://www.fimrc.de/Paperbibliothek/Veroeffentlicht/939/w i-939.pdf

Given, L. M. (2008). *The SAGE encyclopedia of qualitative research methods*. Sage.

GIZ. (2022). *Sector brief Namibia: Renewable energies*. https://www.giz.de/en/downloads/giz2022-en-sector-brief-namibia-renewable-energy.pdf?utm_source=chatgpt.com

Glaser, B. G., & Strauss, A. L. (2006). *The discovery of grounded theory: Strategies for qualitative research*. Routledge.

Government Notice on the Imposition of Levy on Electricity: Electricity Act, 2007 (2016). https://www.lac.org.na/laws/2016/6064.pdf

Green, R. H. (2023). History of Namibia. In *Encyclopaedia Britannica*. Encyclopaedia Britannica. https://www.britannica.com/place/Namibia/History

Habere Güven. (2020, December 9). *Prof. Dr. Kuçuradi: "İnsanların robotlaştırıldığı, robotların da insanlaştırılmaya çalışıldığı bir zamanda yaşıyoruz."*https://www.habereguven.com/prof-dr-kucuradi-insanlarin-robotlastirildigi-robotlarin-da-insanlastirilmaya-calisildigi-bir-zamanda-yasiyoruz

Hackl, D. (n.d.). *Geschichte von Namibia*. Namib. Info.https://www.namib.info/namibia/de/geschichte/

Hancké, B. (2009). *Intelligent research design: A guide for beginning researchers in the social sciences* (1. publ). Oxford University Press.

Harring, S. L., & Odendaal, W. (2002). *"One day we will all be equal". A socio-legal perspective on the Namibian land reform and resettlement process*. Legal Assistance Centre.https://www.lac.org.na/projects/lead/Pdf/oneday.pdf

Heap, J. L., & Roth, P. A. (1973). On phenomenological sociology. *American Sociological Association, American Sociological Review*, *38*(3), 354–367.

Hirmer, S., & Guthrie, P. (2017). The benefits of energy appliances in the off-grid energy sector based on seven off-grid initiatives in rural Uganda. *Renewable and Sustainable Energy Reviews*, *79*, 924–934. https://doi.org/10.1016/j.rser.2017.05.152

Hitchcock, R. K., Babchuk, W. A., & Frost, J. (2021). San Traditional Authorities, communal Conservancies, conflicts, and leadership in Namibia. In A. S. Steinforth & S. Klocke-Daffa (Eds.), *Challenging authorities* (pp. 267–291). Springer International Publishing. https://doi.org/10.1007/978-3-030-76924-6_10

Hoeck, I., Steurer, E., Dolunay, Ö., & Ileka, H. (2022). Challenges for off-grid electrification in rural areas. Assessment of the situation in Namibia using the examples of Gam and Tsumkwe. *Energy, Ecology and Environment, 7*(5), 508–522. https://doi.org/10.1007/s40974-021-00214-5

Honneth, A. (1996). *The struggle for recognition: The moral grammar of social conflicts* (1st MIT Press ed). MIT Press.

Hooghe, L., & Marks, G. (2001, October 12). *Types of multi-Level governance*. European Integration Online Papers (EIoP). http://eiop.or.at/eiop/pdf/2001-011.pdf

Hooghe, L., Marks, G., & Schakel, A. (2020). Multilevel governance (Chapter 11). *Oxford University Press, Comparative Politics*, 193–210.

Hoy, D. C. (2009). *The time of our lives: A critical history of temporality* (The MIT Press). MIT Press.

Humphrey, T. (1992). *An answer to the question: What is enlightenment? Immanuel Kant (1784)*. Hacklett Publishing. https://www.nypl.org/sites/default/files/kant_whatisenlightenment.pdf

Hurmerinta, L., Paavilainen-Mäntymäki, E., & Hassett, M. E. (2016). Tempus fugit: A hermeneutic approach to the internationalisation process. *Management International Review, 56*(6), 805–825. https://doi.org/10.1007/s11575-016-0303-x

Ileka, H., Zongwe, D., & Reuther, K. (2017). Rural electrification with hybrid mini-grids: Finding an efficient and durable ownership model. In *The law reform and development Commission of Namibia at 25: A quarter century of social carpentry* (pp. 351–382). SSRN. https://papers.ssrn.com/sol3/papers.cfm?abstract_id=2937402

International Energy Agency. (2020). *World Energy Outlook 2020*. https://www.iea.org/reports/world-energy-outlook-2020

International Energy Ageny (IEA). (2021). *World Energy Model*. https://www.iea.org/reports/world-energy-model.

International Renewable Energy Agency (IRENA). (2018, November). *Policies and regulations for renewable energy mini-grids*. International Renewable Energy Agency (IRENA). https://www.irena.org/-/media/Files/IRENA/Agency/Publication/2018/Oct/IRENA_mini-grid_policies_2018.pdf

International Renewable Energy Agency (IRENA). (2019). *Off-grid renewable energy solutions to expand electricity access: An opportunity not to be missed*. International Renewable Energy Agency (IRENA). https://www.irena.org/-/media/Files/IRENA/Agency/Publication/2019/Jan/IRENA_Off-grid_RE_Access_2019.pdf

International Renewable Energy Agency (IRENA). (2021). *Off-grid Renewable Energy Statistics 2021* (p. 50). International Renewable Energy Agency.

https://www.gensed.org/assets/attachments/dosyalar/2021-Y%C4%B11%C4%B1-Off-Grid-Yenilenebilir-Enerji-%C4%B0statistik-Raporu.pdf

International Renewable Energy Agency (IRENA). (2022, June 17). *COVID-19 slows progress towards universal energy access*. https://www.irena.org/News/pressreleases/2022/Jun/COVID-19-Slows-Progress-Towards-Universal-Energy-Access

Jami, A., Kouchaki, M., & Gino, F. (2021). I own, so I help out: How psychological ownership increases prosocial behavior. *Journal of Consumer Research, 47*(5), 698–715. https://doi.org/10.1093/jcr/ucaa040

Jinadu, L. A. (1978). Some african theorists of culture and modernization: Fanon, Cabral and some others. *African Studies Review, 21*(1), 121. https://doi.org/10.2307/523767

Juujärvi, S., & Lund, V. (2015). Participatory Action Research as a practice for empowerment in deprived communities. *41st Association for Moral Education Conference, AME2015*. https://www.laurea.fi/globalassets/tki/hankkeen-dokumentteja/participatory-action-research-as-a-practice-of-empowerment.pdf

Kant, I. (1784). *Answer the question: What is enlightenment?* (D. F. Ferrer, Trans.). https://archive.org/details/AnswerTheQuestionWhatIsEnlightenment/KantEnlightmentDanielFidelFerrer2013/page/n3/mode/2up

Kantabutra, S. (2008). What do we know about vision? *Journal of Applied Business Research (JABR), 24*(2). https://doi.org/10.19030/jabr.v24i2.1359

Katjavivi, P. H. (1990). *A history of resistance in Namibia* (1. American ed). Africa World Press [u.a.]. https://www.sahistory.org.za/sites/default/files/archive-files3/peter_h._katjavivi_history_of_resistance_in_namibook4you.pdf

Kenny, A. J. P., & Amadio, A. H. (2023). Philosophy of mind of Aristotle. In *Encyclopaedia Britannica*. Encyclopaedia Britannica. https://www.britannica.com/biography/Aristotle/Philosophy-of-mind

Komter, A. (2010). Solidarity. In H. K. Anheier & S. Toepler (Eds.), *International encyclopedia of civil society* (pp. 1460–1464). Springer US. https://doi.org/10.1007/978-0-387-93996-4_81

Landy, J., & Saler, M. (2009). Introduction: The varieties of modern enchantment. In J. Landy & M. Saler (Eds.), *The Re-enchantment of the world: Secular magic in a rational age* (1st ed.). Stanford University Press. https://doi.org/10.2307/j.ctvr33bqt

Legal Assistance Centre (LAC). (n.d.). *Regional and local government—Factsheet series No. 6 of 8*. Legal Assistance Centre (LAC). https://www.lac.org.na/projects/grap/Pdf/Gov6_Regional_and_Local_Government.pdf

Li, F., & Li, S. (2018). The impact of cross-subsidies on utility service quality in developing countries. *Economic Modelling*, *68*, 217–228. https://doi.org/10.1016/j.econmod.2017.07.013

Li, F., Wang, W., & Yi, Z. (2018). Cross-subsidies and government transfers: Impacts on electricity service quality in Colombia. *Sustainability*, *10*(5), 1599. https://doi.org/10.3390/su10051599

Link, H., McNally, T., Sayre, A., Schmidt, R., & Swap, R. J. (2011). The definition of community: A student perspective. *A Journal of Service Learning & Civic Engagement.*, *2*(2). http://dx.doi.org/10.7253/partj.v0i0.440

Liu, Y. (2020). On the definition of a self-sustaining chemical reaction system and its role in heredity. *Biology Direct*, *15*(1), 15. https://doi.org/10.1186/s13062-020-00269-0

Lo, C. (2018). Going from government to governance. In A. Farazmand (Ed.), *Global encyclopedia of public administration, public policy, and governance* (pp. 1–5). Springer International Publishing. https://doi.org/10.1007/978-3-319-31816-5_3282-1

Local Authorities Act 23 of 1992 (LA Act 23) (1992). http://www.lac.org.na/laws/annoSTAT/Local%20Authorities%20Act%2023%20of%201992.pdf

MacNeill, T. (2020). Culture in development theory. In *Indigenous cultures and sustainable development in Latin America* (pp. 57–83). Springer International Publishing. https://doi.org/10.1007/978-3-030-37023-7

MaCorr Research. (2023). *Sample size calculator*. MaCorr Research Solutions Online. https://www.macorr.com/sample-size-calculator.htm

MacQueen, K. M., McLellan, E., Metzger, D. S., Kegeles, S., Strauss, R. P., Scotti, R., Blanchard, L., & Tortter II, R. T. (2001). What is community? An evidence-based definition for participatory public health. *American Journal of Public Health*, *91*(12), 1929–1938.

Mariscal, C. (2021). Life. In E. N. Zalta (Ed.), *The Standford Encyclopedia of Philosophy* (Winter 2021 Edition). Metaphysics Research Lab, Stanford University. https://plato.stanford.edu/entries/life/

Mark, D. C. (2016). Axel Honneth, The Struggle for Recognition. In J. T. Levy (Ed.), *The Oxford handbook of classics in contemporary political theory* (1st ed.). Oxford University Press. https://doi.org/10.1093/oxfordhb/9780198717133.013.2

Marmot, M. (2013). Self-esteem and health. *BMJ*, *327*(7415), 574–575. https://doi.org/10.1136/bmj.327.7415.574

Marotta, M. (2024). A disenchanted world: Max Weber on magic and modernity. *Journal of Classical Sociology*, *24*(3), 224–242. https://doi.org/10.1177/1468795X231160716

Marshall, L. (1976). *The !Kung of Nyae Nyae*. Harvard Univ. Press.

Marshall, L. J. (Director). (1995). First film [Video recording]. In *!Kung*. Documentary Educational Resources (DER). https://video.alexanderstreet.com/watch/first-film

Maslow, A. H. (1943). A theory of human motivation. *Psychological Review*, *50*(4), 370–396. https://doi.org/10.1037/h0054346

Maslow, A. H. (1980). A theory of human motivation. In H. J. Leavitt, L. R. Pondy, & D. M. Boje (Eds.), *Readings in managerial psychology* (3. ed). Univ. of Chicago Press.

Massey, D. (1991). A global sense of place. *Marxism Today*, 24–29.

Massey, D. (1995). Places and their pasts. *History Workshop Journal*, *39*(1), 182–192. https://doi.org/10.1093/hwj/39.1.182

Mayntz, R. (2001). El estado y la sociedad civil en la gobernanza moderna. *La Revista Del CLAD Reforma y Democracia*, *7–22*(21). https://pure.mpg.de/rest/items/item_1234838/component/file_2060890/content

McDonough, J. K. (2010). Descartes' optics. In *The Cambridge Descartes lexicon*. Cambridge University Press. https://scholar.harvard.edu/files/mcdonough/files/30_descartes_optics_the_cambridge_descartes_lexicon.pdf

Mcgirr, K. (2021, September). *REGIoNAl government in Namibia: Is decentralisation a reality?* Institute for Public Policy and Research (IPPR) and Hanns Seidel Foundation (HSF). https://ippr.org.na/wp-content/uploads/2021/09/IPPR_Decentralisation_WEB.pdf

McLeod, S. (2024, January 24). *Maslow's hierarchy of needs*. Simply Psychology. https://www.simplypsychology.org/maslow.html

McMillan, D. W., & Chavis, D. M. (1986). Sense of community: A definition and theory. *Journal of Community Psychology*, *14*, 6–23.

Mergel, T. (2012, June 21). *Modernization*. The Leibniz Institute of European History (IEG). https://www.ieg-ego.eu/en/threads/models-and-stereotypes/modernization

Meshanko, P. (2013). *The respect effect: Using the science of neuroleadership to inspire a more loyal and productive workplace*. McGraw-Hill Education.

Ministry of Lands, Resettelement and Rehabilitation (MLRR). (1993, April). *Operational strategies for the repatriation of Batswana of Namibian origin*. Ministry of Lands, Resettlement and Rehabilitation; 13-0033A.

Ministry of Mines and Energy. (1998). *White Paper on the Energy Policy of Namibia*. Ministry of Mines and Energy. http://www.mme.gov.na/files/publications/1e3_energy_policy_whitepaper.pdf

Ministry of Mines and Energy (MME). (2000a, August). *Rural Electricity Distribution Master Plan (REDMP) for Namibia—Master Planning Approach and Methodology Report—Volume 16*. Ministry of Mines and Energy.

Ministry of Mines and Energy (MME). (2000b, August). *Rural Electricity Distribution Master Plan (REDMP) for Namibia—National Overview Report—Volume 1: Baseline Report*. Ministry of Mines and Energy (MME).

Ministry of Mines and Energy (MME). (2000c, August). *Rural Electricity Distribution Master Plan (REDMP) for Namibia—Regional Planning Report for the Otjozondjupa Region—Volume 14* (Vol. 14). Ministry of Mines and Energy (MME).

Ministry of Mines and Energy (MME). (2010a). *Rural Electricity Distribution Master Plan for Namibia: Overview & Methodology—Part 1*. Ministry of Mines and Energy (MME). https://www.mme.gov.na/energy/pdf/National/REDMP%20National%20Overview%20&%20Methodology%20-%20Part%201.pdf

Ministry of Mines and Energy (MME). (2010b). *Rural Electricity Distribution Master Plan for Namibia: Overview & Methodology—Part 2*. Ministry of Mines and Energy (MME). https://www.mme.gov.na/energy/pdf/National/REDMP%20National%20Overview%20&%20Methodology%20-%20Part%202.pdf

Ministry of Mines and Energy (MME). (2010c). *Rural Electricity Distribution Master Plan for Namibia: Overview & Methodology—Part 3*. Ministry of Mines and Energy (MME). https://www.mme.gov.na/energy/pdf/National/REDMP%20National%20Overview%20&%20Methodology%20-%20Part%203.pdf

Ministry of Mines and Energy (MME). (2010d). *Rural Electricity Distribution Master Plan for Namibia: Overview & Methodology—Part 4*. Ministry of Mines and Energy (MME). https://www.mme.gov.na/energy/pdf/National/REDMP%20National%20Overview%20&%20Methodology%20-%20Part%204.pdf

Ministry of Mines and Energy (MME). (2013). *Annual Report 2012/13*. Ministry of Mines and Energy. https://www.mme.gov.na/files/publications/MMEAnnualReport.pdf

Ministry of Mines and Energy (MME). (2014). *Lights & Gems*. *1*(1).

Ministry of Mines and Energy (MME). (2016). *National Integrated Resource Plan Review and Update—Final Report (NIRP)*. Ministry of Mines and Energy. https://www.mme.gov.na/files/publications/f74_29a_ELECTRICITY%20SECTOR%20NATIONAL%20INTEGRATED%20RESOURCE%20PLAN%20(NIRP)%202016%20FINAL%20REPORT.pdf

Ministry of Mines and Energy (MME). (2017a, July). *National Energy Policy (NEP)*. Ministry of Mines and Energy. https://www.mme.gov.na

/files/publications/fd8_National%20Energy%20Policy%20-%20July%2020 17.pdf

Ministry of Mines and Energy (MME). (2017b, July). *National Renewable Energy Policy (NREP)*. Ministry of Mines and Energy. https://www.mme.gov.na/files/publications/03f_National%20Renewable%20Energy%20Policy%20-%20July%202017.pdf

Ministry of Mines and Energy (MME). (2018). *National Policy for Independent Power Producers (IPPs)*. Ministry of Mines and Energy. https://www.ecb.org.na/wpcontent/uploads/2022/07/National_IPP_Policy.pdf

Ministry of Mines and Energy (MME). (2019). *(Draft) Off-grid Electrification Policy*.

Ministry of Mines and Energy (MME). (2021). *(Draft) National Electrification Strategy and Implementation Plan (NES)*.

Ministry of Mines and Energy (MME). (2022a, December). *National Electrification Funding Portfolio (NELFP)*.

Ministry of Mines and Energy (MME). (2022b, December). *National Electrification Policy (NELP) 2022-2040*.

Ministry of Mines and Energy (MME). (2023a). *Ministry of Mines and Energy—About us*. Ministry of Mines and Energy - About Us. https://www.mme.gov.na/about-us/mission/

Ministry of Mines and Energy (MME). (2023b). *Ministry of Mines and Energy—History*. Ministry of Mines and Energy - About Us. https://www.mme.gov.na/about-us/history/

Ministry of Mines and Energy (MME). (2023c). *Ministry of Mines and Energy—Renewable energy*. Ministry of Mines and Energy - About Us. https://www.mme.gov.na/directorates/energy/renewable/

Molnar, C., & Gair, J. (2015). *Concepts of biology* (1st Canadian Edition). BCcampus. https://opentextbc.ca/biology/chapter/1-1-themes-and-concepts-of-biology/

Morris, J. (1994). Community care or independent living? *Critical Social Policy*, *14*(40), 24–45. https://doi.org/10.1177/026101839401404002

Namibia Association of CBNRM Support Organizations (NACSO). (2012). *Living with wildlife—The story of Nyae Nyae Conservancy*. Namibia Association of CBNRM Support Organizations (NACSO). http://the-eis.com/elibrary/sites/default/files/downloads/literature/Brochure%20Nyae%20Nyae%20FPis.pdf

Namibia Statistics Agency (NSA). (2013). *Profile of Namibia: Facts, figures, and other fundamental information*.

Namibia Statistics Agency (NSA). (2016). *Namibia Household Income & Expenditure Survey (NHIES) 2015/2016—Key poverty indicators (Preliminary Figures)*. Namibia Statistics Agency.

Namibia Statistics Agency (NSA). (2017). *Namibia Inter-censal Demographic Survey 2016 Report* (p. 136). Namibia Statistics Agency. https://cms.my.na/assets/documents/NIDS_2016.pdf

Namibian Association CBNRM Support Organisation (NACSO). (2023a). *Registered Communal Conservancies*. Namibian Association CBNRM Support Organisation. https://www.nacso.org.na/conservancies

Namibian Association CBNRM Support Organisation (NACSO). (2023b). *Registered Communal Conservancies: Ondjou Conservancy*. Namibian Association CBNRM Support Organisation.https://www.nacso.org.na/cons erveancies/ondjou

NamPower. (2016). Runner refurbishment project at Ruacana completed. *WattsOn- The Official Newsletter of NamPower*, *2*, 8–10.

NamPower. (2019). *NamPower Corporate Strategy and Business Plan 2019-2023*.

NamPower. (2022). *Integrated Annual Report 2022*. NamPower. https://www.nampower.com.na/public/docs/annual-reports/NamPower_2022%20Annual%20Report_210223.pdf

NamPower. (2023). *Welcome to the NamPower Official Website*. Welcome to the NamPower Official Website. https://www.nampower.com.na/About.aspx

NASA. (2024). *About life detection*. Astrobiology at NASA. https://astrobiology.nasa.gov/research/life-detection/about/

National Planning Commission (NPC). (1995). *First National Development Plan (NDP1) 1995/1996—1999/2000*. National Planning Commission (NPC). https://www.npc.gov.na/wp-content/uploads/2022/04/NDP-1.pdf

National Planning Commission (NPC) (Ed.). (2002). *Second National Development Plan (NDP2), 2001/2002-2005/2006*. National Planning Commission (NPC). https://www.npc.gov.na/wp-content/uploads/2022/04/NDP-2-compressed.pdf

National Planning Commission (NPC). (2004). *Preliminary Report—Namibia Household Income & Expenditure Survey (NHIES) 2003/2004*. National Planning Commission.

National Planning Commission (NPC). (2008). *Third National Development Plan (NDP3): 2007/2008 - 2011/12*. National Planning Commission. https://www.npc.gov.na/wp-content/uploads/2021/11/NDP3.pdf

National Planning Commission (NPC). (2012). *Namibia's Fourth National Development Plan NDP4 2012/13 to 2016/17*. National Planning Commission (NPC). https://www.npc.gov.na/wp-content/uploads/2021/11/NDP4.pdf

National Planning Commission (NPC). (2017). *Namibia's 5th National Development Plan (NDP5) 2017/18 to 2021/22*. National Planning Commission (NPC). https://www.npc.gov.na/wp-content/uploads/2021/11/NDP5.pdf

Naughten, J., & Marten, L. (2013). *Conflict and costume: The Herero tribe of Namibia*. Merrell.

NBC Digital News (Director). (2022, April 14). NBC Digital News—Maroelaboom-Mkata-Mangetti Dune electricity power line in Tsumkwe commissioned- NBC [Broadcast]. In *NBC Digital News*. https://www.youtube.com/watch?v=pgOQkeBJR0w

Neal, M. B., Hammer, L. B., & Morgan, D. L. (2015). Using mixed methods in research related to work and family. In M. Pitt-Catsouphes, E. E. Kossek, & S. A. Sweet (Eds.), *The work and family handbook: Multi-disciplinary perspectives, methods, and approaches* (pp. 587–606). Psychology Press.

Norbert Elias Foundation (Director). (2019, March 19). *The task of sociologists* [Video recording]. https://www.youtube.com/watch?v=NYwQSiQA0t0

Nyae Nyae Development Foundation of Namibia (NNDF). (2024). *San*. Nyae Nyae Development Foundation of Namibia (NNDF). https://www.nndfn.org/san

OpenAI. (2023, November). *Response from ChatGPT [Large language model]*. OpenAI.

OpenAI. (2024). *Reflection and refraction of light. AI-generated image*. DALL E. https://openai.com/dall-e.

Ostrom, E. (2010). Beyond markets and states: Polycentric governance of complex economic systems. *American Economic Review*, *100*(3), 641–672. https://doi.org/10.1257/aer.100.3.641

Oxford Learner's Dictionaries. (2023). Wish. In *Oxford Learner's Dictionary*. Oxford University Press. https://www.oxfordlearnersdictionaries.com /definition/english/wish_1

Oxford Reference. (2023). Mantra. In *Oxford Reference*. https://www.oxfordreference.com/display/10.1093/acref/9780198609810.001.0001/acref-9780198609810-e-4333

Panitz, C., Endres, D., Buchholz, M., Khosrowtaj, Z., Sperl, M. F. J., Mueller, E. M., Schubö, A., Schütz, A. C., Teige-Mocigemba, S., & Pinquart, M. (2021). A revised framework for the investigation of expectation update versus maintenance in the context of expectation violations: The ViolEx 2.0 Model. *Frontiers in Psychology*, *12*, 726432. https://doi.org/10.3389 /fpsyg.2021.726432

Parsons, T. (1975). *Gesellschaften: Evolutionäre und komparative Perspektiven* (N. T. Lindquist, Trans.; 1. Aufl). Suhrkamp.

Parsons, T. (1985). *Das System moderner Gesellschaften* (D. Claessens, Trans.). Juventa-Verlag.

Pasqualetti, M. J. (2021). *The thread of energy*. Oxford University Press. https://doi.org/10.1093/oso/9780199394807.001.0001

Peet, R., & Hartwick, E. R. (2009). *Theories of development: Contentions, arguments, alternatives* (2nd ed). Guilford Press.

Perkins, J. H. (2022). Martin J. Pasqualetti. The thread of energy: The ubiquity of energy in modern life. *Springer*, *12*(Journal of Environmental Studies and Sciences), 914–915.

Perret, C., de Villers, T., Carme, F., & Stevenson, J. (2021, March). *Namibia: Geospatial Least Cost Electrification Plan (GLCEP)—Draft Final Report (#7192481).*

Pike, A., Béal, V., Cauchi-Duval, N., Franklin, R., Kinossian, N., Lang, T., Leibert, T., MacKinnon, D., Rousseau, M., Royer, J., Servillo, L., Tomaney, J., & Velthuis, S. (2024). 'Left behind places': A geographical etymology. *Regional Studies*, *58*(6), 1167–1179. https://doi.org/10.1080/00343404.2023.2167972

Pinto, S., Fumincelli, L., Mazzo, A., Caldeira, S., & Martins, J. C. (2017). Comfort, well-being and quality of life: Discussion of the differences and similarities among the concepts. *Porto Biomedical Journal*, *2*(1), 6–12. https://doi.org/10.1016/j.pbj.2016.11.003

PROCEED. (2023). *PROCEED Recommendations for Action*.

PROCEED Final Conference. (2023, September 15). *2nd PROCEED Panel: Quo Vadis Tsumkwe and Gam? Voices from the communities: Reflections, recommendations, visions moderated Q&A session*.

Regional Energy Regulators Association of Southern Africa Authrotirty (RERA). (2023). *RERA: Regional Energy Regulators Association of Southern Africa Authrotirty*. Regional Energy Regulators Association of Southern Africa Authrotirty (RERA). https://rerasadc.com/

Reinhardt, W., Mletzko, C., Sloep, P. B., & Drachsler, H. (2012). Understanding the meaning of awareness in research networks. *Proceedings of 2nd Workshop on Awareness and Reflection in Technology-Enhanced Learning, ARTEL/EC-TEL*, 13–30. https://www.researchgate.net/publication/230640624_Understanding_the_Meaning_of_Awareness_in_Research_Networks

Relph, E. (1970). An inquiry into the relations between phenomenology and geography. *Canadian Geographies / Géographies Canadiennes*, *14*(3), 193–201. https://doi.org/10.1111/j.1541-0064.1970.tb01567.x

Relph, E. (2009). *A pragmatic sense of place*. *20*(3), 24–31. https://ophen.org/pub-115405

Republic of Namibia (Ed.). (2004). *Namibia Vision 2030: Policy Framework for Long-term National Development*. Office of the President. https://www.npc.gov.na/wp-content/uploads/2021/11/vision_2030.pdf

Republic of Namibia. (2009, August 3). *Government Gazette of the Republic of Namibia no.4314*. Government Gazette of the Republic of Namibia. https://www.lac.org.na/laws/2009/4314.pdf

Republic of Namibia. (2016). *Harambee Prosperity Plan I (HPP)*. https://nipdb.com/download/harambee-prosperity-plan-i/

Republic of Namibia. (2019). *Public Enterprises Governance Act 1- PEGA Act*. Government Gazette of the Republic of Namibia. https://www.lac.org.na /laws/2019/6907.pdf

Republic of Namibia. (2021). *Harambee Prosperity Plan II*. http://hppii.gov.na/wp-content/uploads/2021/03/HPP2.pdf

Rhodes, R. A. W. (2007). Understanding governance: Ten years on. *Organization Studies*, *28*(8), 1243–1264. https://doi.org/10.1177/0170840607076586

Ribe, N. M. (1997). Cartesian optics and the mastery of nature. *JSTOR*, *88*(1), 42–61.

Ritchie, C., & Marshall, J. (Directors). (1991). To hold our ground: A field report [Video recording]. In *!Kung*. Documentary Educational Resources (DER). https://video.alexanderstreet.com/watch/to-hold-our-ground-a-field-report

Robson, C. (2002). *Real world research: A resource for social scientists and practitioner-researchers*. Blackwell.

Rojas, M., Méndez, A., & Watkins-Fassler, K. (2023). The hierarchy of needs empirical examination of Maslow's theory and lessons for development. *World Development*, *165*, 106185. https://doi.org/10.1016/j.worlddev.2023.106185

Rothfuss, E., & Dörfler, T. (2023). The geography of the life-world: Spatialising the social theory of Alfred Schütz. *Erdkunde Journal of Human and Physical Geographies*, *77*(2), 149–161. https://doi.org/10.3112/erdkunde.2023.02.04

Royston, S., & Foulds, C. (2021). The making of energy evidence: How exclusions of social sciences and humanities are reproduced (and what researchers can do about it). *Energy Research & Social Science*, *77*, 102084. https://doi.org/10.1016/j.erss.2021.102084

SADC. (1993). *Declaration and Treaty of the Southern African Development Community (SADC)*. SADC.

SADC Centre for Renewable Energy and Energy Efficiency (SACREEE). (2023). *SACREEE - Centre for Renewable Energy and Energy Efficiency*.

SADC Sentre for Renewable Energy and Energy Efficiency (SACREEE) - Vision and Mission. https://www.sacreee.org/vision-mission

Sagan, C., Sagan, D., & Margulis, L. (2023). Life. In *Encyclopaedia Britannica*. Encyclopaedia Britannica. https://www.britannica.com/science/life#ref41332

Saito-Jensen, M. (2015). Multilevel governance theory—Theories and methods for the Study of multilevel environmental governance. *Center for International Forestry Research*, 2–6.

SAPP. (2023). *SAPP Members*. About Us. https://www.sapp.co.zw/about-sapp#members

Sárközy, T. (1982). Problems of social ownership and the proprietory organization. *Acta Oeconomica*, *29*(3/4), 225–257. JSTOR.

Saunders, M., Lewis, P., & Thornhill, A. (2007). *Research methods for business students* (4. ed). Financial Times Prentice Hall.

Sayani, R., Ortega-Arriaga, P., Sandwell, P., Babacan, O., Gambhir, A., Robinson, D., & Nelson, J. (2022). Sizing solar-based mini-grids for growing electricity demand: Insights from rural India. *Journal of Physics: Energy*, *5*(014004). https://iopscience.iop.org/article/10.1088/2515-7655/ac9dc0

Schatzki, T. (2011, November). *Where the action is (on large social phenomena such as sociotechnical regimes)*. Sustainable Practices Research Group. http://www.sprg.ac.uk/uploads/schatzki-wp1.pdf

Schultz, R., & Schumann, C. (2007, January). *Off-Grid Energization Master Plan for Namibia (OGEMP)*. Ministry of Mines and Energy. https://www.mme.gov.na/files/publications/541_off-grid-masterplan.pdf

Schütt, H. (2016). 100% Decentralised, renewable energy for Namibia. In O. C. Ruppel & B. Althusmann (Eds.), *Perspectives on energy security and renewable energies in Sub-Saharan Africa: Practical opportunities and regulatory challenges* (pp. 41–64). Macmillan Education Namibia. https://tlu.nust.na/portal/tlusessions/100%25%20Renewable%20Energy%20and%20Energy%20Efficiency%20in%20Namibia%20FINAL.pdf

Schutz, A. (1962). Phenomenology and the social sciences. In M. Natanson (Ed.), *Collected papers I* (Vol. 11, pp. 118–139). Springer Netherlands. https://doi.org/10.1007/978-94-010-2851-6_5

Seamon, D. (1979). *A Geography of the lifeworld*. St. Martin's Press.

Seamon, D. (2018). *Life takes place: Phenomenology, lifeworlds and place making*. Routledge is an imprint of the Taylor & Francis Group, an Informa Business.

Seckelmann, A. (2001). *Development of urban settlements in independent Namibia*. Namibian Economic Policy Research Unit. Oxford

Shahabi, Z., & Rassi, F. (2015). Husserl's phenomenology and two terms of noema and noesis. *International Letters of Social and Humanistic Sciences*, *53*, 29–34. https://doi.org/10.18052/www.scipress.com/ILSHS.53.29

Shejavali, N. (2015). *Regional governors: The good, the bad and the ugly. Do the benefits of appointment by the President outweigh the loss of democratic representation?* Institute for Public Policy Research (IPPR) Namibia. https://ippr.org.na/wp-content/uploads/2015/11/Briefing%202%20Regional%20Governors_0.pdf

Shove, E., Pantzar, M., & Watson, M. (2012). *The dynamics of social practice: Everyday life and how it changes*. SAGE.

Shove, E., & Walker, G. (2014). What Is energy for? Social practice and energy demand. *Theory, Culture & Society*, *31*(5), 41–58. https://doi.org/10.1177/0263276414536746

Sillah, R. (2015). Africa must be modern: A manifesto. *African Affairs*, *114*(455), 323–325. https://doi.org/10.1093/afraf/adv014

Silvast, A., & Foulds, C. (2022). *Sociology of interdisciplinarity: The dynamics of energy research*. Palgrave Macmillan.

Soja, E. W. (1967). *The geogrpahy of modernization in Kenya* [Syracuse University]. https://www.proquest.com/openview/674f3e3bb720dac8b6dbc542ef7812ab/1?pq-origsite=gscholar&cbl=18750&diss=y

Sørensen, M. P. (2019, November 19). *Second modernity*. Wiley Online Library. https://onlinelibrary.wiley.com/doi/abs/10.1002/9781405165518.wbeos1357

South African Power Pool (SAPP). (2023). *About SAPP*. South African Power Pool (SAPP). https://www.sapp.co.zw/

Southern African Development Community (SADC). (1996, August 24). *Protocol on Energy in the Southern African Development Community (SADC)*. Southern African Development Community (SADC). https://www.sadc.int/sites/default/files/202111/Protocol_on_Energy1996.pdf

Southern African Development Community (SADC). (2024). *History and Treaty*. Southern African Development Community - History and Treaty. https://www.sadc.int/pages/history-and-treaty

Sovacool, B. K. (2011). An international comparison of four polycentric approaches to climate and energy governance. *Energy Policy*, *39*(6), 3832–3844. https://doi.org/10.1016/j.enpol.2011.04.014

Sovacool, B. K., Axsen, J., & Sorrell, S. (2018). Promoting novelty, rigor, and style in energy social science: Towards codes of practice for appropriate methods and research design. *Energy Research & Social Science*, *45*, 12–42. https://doi.org/10.1016/j.erss.2018.07.007

Sovacool, B. K., & Drupady, I. M. (2012). *Energy access, poverty, and development: The governance of small-scale renewable energy in developing Asia*. Ashgate.

Sovacool, B. K., & Martiskainen, M. (2020). Hot transformations: Governing rapid and deep household heating transitions in China, Denmark, Finland and the United Kingdom. *Energy Policy*, *139*(111330), 1–16. https://doi.org/10.1016/j.enpol.2020.111330

Stockmayer, M., Chiguvare, Z., Martanova, L., & Wetzer, W. (2015, June). *Nationally Appropriate Mitigation Action (NAMA): Rural development in Namibia through electrification with renewable energies*. UNDP and Ministry of Environment and Tourism of the Republic of Namibia. https://www.undp.org/sites/g/files/zskgke326/files/publications/NAMA%20Final%20Namibia%202.pdf

Strange, J. M., & Mumford, M. D. (2005). The origins of vision: Effects of reflection, models, and analysis. *The Leadership Quarterly*, *16*(1), 121–148. https://doi.org/10.1016/j.leaqua.2004.07.006

Suzman, J. (2001). *Regional assessment of the status of the San in southern Africa*. Legal Assistance Centre (LAC).

Taiwo, O. (2014). *Africa must be modern: A manifesto* (U.S. edition). Indiana University Press.

Tarr, J. (1998). *A retrospective study of the environmental impacts of emergency borehole supply in the Gam and Khorixas Areas of Namibia* (p. 301). Environmental and Geographical Science Masters Class, University of Cape Town.

Toledano-González, A., Labajos-Manzanares, T., & Romero-Ayuso, D. (2019). Well-Being, self-efficacy and independence in older adults: A randomized trial of occupational therapy. *Archives of Gerontology and Geriatrics*, *83*, 277–284. https://doi.org/10.1016/j.archger.2019.05.002

Tomaselli, K. G. (2005). *Where global contradictions are sharpest—Research stories from the Kalahari*.

United, N. (2024). *Sustainable development goals*. UN Sustainable Development Goals. https://sdgs.un.org/goals/goal7

United Nations. (2015). *Transforming our world: The Agenda 2030 for sustainable development*. United Nations. https://sdgs.un.org/sites/default/files/publications/21252030%20Agenda%20for%20Sustainable%20Development%20web.pdf

United Nations. (2018). *Accelerating SDG7 achievement, Policy brief 08—Interlinkages among energy, poverty and inequalities*. https://sustainabledevelopment.un.org/content/documents/17480PB8.pdf

United Nations. (2023). *Capacity-building*. United Nations-Academic Impact. https://www.un.org/en/academic-impact/capacity-building

United Nations Development Program (UNDP). (2015). *Nationally Appropriate Mitigation Action (NAMA) Sustainable Development Evaluation Tool.* https://www.undp.org/publications/nationally-appropriate-mitigation-action-nama-sustainable-development-evaluationtool

Usenyuk, S., Hyysalo, S., & Whalen, J. (2016). Proximal Design: Users as Designers of Mobility in the Russian North. *Technology and Culture*, *57*(4), 866–908. https://doi.org/10.1353/tech.2016.0110

Uytvanck, M. van. (1993). *Mission Report on the survey on the situation of the Herero repatriation in Gam* (p. 22) [Mission Report by Unicef Consultant]. http://www.nln.gov.na:8080/client/natlib1/search/results;jsessionid=D413731E1CB1C6229C712EB8D2B5BBF6?qu=Namibia.&qf=SUBJECT%09Subject%09Gam%09Gam&qf=PUBDATE%09Publication+Date%091993%091993&ic=true

van der Loo, H., & van Reijen, W. (1992). *Modernisierung: Projekt und Paradox* (M. E. Baumer, Trans.; Deutsche Erstausgabe). Deutscher Taschenbuch Verlag.

van Heerikhuizen, B. (Director). (2015). *Classical sociological theory* [Video recording]. University of Amsterdam. https://www.youtube.com/watch?v=eFzoahZV4eo&list=PLISEtDmihMo0hvvyjFBJhS7sHTxxUzU3A&index=51

Vanheukelom, J. (2017, March). *Understanding the Southern African Development Community. Living with a powerful neighbour*. European Centre for Development Policy Management. https://ecdpm.org/application/files/4916/6133/9639/SADC-Background-Paper-PEDRO-Political-Economy-Dynamics-Regional-Organisations-Africa-ECDPM-2017.pdf

Vitas, M., & Dobovišek, A. (2019). Towards a general definition of life. *Origins of Life and Evolution of the Biosphere : The Journal of the International Society for the Study of the Origin of Life*, *49*(1–2), 77–88. https://doi.org/10.1007/s11084-019-09578-5

Vivelo, F. R. (with American Ethnological Society). (1977). *The Herero of western Botswana: Aspects of change in a group of Bantu-speaking cattle herders*. West Publ.

von Oertzen, D. (2015). *REEE-powering Namibia* (1st edition). Konrad-Adenauer-Stiftung.

von Oertzen, D. (2016). REEE-Powering Namibia. Energizing national development. In O. C. Ruppel & B. Althusmann (Eds.), *Perspectives on energy security and renewable energies in Sub-Saharan Africa: Practical opportunities and regulatory challenges* (Second revised and expanded edition, pp. 65–92). Macmillan Education Namibia.

von Oertzen, D. (2018). *Energy storage systems and their applications in Namibia´s electricity sector*. Konrad-Adenauer-Stiftung.

von Oertzen, D. (2019). *State of the Namibian electricity sector, 2019* (1st edition). Konrad-Adenauer-Stiftung.

Webster, M. (2024). *Governance*. Merriam-Webster Dictionary. https://www.merriam-webster.com/dictionary/governance

Wei, Y., Yang, Y., Zhang, Y., Mu, Z., & Bu, F. (2022). A frame theory of energetic life: A twisting energy solidified on the holographic fractal structure. *Applied Sciences*, *12*(21), 10930. https://doi.org/10.3390/app122110930

Weylandt, M. (2016, November). *Briefing paper: SOE Governance in Namibia: Will a hybrid system work?* Institute for Public Policy Research, Hanns Seidel Foundation. https://ippr.org.na/wp-content/uploads/2017/01/IPPR_SOEs_FINAL.pdf

Wiener, J. M. (1975). The Barrington Moore thesis and its critics. *Theory and Society*, *2*(3), 301–330. JSTOR.

Willow, A. (2022). Visions of transition: Centering the future in engaged sustainability research. *SN Social Sciences*, *2*(5), 56. https://doi.org/10.1007/s43545-022-00356-1

Wootton, D. (2015). *The invention of science: A new history of the scientific revolution* (1st US ed). Allen Lane.

World Bank (WB). (2021). *Access to electricity (% of population)—Namibia*. World Bank - Access to Electricity. https://data.worldbank.org/indicator/EG.ELC.ACCS.ZS?end=2021&locations=NA&name_desc=true&start=1992&view=chart

World Bank (WB). (2023). *Access to electricity (% of population)—Namibia*. World Bank - Access to Electricity. https://data.worldbank.org/indicator/EG.ELC.ACCS.ZS?end=2021&locations=NA&name_desc=true&start=1992&view=chart

World Population Review. (2023). *World Population Review—Namibia*. Namibia Population. https://worldpopulationreview.com/countries/namibia-population

Worldometers. (2023). *Namibia Population (live)*. Worldometers. https://www.worldometers.info/world-population/namibia-population/

Zamfir, I. (2017, April). *Understanding capacity-building/capacity development. A core concept of development policy*. European Parliamentary Research Service (EPRS). https://www.europarl.europa.eu/RegData/etudes/BRIE/2017/599411/EPRS_BRI(2017)599411_EN.pdf

Zubairy, M. S. (2016). A very brief history of light. In M. D. Al-Amri & M. M. El Gomati (Eds.), *Optics in our time* (pp. 3–24). Springer Open. https://doi.org/10.1007/978-3-319-31903-2

# Appendices

## Appendix A. Semi-Structured Survey Questions

**Date, Place**: **No:**

**Type**: ☐ Household ☐ Commercial ☐ ………………

**No. of Residents/Workers**: ☐ Female…………☐ Male………..**Children**:

### PART I: House/Building Infrastructure

1. Do you have power at home? ☐ Yes ☐ No
2. Do you have water at home? ☐ Yes ☐ No
3. Do you have a water shortage? If yes, how often? ☐ Yes ☐ No
   ………………………………..
4. What kind of energy source do you use?

   ☐ Grid ☐ Generator ☐ Battery ☐ Solar Home System
5. Where do you use power most?
   ☐ Lighting ☐ Cooking ☐ Appliances ☐ Other………..
6. From where do you supply your food? …………………………….

### Education:

7. What is your highest degree of education? Primary, High-school, University.
8. How many children do you have? Boy, girl, age
9. Are they going to school? Which grade?
10. Will they pursue higher studies? Where?

### Socio-Economic Data

11. What is your occupation? Do you run a business?
12. If yes, what kind of business do you have? ……………………..
13. What is the occupation of your spouse?
14. What is the income range per household?

    ☐ < 1300 NAD ☐ 1300-3900 NAD ☐ > 3900 NAD
15. Would you think of another occupation for yourself?
16. Do you have any ownership of:
    ☐ Land ☐ Livestock ☐ Other……..
17. If yes, which crops do you cultivate?
18. Which type of livestock do you have?
19. Do you have any need for additional financial source? Family, friends, bank?

    ☐ Yes ☐ No ☐ ……………………………..

**Energy Expenses**

20. What is your payment system?
    □ Monthly Invoice □ Prepaid (Net metering) □ Other…
21. What are your monthly invoice/prepaid power expenses?
    ………………………….. □ I don´t know
22. How do you pay the invoice?
    □ Cash in advance □ cash upon the bill □ other……
23. Who pays the invoice?
24. Is it affordable/expensive/cheap?
25. What happens if you don't pay the invoice?
26. Do you know what you actually pay for in your electricity bill?

**Connection**

27. Why did you want to connect to the mini-grid?
28. Are you satisfied with the mini-grid?
29. How long did it take you to connect to the mini-grid?
    □ 1 week □ 2 weeks □ 3 weeks
30. Was it easy to connect? □ Yes □ No
31. Where/which institutions did you do it?
32. How much did you pay?

**PART II: Technical**

33. Do you have any lighting? If yes, what do you use to illuminate your place?
34. What do you use for cooking?
    □ Firewood □ Gas □ Electricity □ Other…………
35. Which cooking appliance do you use?
    □ Open fireplace □ Gas cooker □ Electric Stove □ Other…………
36. Do you have a heating system? If so, which appliance do you use?
37. Do you have a cooling system? If so, which appliance do you use?
38. What kind of electric appliances do you have in your household? How many?
39. Which electrical appliances would you wish to have in the future?

**PART III: Open Questions:**

1. Can you describe your daily routines before and after the mini-grid? How did they change?
2. Why did you want to connect to the mini-grid?
3. Are there any challenges that you are facing with the mini-grid?

## Appendix B. Guiding Questions of Semi-Structured Interviews

A. For Community Members

- How have you and your family started your life in ……? How long have you been living in ….?
- How has it changed since then?
- (What were the changes with electricity?)
- (What are the challenges you are having with electricity?)
- What would you wish to have for yourself in the future?
- How do you think Tsumkwe will be in the next 5-10 years?

B. For National and Regional Actors

- How have you started working with renewables/solar energy?
- How do you find renewable energy development in Namibia? How was the beginning?
- Would you expect more from the government or ministry in RE sector? What do you think about their role?
- How do you see Namibia´s future in the renewable energy sector?

# Appendix C. Additional Tables of Chapter 4

*All tables are prepared by the author (2020-2023).*

*Table C.1. Distribution of the Number of Employees (Commercial)*

| Number of Employee | | Frequency | Percent | Valid Percent | Cumulative Percent |
|---|---|---|---|---|---|
| Valid | 1 | 5 | 29.4 | 29.4 | 29.4 |
| | 2 | 3 | 17.6 | 17.6 | 47.1 |
| | 3 | 1 | 5.9 | 5.9 | 52.9 |
| | 9 | 2 | 11.8 | 11.8 | 64.7 |
| | 11 | 1 | 5.9 | 5.9 | 70.6 |
| | 15 | 1 | 5.9 | 5.9 | 76.5 |
| | 17 | 1 | 5.9 | 5.9 | 82.4 |
| | 19 | 1 | 5.9 | 5.9 | 88.2 |
| | 22 | 1 | 5.9 | 5.9 | 94.1 |
| | 26 | 1 | 5.9 | 5.9 | 100.0 |
| | Total | 17 | 100.0 | 100.0 | |

*Tale C.2. Average Monthly Income per Household*

| NAD/month | | Frequency | Percent | Valid Percent | Cumulative Percent |
|---|---|---|---|---|---|
| Valid | N/A | 7 | 13.7 | 13.7 | 13.7 |
| | 1-1300 | 13 | 25.5 | 25.5 | 39.2 |
| | 1301-3900 | 14 | 27.5 | 27.5 | 66.7 |
| | 3901-more | 17 | 33.3 | 33.3 | 100.0 |
| | Total | 51 | 100.0 | 100.0 | |

*Table C.3. Ownership of Livestock*

| | | Frequency | Percent | Valid Percent | Cumulative Percent |
|---|---|---|---|---|---|
| Valid | No | 33 | 64.7 | 64.7 | 64.7 |
| | Chickens | 1 | 2.0 | 2.0 | 66.7 |
| | Cattle | 12 | 23.5 | 23.5 | 90.2 |
| | Chickens & cattle | 2 | 3.9 | 3.9 | 94.1 |
| | Goats | 1 | 2.0 | 2.0 | 96.1 |
| | Goats and cattle | 2 | 3.9 | 3.9 | 100.0 |
| | Total | 51 | 100.0 | 100.0 | |

*Table C.4. Ownership of Land*

| | | Frequency | Percent | Valid Percent | Cumulative Percent |
|---|---|---|---|---|---|
| Valid | No | 35 | 68.6 | 68.6 | 68.6 |
| | Plot | 3 | 5.9 | 5.9 | 74.5 |
| | Land | 11 | 21.6 | 21.6 | 96.1 |
| | Garden | 2 | 3.9 | 3.9 | 100.0 |
| | Total | 51 | 100.0 | 100.0 | |

*Table C.5. Distribution of Occupation by the Education*

| | | Highest School Degree | | | | |
|---|---|---|---|---|---|---|
| | | No Education | Primary School | Secondary School | Tertiary School | Total |
| Unem-ployed | Count | 9 | 9 | 0 | 0 | 18 |
| | % Occupation | 50,0% | 50,0% | 0,0% | 0,0% | 100,0% |
| | % within Degree | 56,3% | 39,1% | 0,0% | 0,0% | 35,3% |
| | % of Total | 17,6% | 17,6% | 0,0% | 0,0% | 35,3% |
| Self-Em-ployed | Count | 1 | 2 | 1 | 0 | 4 |
| | % Occupation | 25,0% | 50,0% | 25,0% | 0,0% | 100,0% |
| | % within Degree | 6,3% | 8,7% | 20,0% | 0,0% | 7,8% |
| | % of Total | 2,0% | 3,9% | 2,0% | 0,0% | 7,8% |
| State Em-ployee | Count | 4 | 4 | 1 | 5 | 14 |
| | % Occupation | 28,6% | 28,6% | 7,1% | 35,7% | 100,0% |
| | % within Degree | 25,0% | 17,4% | 20,0% | 71,4% | 27,5% |
| | % of Total | 7,8% | 7,8% | 2,0% | 9,8% | 27,5% |
| Private Sector Em-ployee | Count | 1 | 8 | 3 | 2 | 14 |
| | % Occupation | 7,1% | 57,1% | 21,4% | 14,3% | 100,0% |
| | % within Degree | 6,3% | 34,8% | 60,0% | 28,6% | 27,5% |
| | % of Total | 2,0% | 15,7% | 5,9% | 3,9% | 27,5% |
| Retired | Count | 1 | 0 | 0 | 0 | 1 |
| | % Occupation | 100,0% | 0,0% | 0,0% | 0,0% | 100,0% |
| | % within Degree | 6,3% | 0,0% | 0,0% | 0,0% | 2,0% |
| | % of Total | 2,0% | 0,0% | 0,0% | 0,0% | 2,0% |
| Total | Count | 16 | 23 | 5 | 7 | 51 |
| | % Occupation | 31,4% | 45,1% | 9,8% | 13,7% | 100,0% |
| | % within Degree | 100,0% | 100,0% | 100,0% | 100,0% | 100,0% |
| | % of Total | 31,4% | 45,1% | 9,8% | 13,7% | 100,0% |

*Table C.6. Wish for Children's Future Studies*

| | | Frequency | Percent | Valid Percent | Cumulative Percent |
|---|---|---|---|---|---|
| Valid | N/A | 12 | 23.5 | 23.5 | 23.5 |
| | Yes | 29 | 56.9 | 56.9 | 80.4 |
| | Yes, depends on finances | 3 | 5.9 | 5.9 | 86.3 |
| | Yes, depends on their success | 4 | 7.8 | 7.8 | 94.1 |
| | Yes, depends on circumstances | 3 | 5.9 | 5.9 | 100.0 |
| | Total | 51 | 100.0 | 100.0 | |

*Table C.7. Electricity Supply Satisfaction*

| | | Frequency | Percent | Valid Percent | Cumulative Percent |
|---|---|---|---|---|---|
| Valid | N/A | 4 | 7.8 | 7.8 | 7.8 |
| | Yes | 37 | 72.5 | 72.5 | 80.4 |
| | No | 3 | 5.9 | 5.9 | 86.3 |
| | Yes & No | 7 | 13.7 | 13.7 | 100.0 |
| | Total | 51 | 100.0 | 100.0 | |

*Table C.8. Distribution of Different Appliances by Average Income*

| | | | Average Income per Household (NAD/month) | | | | Total |
|---|---|---|---|---|---|---|---|
| | | | N/A | 1-1300 | 1301-3900 | 3901-more | |
| Quantity of appliances per Household | 0-3 | Count | 2 | 6 | 3 | 0 | 11 |
| | | % Income | 28,6% | 46,2% | 21,4% | 0,0% | 21,6% |
| | | % of Total | 3,9% | 11,8% | 5,9% | 0,0% | 21,6% |
| | 4-6 | Count | 2 | 2 | 8 | 1 | 13 |
| | | % Income | 28,6% | 15,4% | 57,1% | 5,9% | 25,5% |
| | | % of Total | 3,9% | 3,9% | 15,7% | 2,0% | 25,5% |
| | 7-9 | Count | 2 | 2 | 3 | 10 | 17 |
| | | % Income | 28,6% | 15,4% | 21,4% | 58,8% | 33,3% |
| | | % of Total | 3,9% | 3,9% | 5,9% | 19,6% | 33,3% |
| | 10-12 | Count | 1 | 3 | 0 | 6 | 10 |
| | | % Income | 14,3% | 23,1% | 0,0% | 35,3% | 19,6% |
| | | % of Total | 2,0% | 5,9% | 0,0% | 11,8% | 19,6% |
| Total | | Count | 7 | 13 | 14 | 17 | 51 |
| | | % Income | 100,0% | 100,0% | 100,0% | 100,0% | 100,0% |
| | | % of Total | 13,7% | 25,5% | 27,5% | 33,3% | 100,0% |

*Table C.9. Distribution of Different Appliances by Occupation*

| | | Occupation | | | | | |
|---|---|---|---|---|---|---|---|
| | | Unemployed | Self-Employed | State Employee | Private Sector Employee | Retired | Total |
| 0-3 | Count | 6 | 1 | 1 | 2 | 1 | 11 |
| | % Occupation | 33,3% | 25,0% | 7,1% | 14,3% | 100,0% | 21,6% |
| | % of Total | 11,8% | 2,0% | 2,0% | 3,9% | 2,0% | 21,6% |
| 4-6 | Count | 5 | 1 | 3 | 4 | 0 | 13 |
| | % Occupation | 27,8% | 25,0% | 21,4% | 28,6% | 0,0% | 25,5% |
| | % of Total | 9,8% | 2,0% | 5,9% | 7,8% | 0,0% | 25,5% |
| 7-9 | Count | 5 | 1 | 5 | 6 | 0 | 17 |
| | % Occupation | 27,8% | 25,0% | 35,7% | 42,9% | 0,0% | 33,3% |
| | % of Total | 9,8% | 2,0% | 9,8% | 11,8% | 0,0% | 33,3% |
| 10-12 | Count | 2 | 1 | 5 | 2 | 0 | 10 |
| | % Occupation | 11,1% | 25,0% | 35,7% | 14,3% | 0,0% | 19,6% |
| | % of Total | 3,9% | 2,0% | 9,8% | 3,9% | 0,0% | 19,6% |
| Total | Count | 18 | 4 | 14 | 14 | 1 | 51 |
| | % Occupation | 100,0% | 100,0% | 100,0% | 100,0% | 100,0% | 100,0% |
| | % of Total | 35,3% | 7,8% | 27,5% | 27,5% | 2,0% | 100,0% |

*Table C.10. Distribution of Different Appliances by Electricity Expenses*

| | | Electricity Expenses | | | | | Total |
|---|---|---|---|---|---|---|---|
| | | N/A | 0-150 | 151-300 | 301-450 | 451-700 | |
| 0-3 | Count | 0 | 9 | 2 | 0 | 0 | 11 |
| | % Expense | 0,0% | 31,0% | 16,7% | 0,0% | 0,0% | 21,6% |
| | % of Total | 0,0% | 17,6% | 3,9% | 0,0% | 0,0% | 21,6% |
| 4-6 | Count | 1 | 11 | 1 | 0 | 0 | 13 |
| | % Expense | 100,0% | 37,9% | 8,3% | 0,0% | 0,0% | 25,5% |
| | % of Total | 2,0% | 21,6% | 2,0% | 0,0% | 0,0% | 25,5% |
| 7-9 | Count | 0 | 7 | 6 | 3 | 1 | 17 |
| | % Expense | 0,0% | 24,1% | 50,0% | 60,0% | 25,0% | 33,3% |
| | % of Total | 0,0% | 13,7% | 11,8% | 5,9% | 2,0% | 33,3% |
| 10-12 | Count | 0 | 2 | 3 | 2 | 3 | 10 |
| | % Expense | 0,0% | 6,9% | 25,0% | 40,0% | 75,0% | 19,6% |
| | % of Total | 0,0% | 3,9% | 5,9% | 3,9% | 5,9% | 19,6% |
| Total | Count | 1 | 29 | 12 | 5 | 4 | 51 |
| | % Expense | 100,0% | 100,0% | 100,0% | 100,0% | 100,0% | 100,0% |
| | % of Total | 2,0% | 56,9% | 23,5% | 9,8% | 7,8% | 100,0% |

*Table C.11. Distribution of Printers Used in Local Businesses*

| | | Frequency | Percent | Valid Percent | Cumulative Percent |
|---|---|---|---|---|---|
| Valid | 0 | 10 | 58.8 | 58.8 | 58.8 |
| | 1 | 4 | 23.5 | 23.5 | 82.4 |
| | 2 | 1 | 5.9 | 5.9 | 88.2 |
| | 3 | 1 | 5.9 | 5.9 | 94.1 |
| | 15 | 1 | 5.9 | 5.9 | 100.0 |
| | Total | 17 | 100.0 | 100.0 | |

*Table C.12. Distribution of Televisions Used in Local Businesses*

| | | Frequency | Percent | Valid Percent | Cumulative Percent |
|---|---|---|---|---|---|
| Valid | 0 | 9 | 52.9 | 52.9 | 52.9 |
| | 1 | 7 | 41.2 | 41.2 | 94.1 |
| | 10 | 1 | 5.9 | 5.9 | 100.0 |
| | Total | 17 | 100.0 | 100.0 | |

*Table C.13. Distribution of Computers Used in Local Businesses*

| | | Frequency | Percent | Valid Percent | Cumulative Percent |
|---|---|---|---|---|---|
| Valid | 0 | 9 | 52.9 | 52.9 | 52.9 |
| | 1 | 1 | 5.9 | 5.9 | 58.8 |
| | 2 | 3 | 17.6 | 17.6 | 76.5 |
| | 4 | 1 | 5.9 | 5.9 | 82.4 |
| | 5 | 1 | 5.9 | 5.9 | 88.2 |
| | 8 | 1 | 5.9 | 5.9 | 94.1 |
| | 25 | 1 | 5.9 | 5.9 | 100.0 |
| | Total | 17 | 100.0 | 100.0 | |

*Table C.14. Distribution of Digital Payment by Education*

| | | Education | | | | Total |
|---|---|---|---|---|---|---|
| | | No Education | Primary School | Secondary School | Tertiary School | |
| N/A | Count | 13 | 13 | 1 | 3 | 30 |
| | % Education | 81,3% | 56,5% | 20,0% | 42,9% | 58,8% |
| | % of Total | 25,5% | 25,5% | 2,0% | 5,9% | 58,8% |
| Online Banking | Count | 3 | 10 | 4 | 4 | 21 |
| | % Education | 18,8% | 43,5% | 80,0% | 57,1% | 41,2% |
| | % of Total | 5,9% | 19,6% | 7,8% | 7,8% | 41,2% |
| Total | Count | 16 | 23 | 5 | 7 | 51 |
| | % Education | 100,0% | 100,0% | 100,0% | 100,0% | 100,0% |
| | % of Total | 31,4% | 45,1% | 9,8% | 13,7% | 100,0% |

*Table C.15. Distribution of Digital Payment by Occupation*

| | | Occupation | | | | | |
|---|---|---|---|---|---|---|---|
| | | Unem-ployed | Self-Employed | State Employee | Private Sector Employee | Retired | Total |
| N/A | Count | 15 | 2 | 4 | 8 | 1 | 30 |
| | % Occupatic | 83,30% | 50,00% | 28,60% | 57,10% | 100,00% | 58,80% |
| | % of Total | 29,40% | 3,90% | 7,80% | 15,70% | 2,00% | 58,80% |
| Online Bank-ing | Count | 3 | 2 | 10 | 6 | 0 | 21 |
| | % Occupatic | 16,70% | 50,00% | 71,40% | 42,90% | 0,00% | 41,20% |
| | % of Total | 5,90% | 3,90% | 19,60% | 11,80% | 0,00% | 41,20% |
| Total | Count | 18 | 4 | 14 | 14 | 1 | 51 |
| | % Occupatic | 100,00% | 100,00% | 100,00% | 100,00% | 100,00% | 100,00% |
| | % of Total | 35,30% | 7,80% | 27,50% | 27,50% | 2,00% | 100,00% |

*Table C.16. Distribution of Digital Security Cameras*

| | | Frequency | Percent | Valid Percent | Cumulative Percent |
|---|---|---|---|---|---|
| Valid | 0 | 15 | 88.2 | 88.2 | 88.2 |
| | 8 | 1 | 5.9 | 5.9 | 94.1 |
| | 13 | 1 | 5.9 | 5.9 | 100.0 |
| | Total | 17 | 100.0 | 100.0 | |

*Table C.17. Appliances Wished in the Future at Local Businesses*

| Future Appliances / Priority | No. 1 | No. 2 | No. 3 |
|---|---|---|---|
| Fridge | 18% | 6% | |
| Cooling Systems (A/C and fans) | 12% | 6% | |
| Internet servers | 6% | | |
| Broadcasting Equipment (cover-age and main control) | 6% | | |
| Television | 6% | 6% | |
| Computers | 6% | 6% | |
| Heater | | 6% | |
| Electric stove | | 6% | |
| Water pump | | 6% | |
| Kettle | | | 6% |
| Oven | | | 6% |
| Nothing | 47% | 59% | 88% |
| Total | 1 | 1 | 1 |

*Table C.18. Appliances Wished in the Future in Households*

| Future Appliances / Priority | No. 1 | No. 2 | No. 3 |
|---|---|---|---|
| Television | 16% | 5% | 11% |
| Fridge | 9% | 13% | 9% |
| Music Player/System/hi-fi/radio | 9% | 13% | 2% |
| Fan | 9% | 4% | 4% |
| Lights | 5% | 0% | 2% |
| Heater | 5% | 2% | 0 |
| Electrical stove | 4% | 7% | 2% |
| Washing Machine | 5% | 5% | 2% |
| Laptop/IPAD/PC | 4% | 2% | 2% |
| Microwave | 2% | 4% | 2% |
| A/C | 2% | 2% | |
| Sewing Machine | 2% | 0 | |
| Deep-Freezer | | 4% | 2% |
| Kettle | | 2% | 2% |
| Wi-fi | 2% | | |
| Iron | 2% | | |
| Welding Machine | 2% | 2% | |
| Water pressure | 2% | | |
| Geezer | 2% | | |
| Water | | 2% | |
| Grinder | | 2% | |
| Phone | | 2% | |
| Fax | | 2% | |
| Printer | | 2% | 0 |
| Oven | | 0 | 4% |
| Satellite | | 0 | 2% |
| Nothing | 20% | 29% | 57% |
| Total | 1 | 1 | 1 |

*Table C.19. Distribution of Electricity Expenses by Occupation*

| | | Occupation | | | | | |
|---|---|---|---|---|---|---|---|
| | | Unemployed | Self-Employed | State Employee | Private Sector Employee | Retired | Total |
| N/A | Count | 0 | 0 | 0 | 1 | 0 | 1 |
| | % Occupation | 0,0% | 0,0% | 0,0% | 7,1% | 0,0% | 2,0% |
| | % of Total | 0,0% | 0,0% | 0,0% | 2,0% | 0,0% | 2,0% |
| 0-150 | Count | 12 | 3 | 7 | 6 | 1 | 29 |
| | % Occupation | 66,7% | 75,0% | 50,0% | 42,9% | 100,0% | 56,9% |
| | % of Total | 23,5% | 5,9% | 13,7% | 11,8% | 2,0% | 56,9% |
| 151-300 | Count | 2 | 0 | 4 | 6 | 0 | 12 |
| | % Occupation | 11,1% | 0,0% | 28,6% | 42,9% | 0,0% | 23,5% |
| | % of Total | 3,9% | 0,0% | 7,8% | 11,8% | 0,0% | 23,5% |
| 301-450 | Count | 3 | 0 | 1 | 1 | 0 | 5 |
| | % Occupation | 16,7% | 0,0% | 7,1% | 7,1% | 0,0% | 9,8% |
| | % of Total | 5,9% | 0,0% | 2,0% | 2,0% | 0,0% | 9,8% |
| 451-700 | Count | 1 | 1 | 2 | 0 | 0 | 4 |
| | % Occupation | 5,6% | 25,0% | 14,3% | 0,0% | 0,0% | 7,8% |
| | % of Total | 2,0% | 2,0% | 3,9% | 0,0% | 0,0% | 7,8% |
| Total | Count | 18 | 4 | 14 | 14 | 1 | 51 |
| | % Occupation | 100,0% | 100,0% | 100,0% | 100,0% | 100,0% | 100,0% |
| | % of Total | 35,3% | 7,8% | 27,5% | 27,5% | 2,0% | 100,0% |

*Table C.20. Distribution of Electricity Expenses by Rooms*

| | | | Number of Rooms | | | | Total |
|---|---|---|---|---|---|---|---|
| | | | 1 | 2 | 3 | 4 | |
| Monthly Electricity Expense | N/A | Count | 1 | 0 | 0 | 0 | 1 |
| | | % n Rooms | 7,1% | 0,0% | 0,0% | 0,0% | 2,0% |
| | | % of Total | 2,0% | 0,0% | 0,0% | 0,0% | 2,0% |
| | 0-150 | Count | 9 | 13 | 3 | 4 | 29 |
| | | % Rooms | 64,3% | 72,2% | 25,0% | 57,1% | 56,9% |
| | | % of Total | 17,6% | 25,5% | 5,9% | 7,8% | 56,9% |
| | 151-300 | Count | 4 | 3 | 5 | 0 | 12 |
| | | % Rooms | 28,6% | 16,7% | 41,7% | 0,0% | 23,5% |
| | | % of Total | 7,8% | 5,9% | 9,8% | 0,0% | 23,5% |
| | 301-450 | Count | 0 | 0 | 3 | 2 | 5 |
| | | % Rooms | 0,0% | 0,0% | 25,0% | 28,6% | 9,8% |
| | | % of Total | 0,0% | 0,0% | 5,9% | 3,9% | 9,8% |
| | 451-700 | Count | 0 | 2 | 1 | 1 | 4 |
| | | % Rooms | 0,0% | 11,1% | 8,3% | 14,3% | 7,8% |
| | | % of Total | 0,0% | 3,9% | 2,0% | 2,0% | 7,8% |
| Total | | Count | 14 | 18 | 12 | 7 | 51 |
| | | % Rooms | 100,0% | 100,0% | 100,0% | 100,0% | 100,0% |
| | | % of Total | 27,5% | 35,3% | 23,5% | 13,7% | 100,0% |

*Table C.21. Quantity of Different Electric Appliances per Household*

| | | Frequency | Percent | Valid Percent | Cumulative Percent |
|---|---|---|---|---|---|
| Valid | 1 | 2 | 3.9 | 3.9 | 3.9 |
| | 2 | 7 | 13.7 | 13.7 | 17.6 |
| | 3 | 2 | 3.9 | 3.9 | 21.6 |
| | 4 | 3 | 5.9 | 5.9 | 27.5 |
| | 5 | 7 | 13.7 | 13.7 | 41.2 |
| | 6 | 3 | 5.9 | 5.9 | 47.1 |
| | 7 | 6 | 11.8 | 11.8 | 58.8 |
| | 8 | 9 | 17.6 | 17.6 | 76.5 |
| | 9 | 2 | 3.9 | 3.9 | 80.4 |
| | 10 | 7 | 13.7 | 13.7 | 94.1 |
| | 11 | 2 | 3.9 | 3.9 | 98.0 |
| | 12 | 1 | 2.0 | 2.0 | 100.0 |
| | Total | 51 | 100.0 | 100.0 | |

*Table C.22. Quantity of Light Bulbs per Household*

| | | Frequency | Percent | Valid Percent | Cumulative Percent |
|---|---|---|---|---|---|
| Valid | 0 | 1 | 2.0 | 2.0 | 2.0 |
| | 1 | 13 | 25.5 | 25.5 | 27.5 |
| | 2 | 16 | 31.4 | 31.4 | 58.8 |
| | 3 | 3 | 5.9 | 5.9 | 64.7 |
| | 4 | 6 | 11.8 | 11.8 | 76.5 |
| | 5 | 4 | 7.8 | 7.8 | 84.3 |
| | 6 | 1 | 2.0 | 2.0 | 86.3 |
| | 7 | 1 | 2.0 | 2.0 | 88.2 |
| | 8 | 3 | 5.9 | 5.9 | 94.1 |
| | 10 | 2 | 3.9 | 3.9 | 98.0 |
| | 15 | 1 | 2.0 | 2.0 | 100.0 |
| | Total | 51 | 100.0 | 100.0 | |

*Table C.23. Distribution of Mobile Phones by Size of the Households*

| | | Number of Rooms 1 | 2 | 3 | 4 | Total |
|---|---|---|---|---|---|---|
| 0 | Count | 0 | 1 | 0 | 0 | 1 |
| | % within Rooms | 0,0% | 5,6% | 0,0% | 0,0% | 2,0% |
| | % of Total | 0,0% | 2,0% | 0,0% | 0,0% | 2,0% |
| 1 | Count | 9 | 4 | 3 | 1 | 17 |
| | % within Rooms | 64,3% | 22,2% | 25,0% | 14,3% | 33,3% |
| | % of Total | 17,6% | 7,8% | 5,9% | 2,0% | 33,3% |
| 2 | Count | 4 | 7 | 3 | 2 | 16 |
| | % within Rooms | 28,6% | 38,9% | 25,0% | 28,6% | 31,4% |
| | % of Total | 7,8% | 13,7% | 5,9% | 3,9% | 31,4% |
| 3 | Count | 0 | 3 | 2 | 1 | 6 |
| | % within Rooms | 0,0% | 16,7% | 16,7% | 14,3% | 11,8% |
| | % of Total | 0,0% | 5,9% | 3,9% | 2,0% | 11,8% |
| 4 | Count | 0 | 1 | 2 | 0 | 3 |
| | % within Rooms | 0,0% | 5,6% | 16,7% | 0,0% | 5,9% |
| | % of Total | 0,0% | 2,0% | 3,9% | 0,0% | 5,9% |
| 5 | Count | 1 | 2 | 1 | 1 | 5 |
| | % within Rooms | 7,1% | 11,1% | 8,3% | 14,3% | 9,8% |
| | % of Total | 2,0% | 3,9% | 2,0% | 2,0% | 9,8% |
| 6 | Count | 0 | 0 | 1 | 2 | 3 |
| | % within Rooms | 0,0% | 0,0% | 8,3% | 28,6% | 5,9% |
| | % of Total | 0,0% | 0,0% | 2,0% | 3,9% | 5,9% |
| Total | Count | 14 | 18 | 12 | 7 | 51 |
| | % within Rooms | 100,0% | 100,0% | 100,0% | 100,0% | 100,0% |
| | % of Total | 27,5% | 35,3% | 23,5% | 13,7% | 100,0% |

*Table C.24Distribution of Different Appliances per Local Business*

| | | Frequency | Percent | Valid Percent | Cumulative Percent |
|---|---|---|---|---|---|
| Valid | 1 | 1 | 5.9 | 5.9 | 5.9 |
| | 2 | 1 | 5.9 | 5.9 | 11.8 |
| | 3 | 3 | 17.6 | 17.6 | 29.4 |
| | 4 | 1 | 5.9 | 5.9 | 35.3 |
| | 5 | 2 | 11.8 | 11.8 | 47.1 |
| | 6 | 3 | 17.6 | 17.6 | 64.7 |
| | 7 | 1 | 5.9 | 5.9 | 70.6 |
| | 9 | 2 | 11.8 | 11.8 | 82.4 |
| | 11 | 1 | 5.9 | 5.9 | 88.2 |
| | 12 | 1 | 5.9 | 5.9 | 94.1 |
| | 19 | 1 | 5.9 | 5.9 | 100.0 |
| | Total | 17 | 100.0 | 100.0 | |

*Table C.25. Distribution of Cooling System by Type of Business*

| | | | Type | | Total |
|---|---|---|---|---|---|
| | | | Private | Public | |
| Cooling System | No | Count | 8 | 2 | 10 |
| | | % within Type | 80.0% | 28.6% | 58.8% |
| | | % of Total | 47.1% | 11.8% | 58.8% |
| | Yes | Count | 2 | 5 | 7 |
| | | % within Type | 20.0% | 71.4% | 41.2% |
| | | % of Total | 11.8% | 29.4% | 41.2% |
| Total | | Count | 10 | 7 | 17 |
| | | % within Type | 100.0% | 100.0% | 100.0% |
| | | % of Total | 58.8% | 41.2% | 100.0% |

*Table C.26. Distribution of Digital Payment by Average Income*

| | | Average Income | | | | Total |
|---|---|---|---|---|---|---|
| | | N/A | 0-1300 | 1301-3900 | 3901-more | |
| N/A | Count | 6 | 11 | 9 | 4 | 30 |
| | % Income | 85,7% | 84,6% | 64,3% | 23,5% | 58,8% |
| | % of Total | 11,8% | 21,6% | 17,6% | 7,8% | 58,8% |
| Online Banking | Count | 1 | 2 | 5 | 13 | 21 |
| | % Income | 14,3% | 15,4% | 35,7% | 76,5% | 41,2% |
| | % of Total | 2,0% | 3,9% | 9,8% | 25,5% | 41,2% |
| Total | Count | 7 | 13 | 14 | 17 | 51 |
| | % Income | 100,0% | 100,0% | 100,0% | 100,0% | 100,0% |
| | % of Total | 13,7% | 25,5% | 27,5% | 33,3% | 100,0% |

# Appendix D. Additional Tables of Chapter 5

*Table D.1. Distribution of Rooms*

| | | Frequency | Percent | Valid Percent | Cumulative Percent |
|---|---|---|---|---|---|
| Valid | 1 | 14 | 50.0 | 50.0 | 50.0 |
| | 2 | 9 | 32.1 | 32.1 | 82.1 |
| | 3 | 3 | 10.7 | 10.7 | 92.9 |
| | 4 | 1 | 3.6 | 3.6 | 96.4 |
| | 6 | 1 | 3.6 | 3.6 | 100.0 |
| | Total | 28 | 100.0 | 100.0 | |

*Table D.2. Distribution of Employment by Education*

| | | Education | | | | |
|---|---|---|---|---|---|---|
| | | No education | Primary | Secondary | Tertiary | Total |
| Unemployed | Count | 2 | 8 | 2 | 1 | 13 |
| | % Education | 33,3% | 61,5% | 40,0% | 25,0% | 46,4% |
| | % of Total | 7,1% | 28,6% | 7,1% | 3,6% | 46,4% |
| State Employee | Count | 1 | 1 | 0 | 1 | 3 |
| | % Education | 16,7% | 7,7% | 0,0% | 25,0% | 10,7% |
| | % of Total | 3,6% | 3,6% | 0,0% | 3,6% | 10,7% |
| Private Sector Employee | Count | 0 | 0 | 1 | 2 | 3 |
| | % Education | 0,0% | 0,0% | 20,0% | 50,0% | 10,7% |
| | % of Total | 0,0% | 0,0% | 3,6% | 7,1% | 10,7% |
| Self-Employed | Count | 0 | 4 | 2 | 0 | 6 |
| | % Education | 0,0% | 30,8% | 40,0% | 0,0% | 21,4% |
| | % of Total | 0,0% | 14,3% | 7,1% | 0,0% | 21,4% |
| Retired | Count | 3 | 0 | 0 | 0 | 3 |
| | % Education | 50,0% | 0,0% | 0,0% | 0,0% | 10,7% |
| | % of Total | 10,7% | 0,0% | 0,0% | 0,0% | 10,7% |
| Total | Count | 6 | 13 | 5 | 4 | 28 |
| | % Education | 100,0% | 100,0 | 100,0% | 100,0 | 100,0% |
| | % of Total | 21,4% | 46,4% | 17,9% | 14,3% | 100,0% |

*Table D.3. Financial Need from Family, Friends, or Financial Institutions*

| | | Frequency | Percent | Valid Percent | Cumulative Percent |
|---|---|---|---|---|---|
| Valid | No | 18 | 64.3 | 64.3 | 64.3 |
| | Yes | 10 | 35.7 | 35.7 | 100.0 |
| | Total | 28 | 100.0 | 100.0 | |

*Table D.4. Distribution of Mobile Phone by Education*

| | | Education | | | | Total |
|---|---|---|---|---|---|---|
| | | No Education | Primary | Secondary | Tertiary | |
| 0 | Count | 0 | 1 | 0 | 1 | 2 |
| | % Education | 0,0% | 7,7% | 0,0% | 25,0% | 7,1% |
| | % of Total | 0,0% | 3,6% | 0,0% | 3,6% | 7,1% |
| 1 | Count | 3 | 3 | 2 | 2 | 10 |
| | % Education | 50,0% | 23,1% | 40,0% | 50,0% | 35,7% |
| | % of Total | 10,7% | 10,7% | 7,1% | 7,1% | 35,7% |
| 2 | Count | 2 | 3 | 1 | 1 | 7 |
| | % Education | 33,3% | 23,1% | 20,0% | 25,0% | 25,0% |
| | % of Total | 7,1% | 10,7% | 3,6% | 3,6% | 25,0% |
| 3 | Count | 1 | 2 | 0 | 0 | 3 |
| | % Education | 16,7% | 15,4% | 0,0% | 0,0% | 10,7% |
| | % of Total | 3,6% | 7,1% | 0,0% | 0,0% | 10,7% |
| 4 | Count | 0 | 3 | 1 | 0 | 4 |
| | % Education | 0,0% | 23,1% | 20,0% | 0,0% | 14,3% |
| | % of Total | 0,0% | 10,7% | 3,6% | 0,0% | 14,3% |
| 6 | Count | 0 | 1 | 0 | 0 | 1 |
| | % Education | 0,0% | 7,7% | 0,0% | 0,0% | 3,6% |
| | % of Total | 0,0% | 3,6% | 0,0% | 0,0% | 3,6% |
| 7 | Count | 0 | 0 | 1 | 0 | 1 |
| | % Education | 0,0% | 0,0% | 20,0% | 0,0% | 3,6% |
| | % of Total | 0,0% | 0,0% | 3,6% | 0,0% | 3,6% |
| Total | Count | 6 | 13 | 5 | 4 | 28 |
| | % Education | 100,0% | 100,0% | 100,0% | 100,0% | 100,0% |
| | % of Total | 21,4% | 46,4% | 17,9% | 14,3% | 100,0% |

*Table D.5. Distribution of Mobile Phones by Income*

| Mobile Phone | | Income | | | | Total |
|---|---|---|---|---|---|---|
| | | No income | 1-1300 | 1301-3900 | 3901-more | |
| 0 | Count | 1 | 1 | 0 | 0 | 2 |
| | % Income | 16,7% | 7,1% | 0,0% | 0,0% | 7,1% |
| | % of Total | 3,6% | 3,6% | 0,0% | 0,0% | 7,1% |
| 1 | Count | 2 | 6 | 1 | 1 | 10 |
| | % Income | 33,3% | 42,9% | 33,3% | 20,0% | 35,7% |
| | % of Total | 7,1% | 21,4% | 3,6% | 3,6% | 35,7% |
| 2 | Count | 1 | 3 | 1 | 2 | 7 |
| | % Income | 16,7% | 21,4% | 33,3% | 40,0% | 25,0% |
| | % of Total | 3,6% | 10,7% | 3,6% | 7,1% | 25,0% |
| 3 | Count | 0 | 1 | 0 | 2 | 3 |
| | % Income | 0,0% | 7,1% | 0,0% | 40,0% | 10,7% |
| | % of Total | 0,0% | 3,6% | 0,0% | 7,1% | 10,7% |
| 4 | Count | 1 | 2 | 1 | 0 | 4 |
| | % Income | 16,7% | 14,3% | 33,3% | 0,0% | 14,3% |
| | % of Total | 3,6% | 7,1% | 3,6% | 0,0% | 14,3% |
| 6 | Count | 0 | 1 | 0 | 0 | 1 |
| | % Income | 0,0% | 7,1% | 0,0% | 0,0% | 3,6% |
| | % of Total | 0,0% | 3,6% | 0,0% | 0,0% | 3,6% |
| 7 | Count | 1 | 0 | 0 | 0 | 1 |
| | % Income | 16,7% | 0,0% | 0,0% | 0,0% | 3,6% |
| | % of Total | 3,6% | 0,0% | 0,0% | 0,0% | 3,6% |
| Total | Count | 6 | 14 | 3 | 5 | 28 |
| | % Income | 100,0% | 100,0% | 100,0% | 100,0% | 100,0% |
| | % of Total | 21,4% | 50,0% | 10,7% | 17,9% | 100,0% |

*Table D.6. Distribution of Mobile Phones by Occupation*

| | | Occupation | | | | | Total |
|---|---|---|---|---|---|---|---|
| | | Unemployed | State Employee | Private Sector Employee | Self-Employed | Retired | |
| 0 | Count | 2 | 0 | 0 | 0 | 0 | 2 |
| | % Occupation | 15,4% | 0,0% | 0,0% | 0,0% | 0,0% | 7,1% |
| | % of Total | 7,1% | 0,0% | 0,0% | 0,0% | 0,0% | 7,1% |
| 1 | Count | 5 | 1 | 1 | 0 | 3 | 10 |
| | % Occupation | 38,5% | 33,3% | 33,3% | 0,0% | 100,0% | 35,7% |
| | % of Total | 17,9% | 3,6% | 3,6% | 0,0% | 10,7% | 35,7% |
| 2 | Count | 3 | 0 | 1 | 3 | 0 | 7 |
| | % Occupation | 23,1% | 0,0% | 33,3% | 50,0% | 0,0% | 25,0% |
| | % of Total | 10,7% | 0,0% | 3,6% | 10,7% | 0,0% | 25,0% |
| 3 | Count | 0 | 1 | 0 | 2 | 0 | 3 |
| | % Occupation | 0,0% | 33,3% | 0,0% | 33,3% | 0,0% | 10,7% |
| | % of Total | 0,0% | 3,6% | 0,0% | 7,1% | 0,0% | 10,7% |
| 4 | Count | 2 | 1 | 0 | 1 | 0 | 4 |
| | % Occupation | 15,4% | 33,3% | 0,0% | 16,7% | 0,0% | 14,3% |
| | % of Total | 7,1% | 3,6% | 0,0% | 3,6% | 0,0% | 14,3% |
| 6 | Count | 1 | 0 | 0 | 0 | 0 | 1 |
| | % Occupation | 7,7% | 0,0% | 0,0% | 0,0% | 0,0% | 3,6% |
| | % of Total | 3,6% | 0,0% | 0,0% | 0,0% | 0,0% | 3,6% |
| 7 | Count | 0 | 0 | 1 | 0 | 0 | 1 |
| | % Occupation | 0,0% | 0,0% | 33,3% | 0,0% | 0,0% | 3,6% |
| | % of Total | 0,0% | 0,0% | 3,6% | 0,0% | 0,0% | 3,6% |
| Total | Count | 13 | 3 | 3 | 6 | 3 | 28 |
| | % Occupation | 100,0% | 100,0% | 100,0% | 100,0% | 100,0% | 100,0% |
| | % of Total | 46,4% | 10,7% | 10,7% | 21,4% | 10,7% | 100,0% |

*Table D.7. Descriptive Statistics of Different Appliances at Local Businesses*

| | N | Minimum | Maximum | Mean | Std. Deviation |
|---|---|---|---|---|---|
| Total Different Appliances | 9 | 3 | 10 | 6.67 | 2.000 |
| Lights (Fluorescent) | 9 | 0 | 50 | 7.56 | 16.394 |
| Lights (Bulbs) | 9 | 1 | 23 | 7.56 | 7.485 |
| Fridge | 9 | 0 | 5 | 1.56 | 1.944 |
| Deep Freezer | 9 | 0 | 2 | 0.33 | 0.707 |
| Electric Cash Box | 9 | 0 | 1 | 0.33 | 0.500 |
| Hot Plate | 9 | 0 | 1 | 0.11 | 0.333 |
| Radio | 9 | 0 | 1 | 0.22 | 0.441 |
| Computer - Laptop | 9 | 0 | 40 | 8.11 | 13.824 |
| Fan | 9 | 0 | 2 | 0.67 | 0.707 |
| Electric Stove | 9 | 0 | 1 | 0.11 | 0.333 |
| Copy Machine | 9 | 0 | 3 | 0.44 | 1.014 |
| Electric Heater | 9 | 0 | 1 | 0.11 | 0.333 |
| Printer | 9 | 0 | 2 | 0.89 | 0.601 |
| Digital Security Camera | 9 | 0 | 5 | 1.11 | 2.205 |
| Monitor | 9 | 0 | 6 | 1.89 | 2.261 |
| Air Conditioner | 9 | 0 | 1 | 0.22 | 0.441 |
| Television | 9 | 0 | 2 | 0.22 | 0.667 |
| Iron | 9 | 0 | 1 | 0.11 | 0.333 |
| Valid N (listwise) | 9 | | | | |

*Table D.8. Distribution of Light Bulb or Fluorescent Light*

| | | Frequency | Percent | Valid Percent | Cumulative Percent |
|---|---|---|---|---|---|
| Valid | 1 | 3 | 33.3 | 33.3 | 33.3 |
| | 2 | 1 | 11.1 | 11.1 | 44.4 |
| | 7 | 1 | 11.1 | 11.1 | 55.6 |
| | 9 | 1 | 11.1 | 11.1 | 66.7 |
| | 10 | 1 | 11.1 | 11.1 | 77.8 |
| | 14 | 1 | 11.1 | 11.1 | 88.9 |
| | 23 | 1 | 11.1 | 11.1 | 100.0 |
| | Total | 9 | 100.0 | 100.0 | |

*Table D.9. Distribution of Digital Payment by Education*

| | | Education | | | | Total |
|---|---|---|---|---|---|---|
| | | No Education | Primary | Secondary | Tertiary | |
| Cash | Count | 5 | 12 | 4 | 2 | 23 |
| | % within Education | 83,3% | 92,3% | 80,0% | 50,0% | 82,1% |
| | % of Total | 17,9% | 42,9% | 14,3% | 7,1% | 82,1% |
| Cash and Digital Payment | Count | 1 | 1 | 1 | 2 | 5 |
| | % within Education | 16,7% | 7,7% | 20,0% | 50,0% | 17,9% |
| | % of Total | 3,6% | 3,6% | 3,6% | 7,1% | 17,9% |
| Total | Count | 6 | 13 | 5 | 4 | 28 |
| | % within Education | 100,0% | 100,0% | 100,0% | 100,0% | 100,0% |
| | % of Total | 21,4% | 46,4% | 17,9% | 14,3% | 100,0% |

*Table D.10. Distribution of Digital Payment by Income*

| | | Income | | | | Total |
|---|---|---|---|---|---|---|
| | | No Income | 1-1300 | 1301-3900 | 3901-more | |
| Cash | Count | 6 | 14 | 1 | 2 | 23 |
| | % within Income | 100,0% | 100,0% | 33,3% | 40,0% | 82,1% |
| | % of Total | 21,4% | 50,0% | 3,6% | 7,1% | 82,1% |
| Cash and Digital Payment | Count | 0 | 0 | 2 | 3 | 5 |
| | % within Income | 0,0% | 0,0% | 66,7% | 60,0% | 17,9% |
| | % of Total | 0,0% | 0,0% | 7,1% | 10,7% | 17,9% |
| Total | Count | 6 | 14 | 3 | 5 | 28 |
| | % within Income | 100,0% | 100,0% | 100,0% | 100,0% | 100,0% |
| | % of Total | 21,4% | 50,0% | 10,7% | 17,9% | 100,0% |

*Table D.11. Distribution of Digital Payment by Occupation*

| | | Occupation | | | | | Total |
|---|---|---|---|---|---|---|---|
| | | Unem-ployed | State Employe | Private Sector Employee | Self-Employed | Retired | |
| Cash | Count | 13 | 0 | 2 | 5 | 3 | 23 |
| | % Occupation | 100,0% | 0,0% | 66,7% | 83,3% | 100,0% | 82,1% |
| | % of Total | 46,4% | 0,0% | 7,1% | 17,9% | 10,7% | 82,1% |
| Cash and Digital Payment | Count | 0 | 3 | 1 | 1 | 0 | 5 |
| | % Occupation | 0,0% | 100,0% | 33,3% | 16,7% | 0,0% | 17,9% |
| | % of Total | 0,0% | 10,7% | 3,6% | 3,6% | 0,0% | 17,9% |
| Total | Count | 13 | 3 | 3 | 6 | 3 | 28 |
| | % Occupation | 100,0% | 100,0% | 100,0% | 100,0% | 100,0% | 100,0% |
| | % of Total | 46,4% | 10,7% | 10,7% | 21,4% | 10,7% | 100,0% |

*Table D.12. Distribution of Ownership by Education*

| | | Education | | | | Total |
|---|---|---|---|---|---|---|
| | | No Education | Primary | Secondary | Tertiary | |
| Yes | Count | 3 | 7 | 3 | 3 | 16 |
| | % within Education | 50,0% | 53,8% | 60,0% | 75,0% | 57,1% |
| | % of Total | 10,7% | 25,0% | 10,7% | 10,7% | 57,1% |
| No | Count | 3 | 6 | 2 | 1 | 12 |
| | % within Education | 50,0% | 46,2% | 40,0% | 25,0% | 42,9% |
| | % of Total | 10,7% | 21,4% | 7,1% | 3,6% | 42,9% |
| Total | Count | 6 | 13 | 5 | 4 | 28 |
| | % within Education | 100,0% | 100,0% | 100,0% | 100,0% | 100,0% |
| | % of Total | 21,4% | 46,4% | 17,9% | 14,3% | 100,0% |

*Table D.13. Distribution of Ownership by Occupation*

| | | Occupation | | | | | |
|---|---|---|---|---|---|---|---|
| | | Unem-ployed | State Employee | Private Sector Employee | Self-Employed | Retired | Total |
| Yes | Count | 5 | 2 | 3 | 5 | 1 | 16 |
| | % Occupation | 38,5% | 66,7% | 100,0% | 83,3% | 33,3% | 57,1% |
| | % of Total | 17,9% | 7,1% | 10,7% | 17,9% | 3,6% | 57,1% |
| No | Count | 8 | 1 | 0 | 1 | 2 | 12 |
| | % Occupation | 61,5% | 33,3% | 0,0% | 16,7% | 66,7% | 42,9% |
| | % of Total | 28,6% | 3,6% | 0,0% | 3,6% | 7,1% | 42,9% |
| Total | Count | 13 | 3 | 3 | 6 | 3 | 28 |
| | % Occupation | 100,0% | 100,0% | 100,0% | 100,0% | 100,0% | 100,0% |
| | % of Total | 46,4% | 10,7% | 10,7% | 21,4% | 10,7% | 100,0% |

*Table D.14. Distribution of Ownership by Income*

| | | | Average Income (NAD/month) | | | | |
|---|---|---|---|---|---|---|---|
| | | | No Income | 1-1300 | 1301-3900 | 3901-more | Total |
| Ownership of land and/or livestock | Yes | Count | 3 | 7 | 3 | 3 | 16 |
| | | % within Income | 50,0% | 50,0% | 100,0% | 60,0% | 57,1% |
| | | % of Total | 10,7% | 25,0% | 10,7% | 10,7% | 57,1% |
| | No | Count | 3 | 7 | 0 | 2 | 12 |
| | | % within Income | 50,0% | 50,0% | 0,0% | 40,0% | 42,9% |
| | | % of Total | 10,7% | 25,0% | 0,0% | 7,1% | 42,9% |
| Total | | Count | 6 | 14 | 3 | 5 | 28 |
| | | % within Income | 100,0% | 100,0% | 100,0% | 100,0% | 100,0% |
| | | % of Total | 21,4% | 50,0% | 10,7% | 17,9% | 100,0% |

*Table D.15. Distribution of Electricity Expenses by Occupation*

| | | Occupation | | | | | Total |
|---|---|---|---|---|---|---|---|
| | | Unemployed | State Employee | Private Sector Employee | Self-Employed | Retired | |
| N/A | Count | 7 | 0 | 0 | 1 | 2 | 10 |
| | % Occupation | 53,8% | 0,0% | 0,0% | 16,7% | 66,7% | 35,7% |
| | % of Total | 25,0% | 0,0% | 0,0% | 3,6% | 7,1% | 35,7% |
| 0-150 | Count | 4 | 1 | 1 | 1 | 0 | 7 |
| | % Occupation | 30,8% | 33,3% | 33,3% | 16,7% | 0,0% | 25,0% |
| | % of Total | 14,3% | 3,6% | 3,6% | 3,6% | 0,0% | 25,0% |
| 151-300 | Count | 1 | 2 | 2 | 1 | 1 | 7 |
| | % Occupation | 7,7% | 66,7% | 66,7% | 16,7% | 33,3% | 25,0% |
| | % of Total | 3,6% | 7,1% | 7,1% | 3,6% | 3,6% | 25,0% |
| 451-700 | Count | 1 | 0 | 0 | 3 | 0 | 4 |
| | % Occupation | 7,7% | 0,0% | 0,0% | 50,0% | 0,0% | 14,3% |
| | % of Total | 3,6% | 0,0% | 0,0% | 10,7% | 0,0% | 14,3% |
| Total | Count | 13 | 3 | 3 | 6 | 3 | 28 |
| | % Occupation | 100,0% | 100,0% | 100,0% | 100,0% | 100,0% | 100,( |
| | % of Total | 46,4% | 10,7% | 10,7% | 21,4% | 10,7% | 100,( |

*Table D.16. Distribution of Different Appliances by Rooms*

| | | Rooms | | | | | Total |
|---|---|---|---|---|---|---|---|
| | | 1 | 2 | 3 | 4 | 6 | |
| 1-3 | Count | 8 | 0 | 0 | 0 | 0 | 8 |
| | % Rooms | 57,1% | 0,0% | 0,0% | 0,0% | 0,0% | 28,6% |
| | % of Total | 28,6% | 0,0% | 0,0% | 0,0% | 0,0% | 28,6% |
| 4-6 | Count | 3 | 5 | 1 | 1 | 0 | 10 |
| | % Rooms | 21,4% | 55,6% | 33,3% | 100,0' | 0,0% | 35,7% |
| | % of Total | 10,7% | 17,9% | 3,6% | 3,6% | 0,0% | 35,7% |
| 7-9 | Count | 3 | 4 | 2 | 0 | 1 | 10 |
| | % Rooms | 21,4% | 44,4% | 66,7% | 0,0% | 100,0% | 35,7% |
| | % of Total | 10,7% | 14,3% | 7,1% | 0,0% | 3,6% | 35,7% |
| Total | Count | 14 | 9 | 3 | 1 | 1 | 28 |
| | % Rooms | 100,0' | 100,0' | 100,0' | 100,0' | 100,0% | 100,0% |
| | % of Total | 50,0% | 32,1% | 10,7% | 3,6% | 3,6% | 100,0% |

*Table D.17. Distribution of Monthly Electricity Expenses by Rooms*

| | | Rooms | | | | | Total |
|---|---|---|---|---|---|---|---|
| | | 1 | 2 | 3 | 4 | 6 | |
| N/A | Count | 7 | 3 | 0 | 0 | 0 | 10 |
| | % Rooms | 50,0% | 33,3% | 0,0% | 0,0% | 0,0% | 35,7% |
| | % of Total | 25,0% | 10,7% | 0,0% | 0,0% | 0,0% | 35,7% |
| 0-150 | Count | 2 | 2 | 1 | 1 | 1 | 7 |
| | % Rooms | 14,3% | 22,2% | 33,3% | 100,0% | 100,0% | 25,0% |
| | % of Total | 7,1% | 7,1% | 3,6% | 3,6% | 3,6% | 25,0% |
| 151-300 | Count | 4 | 3 | 0 | 0 | 0 | 7 |
| | % Rooms | 28,6% | 33,3% | 0,0% | 0,0% | 0,0% | 25,0% |
| | % of Total | 14,3% | 10,7% | 0,0% | 0,0% | 0,0% | 25,0% |
| 451-700 | Count | 1 | 1 | 2 | 0 | 0 | 4 |
| | % Rooms | 7,1% | 11,1% | 66,7% | 0,0% | 0,0% | 14,3% |
| | % of Total | 3,6% | 3,6% | 7,1% | 0,0% | 0,0% | 14,3% |
| Total | Count | 14 | 9 | 3 | 1 | 1 | 28 |
| | % Rooms | 100,0% | 100,0% | 100,0% | 100,0% | 100,0% | 100,0% |
| | % of Total | 50,0% | 32,1% | 10,7% | 3,6% | 3,6% | 100,0% |

*Table D.18. Distribution of Hot Plate/Electric Stove by Education*

| | | | Education | | | | Total |
|---|---|---|---|---|---|---|---|
| | | | No Education | Primary | Secondary | Tertiary | |
| Hot Plate/ Electric Stove | 0 | Count | 2 | 7 | 2 | 2 | 13 |
| | | % Education | 33,3% | 53,8% | 40,0% | 50,0% | 46,4% |
| | | % of Total | 7,1% | 25,0% | 7,1% | 7,1% | 46,4% |
| | 1 | Count | 4 | 6 | 3 | 1 | 14 |
| | | % Education | 66,7% | 46,2% | 60,0% | 25,0% | 50,0% |
| | | % of Total | 14,3% | 21,4% | 10,7% | 3,6% | 50,0% |
| | 5 | Count | 0 | 0 | 0 | 1 | 1 |
| | | % Education | 0,0% | 0,0% | 0,0% | 25,0% | 3,6% |
| | | % of Total | 0,0% | 0,0% | 0,0% | 3,6% | 3,6% |
| Total | | Count | 6 | 13 | 5 | 4 | 28 |
| | | % Education | 100,0% | 100,0% | 100,0% | 100,0% | 100,0% |
| | | % of Total | 21,4% | 46,4% | 17,9% | 14,3% | 100,0% |

*Table D.19. Distribution of Different Energy Appliances*

| | | Frequency | Percent | Valid Percent | Cumulative Percent |
|---|---|---|---|---|---|
| Valid | 1 | 1 | 3.6 | 3.6 | 3.6 |
| | 2 | 3 | 10.7 | 10.7 | 14.3 |
| | 3 | 4 | 14.3 | 14.3 | 28.6 |
| | 4 | 3 | 10.7 | 10.7 | 39.3 |
| | 5 | 4 | 14.3 | 14.3 | 53.6 |
| | 6 | 3 | 10.7 | 10.7 | 64.3 |
| | 7 | 4 | 14.3 | 14.3 | 78.6 |
| | 8 | 2 | 7.1 | 7.1 | 85.7 |
| | 9 | 4 | 14.3 | 14.3 | 100.0 |
| | Total | 28 | 100.0 | 100.0 | |

*Table D.20. Distribution of Different Appliances by Income*

| | | Income | | | | Total |
|---|---|---|---|---|---|---|
| | | No income | 1-1300 | 1301-390( | 3901-mo | |
| 1-3 | Count | 1 | 6 | 0 | 1 | 8 |
| | % within Income | 16,7% | 42,9% | 0,0% | 20,0% | 28,6% |
| | % of Total | 3,6% | 21,4% | 0,0% | 3,6% | 28,6% |
| 4-6 | Count | 3 | 5 | 1 | 1 | 10 |
| | % within Income | 50,0% | 35,7% | 33,3% | 20,0% | 35,7% |
| | % of Total | 10,7% | 17,9% | 3,6% | 3,6% | 35,7% |
| 7-9 | Count | 2 | 3 | 2 | 3 | 10 |
| | % within Income | 33,3% | 21,4% | 66,7% | 60,0% | 35,7% |
| | % of Total | 7,1% | 10,7% | 7,1% | 10,7% | 35,7% |
| Total | Count | 6 | 14 | 3 | 5 | 28 |
| | % within Income | 100,0% | 100,0% | 100,0% | 100,0% | 100,0% |
| | % of Total | 21,4% | 50,0% | 10,7% | 17,9% | 100,0% |

*Table D.21. Distribution of Iron by Education*

| | | | Education | | | | Total |
|---|---|---|---|---|---|---|---|
| | | | No Education | Primary | Secondary | Tertiary | |
| Iron | 0 | Count | 4 | 1 | 3 | 2 | 10 |
| | | % Education | 66,7% | 7,7% | 60,0% | 50,0% | 35,7% |
| | | % of Total | 14,3% | 3,6% | 10,7% | 7,1% | 35,7% |
| | 1 | Count | 2 | 11 | 2 | 2 | 17 |
| | | % Education | 33,3% | 84,6% | 40,0% | 50,0% | 60,7% |
| | | % of Total | 7,1% | 39,3% | 7,1% | 7,1% | 60,7% |
| | 2 | Count | 0 | 1 | 0 | 0 | 1 |
| | | % Education | 0,0% | 7,7% | 0,0% | 0,0% | 3,6% |
| | | % of Total | 0,0% | 3,6% | 0,0% | 0,0% | 3,6% |
| Total | | Count | 6 | 13 | 5 | 4 | 28 |
| | | % Education | 100,0% | 100,0% | 100,0% | 100,0% | 100,0% |
| | | % of Total | 21,4% | 46,4% | 17,9% | 14,3% | 100,0% |

*Table D.22. Distribution of Iron by Income*

| | | | Income | | | | Total |
|---|---|---|---|---|---|---|---|
| | | | No Income | 1-1300 | 1301-3900 | 3901-more | |
| Iron | 0 | Count | 1 | 7 | 1 | 1 | 10 |
| | | % Income | 16,7% | 50,0% | 33,3% | 20,0% | 35,7% |
| | | % of Total | 3,6% | 25,0% | 3,6% | 3,6% | 35,7% |
| | 1 | Count | 5 | 6 | 2 | 4 | 17 |
| | | % Income | 83,3% | 42,9% | 66,7% | 80,0% | 60,7% |
| | | % of Total | 17,9% | 21,4% | 7,1% | 14,3% | 60,7% |
| | 2 | Count | 0 | 1 | 0 | 0 | 1 |
| | | % Income | 0,0% | 7,1% | 0,0% | 0,0% | 3,6% |
| | | % of Total | 0,0% | 3,6% | 0,0% | 0,0% | 3,6% |
| Total | | Count | 6 | 14 | 3 | 5 | 28 |
| | | % Income | 100,0% | 100,0% | 100,0% | 100,0% | 100,0% |
| | | % of Total | 21,4% | 50,0% | 10,7% | 17,9% | 100,0% |

*Table D.23. Distribution of Cooling System by Type of Business*

| | | | Type of Business | | Total |
|---|---|---|---|---|---|
| | | | Private | Public | |
| Space Cooling System | No | Count | 1 | 2 | 3 |
| | | % within Type | 33.3% | 33.3% | 33.3% |
| | | % of Total | 11.1% | 22.2% | 33.3% |
| | Yes | Count | 2 | 4 | 6 |
| | | % within Type | 66.7% | 66.7% | 66.7% |
| | | % of Total | 22.2% | 44.4% | 66.7% |
| Total | | Count | 3 | 6 | 9 |
| | | % within Type | 100.0% | 100.0% | 100.0% |
| | | % of Total | 33.3% | 66.7% | 100.0% |

**Appendix E. List of Interviews**

Interview 1, Tsumkwe, August, 2020.

Interview 2, Tsumkwe, August, 2020.

Interview 3, Tsumkwe, August, 2020.

Interview 4, Tsumkwe, August, 2020.

Interview 5, Tsumkwe, August, 2020.

Interview 6, Tsumkwe, August, 2020.

Interview 7, Tsumkwe, August, 2020.

Interview 8, Tsumkwe, August, 2020.

Interview 9, Tsumkwe, August, 2020.

Interview 10, Tsumkwe, August, 2020.

Interview 11, Tsumkwe, August, 2020.

Interview 12, Tsumkwe, August, 2020.

Interview 13, Tsumkwe, April, 2022.

Interview 14, Tsumkwe, April, 2022.

Interview 1 - G, Gam, April, 2022.

Interview 2 - G, Gam, April, 2022.

Interview 3 - G, Gam, April, 2022.

Interview 4 - G, Gam, April, 2022.

Interview 5 - G, Gam, April, 2022.

Interview 6 - G, Gam, April, 2022.

Interview 7 - G, Gam, April, 2022.

Interview 8 - G, Gam, April, 2022.

Interview 9 - G, Gam, April, 2022.

Interview 10 - G, Gam, April, 2022.

National/Regional Actor (NRA) 1, September, 2020.

National/Regional Actor (NRA) 2, September, 2020.

National/Regional Actor (NRA) 3, September, 2020.

National/Regional Actor (NRA) 4, September, 2020.

National/Regional Actor (NRA) 5, September, 2020.

National/Regional Actor (NRA) 6, September, 2020.

Mangetti Dune Inauguration – NBC Digital News, April 2022.

Mangetti Dune Inauguration – Inauguration Speeches, April, 2022.